ROWAN UNIVERSITY
CAMPBELL LIBRARY
201 MULLICA HILL RD.
GLASSBORO, NJ 08028-1701

SYSTEM THEORY AND PRACTICAL APPLICATIONS OF BIOMEDICAL SIGNALS

IEEE Press Series on Biomedical Engineering

The focus of our series is to introduce current and emerging technologies to biomedical and electrical engineering practitioners, researchers, and students. This series seeks to foster interdisciplinary biomedical engineering education to satisfy the needs of the industrial and academic areas. This requires an innovative approach that overcomes the difficulties associated with the traditional textbooks and edited collections.

Metin Akay, *Series Editor*
Dartmouth College

Advisory Board

Thomas Budinger	Simon Haykin	Richard Robb
Ingrid Daubechies	Murat Kunt	Richard Satava
Andrew Daubenspeck	Paul Lauterbur	Malvin Teich
Murray Eden	Larry McIntire	Herbert Voigt
James Greenleaf	Robert Plonsey	Lotfi Zadeh

Editorial Board

Eric W. Abel	Gabor Herman	Kris Ropella
Dan Adam	Helene Hoffman	Joseph Rosen
Peter Adlassing	Donna Hudson	Christian Roux
Berj Bardakjian	Yasemin Kahya	Janet Rutledge
Erol Basar	Michael Khoo	Wim L. C. Rutten
Katarzyna Blinowska	Yongmin Kim	Alan Sahakian
Bernadette Bouchon-Meunier	Andrew Laine	Paul S. Schenker
Tom Brotherton	Rosa Lancini	G. W. Schmid-Schönbein
Eugene Bruce	Swamy Laxminarayan	Ernest Stokely
Jean-Louis Coatrieux	Richard Leahy	Ahmed Tewfik
Sergio Cerutti	Zhi-Pei Liang	Nitish Thakor
Maurice Cohen	Jennifer Linderman	Michael Unser
John Collier	Richard Magin	Eugene Veklerov
Steve Cowin	Jaakko Malmivuo	Al Wald
Jerry Daniels	Jorge Monzon	Bruce Wheeler
Jaques Duchene	Michael Neuman	Mark Wiederhold
Walter Greenleaf	Banu Onaral	William Williams
Daniel Hammer	Keith Paulsen	Andy Yagle
Dennis Healy	Peter Richardson	Yuan-Ting Zhang

Books in the IEEE Press Series on Biomedical Engineering

Akay, M., *Nonlinear Biomedical Signal Processing: Volume I, Fuzzy Logic, Neural Networks, and New Algorithms*

Akay, M., *Nonlinear Biomedical Signal Processing: Volume II Dynamic Analysis and Modeling*

Akay, M., *Time Frequency and Wavelets in Biomedical Signal Processing*

Baura, G. D., *System Theory and Practical Applications of Biomedical Signals*

Hudson, D. L. and M. E. Cohen, *Neural Networks and Artificial Intelligence for Biomedical Engineering*

Khoo, M. C. K., *Physiological Control Systems: Analysis, Simulation, and Estimation*

Liang, Z-P. and P. C. Lauterbur, *Principles of Magnetic Resonance Imaging: A Signal Processing Perspective*

Ying, H., *Fuzzy Control and Modeling: Analytical Foundations and Applications*

SYSTEM THEORY AND PRACTICAL APPLICATIONS OF BIOMEDICAL SIGNALS

Gail D. Baura
CardioDynamics
San Diego, CA

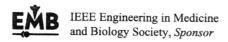
IEEE Engineering in Medicine
and Biology Society, *Sponsor*

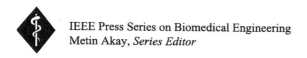
IEEE Press Series on Biomedical Engineering
Metin Akay, *Series Editor*

IEEE Press

A JOHN WILEY & SONS, INC., PUBLICATION

This text is printed on acid-free paper. ∞

Copyright © 2002 by the Institute of Electrical and Electronics Engineers, Inc. All rights reserved.

No part of this publication may be reproduced, stored in a retrieval system or transmitted in any form or by any means, electronic, mechanical, photocopying, recording, scanning or otherwise, except as permitted under Section 107 or 108 of the 1976 United States Copyright Act, without either the prior written permission of the Publisher, or authorization through payment of the appropriate per-copy fee to the Copyright Clearance Center, Inc., 222 Rosewood Drive, Danvers, MA 01923, (978) 750-8400, fax (978) 750-4744, or on the web at www.copyright.com. Requests to the Publisher for permission should be addressed to the Permissions Department, John Wiley & Sons, Inc., 111 River Street, Hoboken, NJ 07030, (201) 748-6011, fax (201) 748-6008, e-mail: permcoordinator@wiley.com.

Limit of Liability/Disclaimer of Warranty: While the publisher and author have used their best efforts in preparing this book, they make no representation or warranties with respect to the accuracy or completeness of the contents of this book and specifically disclaim any implied warranties of merchantability or fitness for a particular purpose. No warranty may be created or extended by sales representatives or written sales materials. The advice and strategies contained herein may not be suitable for your situation. You should consult with a professional where appropriate. Neither the publisher nor author shall be liable for any loss of profit or any other commercial damages, including but not limited to special, incidental, consequential, or other damages.

For general information on our other products and services please contact our Customer Care Department within the U.S. at 877-762-2974, outside the U.S. at 317-572-3993 or fax 317-572-4002.

Wiley also publishes its books in a variety of electronic formats. Some content that appears in print, however, may not be available in electronic format.

Library of Congress Cataloging in Publication Data is available.

ISBN 0-471-23653-5

Printed in the United States of America.

10 9 8 7 6 5 4 3 2

IEEE Press
445 Hoes Lane, P.O. Box 1331
Piscataway, NJ 08855-1331

IEEE Press Editorial Board
Stamatios V. Kartalopoulos, *Editor in Chief*

M. Akay	M. E. El-Hawary	M. Padgett
J. B. Anderson	R. J. Herrick	W. D. Reeve
R. J. Baker	D. Kirk	S. Tewksbury
J. E. Brewer	M. S. Newman	G. Zobrist

Kenneth Moore, *Director of IEEE Press*
Catherine Faduska, *Senior Acquisitions Editor*
John Griffin, *Acquisitions Editor*
Christina Kuhnen, *Associate Acquisitions Editor*

IEEE Engineering in Medicine and Biology Society, *Sponsor*
EMB-S Liaison to IEEE Press, Metin Akay

Technical Reviewers

Kris Ropella, *Marquette University*
Rich Shiavi, *Vanderbilt University*

Books of Related Interest from IEEE Press

BIOMEDICAL SIGNAL ANALYSIS: A Case-Study Approach
Rangaraj M. Rangayyan
2001   Hardcover   516 pp   IEEE Order No. PC5799   ISBN 0-471-20811-6

NONLINEAR BIOMEDICAL SIGNAL PROCESSING VOL. I: Fuzzy Logic, Neural Networks, and New Algorithms
Edited by Metin Akay
2000   Hardcover   259 pp   IEEE Order No. PC5861   ISBN 0-7803-6011-7

NONLINEAR BIOMEDICAL SIGNAL PROCESSING VOL. II: Dynamic Analysis and Modeling
Edited by Metin Akay
2001   Hardcover   341 pp   IEEE Order No. PC5862   ISBN 0-7803-6012-5

UNDERSTANDING NEURAL NETWORKS AND FUZZY LOGIC: Basic Concepts & Applications
Stamatios V. Kartalopoulos
A volume in the IEEE Press Understanding & Technology Series
1996   Softcover   232 pp   IEEE Order No. PP5591   ISBN 0-7803-1128-0

To Larry Spiro, my bon vivant, *without whose infinite patience and love this book could not have been written.*

CONTENTS

Preface xv

Nomenclature xix

I Filters

1 System Theory and Frequency-Selective Filters 3
1.1 Input-Output Description 3
1.2 Linear Constant Coefficient Difference Equations 6
1.3 Basic Frequency-Selective Filter Concepts 9
1.4 Design of IIR Digital Filters from Analog Filters 13
1.5 Design of FIR Filters by Windowing 20
1.6 Pseudorandom Binary Sequence Filter 24
1.7 Summary 26
1.8 References 27
1.9 Recommended Exercises 28

2 Low Flow Rate Occlusion Detection Using Resistance Monitoring 29
2.1 Physiology of Intravenous Drug Administration 29
2.2 Intravenous Infusion Devices 32
2.3 Problem Significance 36
2.4 Resistance Monitoring in the IVAC Signature Edition Pump 38
2.5 Summary 42
2.6 References 42
2.7 Matlab Exercises 43
2.8 Intraarterial Blood Pressure Exercises 43

3 Adaptive Filters 46
3.1 Adaptive Noise Cancellation Proof 46
3.2 Optimization Concepts 48

3.3	Least Mean Squares Algorithm for Finite Impulse Response Filters	49
3.4	Infinite Impulse Response Filters	52
3.5	Adaptive Noise Cancellation	60
3.6	Summary	63
3.7	References	65
3.8	Recommended Exercises	65

4 Improved Pulse Oximetry — 66

4.1	Physiology of Oxygen Transport	66
4.2	In Vitro Oxygen Measurements	71
4.3	Problem Significance	75
4.4	Adaptive Noise Cancellation in Masimo Software	78
4.5	Summary	84
4.6	References	84
4.7	Noninvasive Blood Pressure Exercises	85

5 Time-Frequency and Time-Scale Analysis — 87

5.1	Time-Frequency Representations	91
5.2	Spectrogram	94
5.3	Wigner Distribution	97
5.4	Kernel Method	100
5.5	Time-Scale Representations	104
5.6	Scalograms	105
5.7	Summary	109
5.8	References	109
5.9	Recommended Exercises	111

6 Improved Impedance Cardiography — 112

6.1	Physiology of Cardiac Output	112
6.2	In Vivo and In Vitro Cardiac Output Measurements	116
6.3	Problem Significance	122
6.4	Spectrogram Processing in Drexel Patents	124
6.5	Wavelet Processing in CardioDynamics Software	127
6.6	Summary	133
6.7	References	134
6.8	Electrocardiogram QRS Detection Exercises	136

II Models for Real Time Processing

7 Linear System Identification — 141

7.1	The ARMAX Model and Variations	141
7.2	Uniqueness Properties	144
7.3	Model Identifiability	145
7.4	Prediction Error Methods	145
7.5	Instrumental Variable Methods	150
7.6	Recursive Least Squares Algorithm	152

7.7	Model Validation	157
7.8	Summary	159
7.9	References	161
7.10	Recommended Exercises	162

8 External Defibrillation Waveform Optimization — 163

8.1	Physiology	163
8.2	External Defibrillation Waveforms	167
8.3	Problem Significance	171
8.4	Previous Studies	173
8.5	Application of the ARX Model to Prediction of Transthoracic Impedance	174
8.6	Transthoracic Impedance as the Basis of External Defibrillation Waveform Optimization	185
8.7	Summary	188
8.8	References	189
8.9	Digital Thermometry Exercises	192

9 Nonlinear System Identification — 195

9.1	Historical Review	195
9.2	Supervised Multilayer Networks	199
9.3	Unsupervised Neural Networks: Kohonen Network	205
9.4	Unsupervised Networks: Adaptive Resonance Theory Network	210
9.5	Model Validation	212
9.6	Summary	214
9.7	References	215
9.8	Recommended Exercises	217

10 Improved Screening for Cervical Cancer — 218

10.1	Physiology	218
10.2	Pap Smear	220
10.3	Problem Significance	224
10.4	Semiautomation of Cervical Cancer Screening	228
10.5	Cervical Cancer Screening Using Neural Networks	231
10.6	Summary	234
10.7	References	235
10.8	Cardiac Output Exercises	237

11 Fuzzy Models — 239

11.1	Historical Review	239
11.2	Fuzzification	242
11.3	Rule Base Inference	244
11.4	Defuzzification	246
11.5	Knowledge Base	248
11.6	Model Validation	251
11.7	Fuzzy Control	251

11.8	Fuzzy Pattern Recognition	255
11.9	Summary	257
11.10	References	260
11.11	Recommended Exercises	261

12 Continuous Noninvasive Blood Pressure Monitoring: Proof of Concept — 262

12.1	Physiology	262
12.2	In Vivo and In Vitro Blood Pressure Measurements	264
12.3	Problem Significance	266
12.4	Previous Studies	267
12.5	Work Based on Digital Signal Processing	273
12.6	Continuous Blood Pressure Measurement	282
12.7	Summary	284
12.8	References	287
12.9	Infusion Pump Occlusion Alarm Exercises	289

III Compartmental Models

13 The Linear Compartmental Model — 295

13.1	Protein Structure	295
13.2	Experimental Design	297
13.3	Kinetic Models	299
13.4	Model Identifiability	304
13.5	Nonlinear Least Squares Estimation	306
13.6	Sampling Schedules	308
13.7	Model Validation	312
13.8	Summary	313
13.9	References	314
13.10	Recommended Exercises	315

14 Pharmacologic Stress Testing Using Closed-Loop Drug Delivery — 316

14.1	Pharmacokinetics and Pharmacodynamics	316
14.2	Control Theory	320
14.3	Problem Significance	325
14.4	Closed-Loop Drug Infusion in Pharmacological Stress Tests	328
14.5	Summary	334
14.6	References	335
14.7	Peripheral Insulin Kinetics Exercises	337

15 The Nonlinear Compartmental Model — 340

15.1	Michaelis–Menten Dynamics	340
15.2	Bilinear Relation	347
15.3	Summary	351
15.4	Recommended References	353
15.5	Recommended Exercises	353

16 The Role of Nonlinear Compartmental Models in Development of Antiobesity Drugs — 354

- 16.1 Body Weight Regulation — 354
- 16.2 Receptor-Mediated Transport Across The Blood-Brain Barrier — 357
- 16.3 Problem Significance — 361
- 16.4 Previous Blood-Brain Barrier Insulin Studies — 363
- 16.5 Saturable Transport of Insulin from Plasma into the CNS — 368
- 16.6 Summary — 373
- 16.7 References — 375
- 16.8 Central Insulin Kinetics Exercises — 377

IV System Theory Implementation

17 Algorithm Implementation — 383

- 17.1 Data Types — 383
- 17.2 Digital Signal Processors — 385
- 17.3 Embedded Systems — 387
- 17.4 FDA Review of Medical Device Software — 389
- 17.5 Summary — 393
- 17.6 References — 394

18 The Need for More System Theory in Low-Cost Medical Applications — 395

- 18.1 Future Employment for Biomedical Engineering Graduate Students — 395
- 18.2 The Loss of Innovation in the Medical Device Industry — 396
- 18.3 Low-Cost Medical Monitoring and System Theory — 398
- 18.4 Addressing the Need for Innovation in a Cost-Conscious Environment — 401
- 18.5 References — 403

Glossary — 407

Index — 431

PREFACE

The impetus for this book was a conversation I had with my friend, Dr. Jenq-Neng Hwang, during the 1996 ICASSP conference. When I asked why few applications were presented, Jenq-Neng replied that most professors do not have access to data, especially the physiologic data in which I am interested. Later that year, I saw an advertisement in *EMBS Magazine* soliciting book proposals and decided to write a book that could serve to bridge the gap between industrial medical instrumentation applications and academic system theory.

This text is divided into four parts. In Part I, classic and current filtering techniques for real-time applications are discussed. These include frequency-selective filters, the pseudorandom binary sequence, adaptive filters, time-frequency representations, and time-scale representations. In Part II, modeling techniques for real-time applications are discussed. These include the autoregressive moving average with exogenous input model, the artificial neural network model, and the fuzzy model. In Part III, linear and nonlinear compartmental models are discussed. These models have been applied to physiologic data such as metabolite and drug transport with uneven sampling intervals. In Part IV, algorithmic implementations and the need for more system theory in the medical instrumentation industry are highlighted.

RECOMMENDED READING STRATEGIES

This book is intended as a textbook for a system theory applications course, within the medical instrumentation course series of a biomedical engineering/bioengineering graduate program. It may also be used as a reference book for industrial medical instrumentation. The chapters are intentionally organized in groups of two chapters, with the first chapter describing a system theory technology, and the second chapter describing an industrial application of this technology. Although this organization is somewhat unorthodox, it is designed to sustain the interest of graduate students.

Each theory chapter contains a general overview of a system theory technology, which is intended as background material for the application chapter, rather than as a comprehensive review. Textbooks that may serve as references for each technology are recommended at the end of each theory chapter. Each application chapter contains a history of the highlighted medical instrument, summary of appropriate physiology, discussion of the problem of interest and previous empirical solutions, and review of a solution that utilizes the theory in the previous chapter. When a new term is first introduced, it is set in boldface, and defined in the Glossary. Depending on the reader's background, it is recommended that the chapters be read in the following order:

Biomedical engineering researchers and graduate students with system theory background:

    Chapters 1–18—original order

Medical instrumentation engineers:

    Chapter 18—summary chapter
    Even numbered chapters—application chapters
    Odd numbered chapters—background theories corresponding to applications of interest
    Chapter 17—optional; implementation chapter

Other biomedical engineering researchers and graduate students:

    Even numbered chapters—application chapters
    Odd numbered chapters—background theories corresponding to applications of interest
    Chapters 17–18—implementation chapters

Electrical engineering researchers and graduate students with system theory background:

    Even numbered chapters—background applications corresponding to theories of interest (except Chapters 14, 16)
    Chapters 17–18—implementation chapters
    Chapters 1, 3, 5, 7, 9, 11—optional; theory chapters
    Chapters 13–16—optional; contains advanced physiology and biochemistry

ACKNOWLEDGMENTS

These 18 chapters were written in 26 months from late 1998 to 2000, in the midst of working full time in industry. I would like to thank my colleagues for reviewing groups of chapters: Dr. Bill Barnes, Dr. Leon Cohen, Joe Elf, Dr. David Foster, Dr. Moritz Harteneck, Dr. Sandy Ng, Dr. Shankar Reddy, Dr. Alfredo Ruggeri, and Dr. Xiang Wang. My Ph.D. advisor David Foster was especially supportive of this project. I would also like to recognize these colleagues and corporations for sharing data used in chapter exercises: Stuart Gallant of Tensys Corporation, Dr. Michael Schwartz, Dr. Masaru Sugimachi, the SAAM Institute, and Welch Allyn Inc. Chapter 17 is dedicated to Stuart Gallant, in recognition of his love of integer mathematics.

    My friend Alan Davison served as the perfect editor for this project. In his own endearing, anal-retentive fashion, he pointed out endless improvements that increased each chapter's readability and applicability.

    I received incredible emotional support for this project from my coworkers at VitalWave Corporation: Fred Bacher, Kurt Blessinger, Dave Eshbaugh, Simon Finburgh,

Mano Goharla'ee, and Kevin Woolf. My dear friends Dulce Capadocia and Maddy Ramirez also were extremely supportive during these lost years. My husband Larry encouraged this project and patiently endured its duration.

The following people were not directly involved, but influenced this text: To my undergraduate professors at Loyola Marymount University, thank you for providing such a strong engineering foundation upon which to build: Cliff d'Autremont, Dr. Joe Callanan, Dr. Tai-Wu Kao, Dr. John Page, Bob Ritter, Dr. Paul Rude. Thanks also to my supervisors at St. Mary Medical Center who gave me my first (student) experience as a biomedical engineer: Bob Ward and Chris Wentzel. I became a physiologist as well as engineer thanks to Ph.D. coadvisor, Dr. Dan Porte, Jr., with interpretative translation from Dr. Michael Schwartz and biomedical engineer/physiologist Dr. Jay Taborsky. Through my industrial mentors Dennis Hepp and Ron Bromfield, I have been given the rare industrial opportunity and freedom to explore physiologic mechanisms underlying various signals.

Thank you, IEEE Press editors John Griffin and Chrissy Kuhnen, for moving this manuscript toward publication. I welcome comments to this text at *www.gailbaura.com*.

GAIL D. BAURA

San Diego, 2001

NOMENCLATURE

THEORY CHAPTERS

$u(k)$	scalar
$\mathbf{u}(k)$	vector
$\underline{\mathbf{U}}(k)$	matrix
μ_n, ρ_n	constants
a	scale, scaling constant
$a_n, \tilde{a}_n$	feedback coefficients
$ac(k)$	acceleration
α	IIR filter parameter formula, coefficient in numerator polynomial
$\underline{\mathbf{A}}(k)$	feedback matrix of rate constants
$A(\theta, \tau)$	symmetric ambiguity function
AIC	Akaike's information criterion
AUC	area under curve of drug concentration in blood or plasma versus time
$b(t)$	glucose balance
$b_n, \tilde{b}_n$	feedforward coefficients
$\underline{\mathbf{B}}(k)$	feedforward matrix of rate constants
B_0	net glucose balance
β	coefficient in denominator polynomial
$\mathbf{c}(t)$	concentration vector
$c_i(t)$	concentration in compartment i
c_n	moving average coefficients
$\underline{\mathbf{C}}(k)$	relationship between concentration and desired output vectors
$C(\tau, \omega)$	Cohen's class
Cl	clearance rate
CV	coefficient of variation
d	delay
$d(k)$	PRBS decoder
$de(k)$	distance error
$d/du(k)$	derivative

$\delta(k)$	impulse
δ	error
$\partial/\partial u(k)$	partial derivative
$\nabla \mathbf{u}[\mathbf{w}(k)]$	gradient
e	minimum value
$e(k)$	white noise, PRBS encoder
$\mathbf{e}(t)$	error vector
E	enzyme
$E\{\ \}$	expected value
$E(t)$	glucose effectiveness
ES	enzyme–substrate complex
E_T	total enzyme
ε	allowable Chebyshev passband ripple
$\boldsymbol{\varepsilon}(k)$	error vector
$\varepsilon_0(k)$	error sequence
$f(k)$	auxiliary process of HARF adaptive filter
f_n	zeros of transfer function
FSD	fractional standard deviation
$\Im\{\ \}$	Fourier transform
$\Im^{-1}\{\ \}$	Inverse Fourier transform
$\boldsymbol{\phi}(k)$	regression vector
$\phi(\theta, \tau)$	time-frequency kernel
$\phi[a\theta, (\tau/a)]$	time-scale kernel
$\phi_{uu}(k)$	autocorrelation of $u(k)$
$\phi_{uy}(k)$	cross-correlation of $u(k)$ and $y(k)$
$\Phi_{uu}(e^{j\omega})$	power density spectrum (or power spectrum)
$\Phi_{uy}(e^{j\omega})$	cross power density spectrum
g_n	poles of transfer function
$g(D)$	primitive polynomial
$g(t)$	glucose concentration, activation function
G	downsampling constant
$\mathbf{G}(k)$	Jacobian matrix
$G(z)$	white noise Gaussian process
$grd[H(e^{j\omega})]$	frequency response group delay
$h(k)$	impulse response
$h_d(k)$	desired impulse response
$h_s(k)$	scaled impulse response
$h_{\text{lowpass}}(k)$	desired lowpass filter
$h_{\text{highpass}}(k)$	desired highpass filter
$h_{\text{bandpass}}(k)$	desired bandpass filter
$H(e^{j\omega})$	Fourier transfer function
$H_d(e^{j\omega})$	desired Fourier transfer function
$H_c(s)$	Laplace transfer function
$H(z)$	z-domain transfer function
$\mathbf{H}(k)$	system operator
$|H(e^{j\omega})|$	frequency response magnitude
$|H_c(j\Omega)|$	continuous frequency response magnitude
$|H_c(j\Omega)|^2$	continuous magnitude squared function

$\angle H(e^{j\omega})$	phase response
HPLC	high-performance liquid chromatography
i	compartment to which flux is transported
$i(t)$	insulin concentration
$\mathbf{I}(t)$	identity matrix
ISF	interstitial fluid
IVGTT	intravenous glucose tolerance test
$Im\{\ \}$	imaginary part
j	compartment from which flux is transported
$\mathbf{J}(\mathbf{k}, \sigma^2, SS)$	Fisher information matrix
ϑ	margin
k	discrete time (samples)
$\mathbf{k}$	rate constant vector
k_{cat}	catalyst rate constant
k_{ij}	rate constant to compartment i from compartment j
K	dimension of parameter vector
K_M	Michaelis constant
λ	eigenvalue
$\lambda(k)$	forgetting factor
$\mathbf{L}(k)$	$\mathbf{P}(k)\boldsymbol{\phi}(k)$
M	order of feedforward coefficients, maximal length of PRBS
$M(\Omega, \tau)$	joint time-frequency characteristic function
$M(\Omega, \sigma)$	joint time-scale characteristic function
max	maximum subscript
min	minimum subscript
μ	step size
$\mu(\mathbf{z})$	membership function
n	summation index
$n_0(k)$	noise associated with input signal for adaptive filter
$n_1(k)$	reference noise source for adaptive filter
N	order of feedback coefficients
$N_c(k)$	neighborhood of Kohonen network
NN	number of data points
OSS	optimum sampling schedule
p	number of parameters
$\mathbf{p}$	parameter vector
$\mathbf{p}(k)$	$E\{y(k)\mathbf{u}(k)\}$
P	order of moving average coefficients, product, number of input–output pairs, total number of membership functions
$\mathbf{P}(k)$	$(1/k)\mathfrak{R}^{-1}(k)$
$P(t, a)$	time-scale distribution, affine class
$P(t, \Omega)$	joint energy density, time-frequency distribution, probability distribution
$P_S(t, s)$	scalogram
$P_{CW}(t, \Omega)$	Choi–Williams distribution
$P_{PW}(t, \Omega)$	pseudo-Wigner distribution
$P_{STFT}(t, \Omega)$	short-time Fourier transform
$P_W(t, \Omega)$	Wigner distribution, Wigner–Ville distribution
$P_{ZAM}(t, \Omega)$	Zhao–Atlas–Marks distribution

$\theta(k)$	regression vector
$q(k)$	normalizing factor
$q_i(k)$	eigenvector i
$q_i(t)$	mass in compartment I
q^{-1}	backwards shift operator
Q	quality factor
θ	neuron threshold
$\boldsymbol{\theta}$	parameter vector
r	ratio, real number
$r(t)$	insulin concentration in remote compartment
$rs(k)$	relative speed
$\underline{\mathbf{R}}(k)$	$E\{\mathbf{u}(k)\mathbf{u}^T(k)\}$, $(1/k)\mathfrak{R}(k)$
$\underline{\mathbf{R}}(t)$	output correlation matrix
$\underline{\mathbf{R}}_\mathbf{h}(k)$	hidden neuron correlation matrix
$\underline{\mathbf{R}}_\mathbf{i}(k)$	input correlation matrix
$R_{ij}(t)$	flux of molecules entering compartment i from compartment j
$Re\{\ \}$	real part
$\mathfrak{R}$	set of real numbers
$\mathfrak{R}(k)$	$\Sigma_{i=1}^{k} \phi(i)\, \phi^T(i)$
ρ	learning rate
RIA	radioimmunoassay
s	Laplace frequency
sc	scaled subscript
s_n	Laplace poles
S	substrate
$\underline{\mathbf{S}}(\hat{k})$	sensitivity matrix
S_I	insulin sensitivity
SS	sampling schedule
$\sigma^2(k)$	system variance
$\sigma^2_{i|j}$	conditional standard deviation
t	continuous time
$t_{1/2}$	molecule half-life
$t_{1/2\alpha}$	distribution half-life
$t_{1/2\beta}$	elimination half-life
$\underline{\mathbf{T}}(k)$	transformation matrix
T_d	sampling interval
τ	signal period
$u(k)$	input sequence
$u(t)$	exogenous input to a compartment
$|u(t)^2|$	instantaneous power of continuous signal
$u_p(t)$	disappearance of glucose into peripheral tissues
ν	vigilance parameter
$U(e^{j\omega})$	Fourier transform of input sequence
$U(e^{j\Omega})$	Fourier transform of continuous input
$U(\Omega)$	abbreviated form of $U(e^{j\Omega})$
$\mathbf{U}(k)$	fuzzy input vector
$U(z)$	Fourier transform of input sequence
$U_{mi}(k)$	fuzzy input

$\|U(\Omega)^2\|$	power density spectrum
$\mathcal{U}$	universal set
$v(k)$	disturbance sequence, SHARF moving average
$v_i(t)$	volume of compartment i
$v_{ii}(t,\hat{k})$	diagonal element of covariance matrix
V	velocity
$V(\theta)$	covariance
V_D	apparent volume of distribution
V_{max}	maximum velocity
$\underline{\mathbf{V}}(k)$	covariance matrix
$V_N(x)$	Nth order Chebyshev polynomial
$w(k)$	finite duration window
$\mathbf{w}(k)$	adaptive filter weight vector
$w_{lm}(k)$	weight between neural input and hidden layers
$w_s(k)$	membership function support value
$\underline{\mathbf{W}}(t)$	weighting matrix
$W_{nl}(k)$	weight between neural hidden and output layers
$WT(t, a)$	continuous wavelet transform
$WT(k, 2^j)$	dyadic discrete wavelet transform
ω	discrete frequency (radians/sample)
ω_c	discrete cutoff frequency
ω_p	passband frequency
ω_s	stopband frequency
Ω	continuous frequency
Ω_c	continuous cutoff frequency
Ω_N	Nyquist frequency
$x(t)$	state variable
$x_l(k)$	neural hidden layer response
$\xi[\mathbf{w}(k)]$	performance function
$y(k)$	output sequence, desired output sequence
$\hat{y}_c(k)$	central neuron
$Y_{ni}(k)$	fuzzy input
z	z frequency
$\mathbf{z}(k)$	instrument vector
z^{-1}	unit delay
$\mathbf{z}(t, \mathbf{k})$	vector of noisy measurements of outputs
$Z\{\ \}$	z-transform
$Z^{-1}\{\ \}$	inverse z-transform
$\psi(k), \psi_a(k)$	wavelet
$\psi_b(k)$	scaling function
$\hat{\psi}(\Omega)$	Fourier transform of $\psi(t)$
$2\Omega_N$	Nyquist rate
$[\ ]$	concentration
$\|\|\ \|\|$	Euclidean distance
∞	infinity
$\cap$	intersection
$\cup$	union
$\cdot$	derivative superscript

^	estimate superscript
—	mean superscript, normalized subscript
~	measured superscript
*	optimal superscript
T	transpose superscript

APPLICATION CHAPTERS

a, b, c, d	constants
A	area under thermodilution curve, cross-sectional area of aorta
$A_i(\lambda)$	unscattered absorbance of i
$A_t(\lambda)$	total unscattered absorbance
AED	automatic external defibrillator
AV	atrioventricular
α	receptor subtype, constant
BTE	biphasic truncated exponential
BV	blood velocity
β	receptor subtype, constant
c	concentration, velocity of sound in blood
C_{AO_2}	oxygen content in arterial blood
C_d	defibrillator capacitance
C_t	transthoracic capacitance
C_{VO_2}	oxygen content in pulmonary artery blood
C_{tubing}	tubing compliance
CAD	coronary artery disease
CLDD	closed-loop drug delivery
CO	cardiac output
COHb	carboxyhemoglobin
CO_2	carbon dioxide
d	onset delay
d	arterial deoxyhemoglobin subscript
d/dt	derivative
D	narrowest aortic diameter, drug molecule
DR	drug-receptor complex
Δ	change
$E(s)$	error transfer function
E_{50}	total energy required for 50% successful defibrillation
EC_{50}	drug concentration that elicits half-maximal response
ECG	electrocardiogram
ε_i	convergence parameter
$\varepsilon_i(\lambda)$	extinction coefficient of i
f	function root
f_d	Doppler shift frequency
f_r	respiratory frequency
f_t	transmitted frequency
F	correction factor
F_{gas}	total gas volume

FDA	Food and Drug Administration
Fe^{++}	iron (ferrous) ion
FFT	fast Fourier transform
$G_c(s)$	controller transfer function
$G_p(s)$	prefilter transfer function
$H(s)$	plant transfer function
Hb	deoxyhemoglobin
HbO_2	oxyhemoglobin
$HRa(t)$	measured heart rate increase above baseline
η	fluid viscosity
inf	infusion rate
I	transmitted light
$I(k)$	current
I_0	incident light
$I_T(t)$	thoracic current
k	discrete time
k_1	association rate constant
k_2	dissociation rate constant
K	number of samples in waveform period
K_d	derivative term of PID controller
K_i	integral term of PID controller
K_p	proportional gain of PID controller
K_D	equilibrium constant
K_1	density factor
K_2	manufacturer's computation constant
l	path length
L	tubing length, cylindrical length, inductance
LVET	left ventricular ejection time
λ	wavelength
m	eigenvalue
max	maximum subscript
MetHb	methemoglobin
min	minimum subscript
n	number of samples
$n(k)$	noise signal
$N(s)$	sensor noise
o	arterial oxyhemoglobin subscript
O_2	oxygen
p	probability
P_a	arterial partial pressure of gas
P_A	alveolar partial pressure of gas
P_{alarm}	pressure alarm setting
P_{gas}	partial pressure of gas
P_h	hydrostatic pressure gradient
P_v	venous partial pressure of gas, mean pressure in vein
P_0	pressure in peripheral vein without flow
P_1	pressure at tubing inlet
P_2	pressure at tubing output

P_{50}	half-saturation pressure
PI	proportional-integral control
PID	proportional-integral-derivative control
PMA	premarket approval
$\tilde{Q}$	flow
$\dot{Q}$	lung perfusion
QRS	electrocardiogram feature composed of Q, R, and S waves
r	tubing radius, correlation coefficient
R	ratio of two absorbance derivatives, drug receptor
R_d	defibrillator resistor
R_l	load resistor
R_{mfr}	resistance through mechanical flow regulator
R_n	resistance through delivery system excluding mechanical flow regulator
R_{sys}	resistance for all tubing and catheter combinations
R_t	transthoracic resistance
R_{total}	total resistance
R_{vein}	median resistance of patient vein
RC	resistance–capacitance circuit
RLC	resistance–inductance–capacitance circuit
ρ	resistivity
SA	sinoatrial
SCA	sudden cardiac arrest
S_aO_2	functional arterial oxygen saturation
S_pO_2	S_aO_2 measured by pulse oximetry
SV	stroke volume
T_B	blood temperature in pulmonary artery
T_I	injectate temperature
T_S	settling time
Target_D	maximum desired heart rate increase
TTA	time to alarm
τ	time constant
θ	angle
$u(k)$	input signal
$V(s)$	unpredicted disturbance input
V_A	alveolar space tidal volume
V_D	dead space tidal volume
V_E	expiratory tidal volume
V_I	injectate volume
$V_{in}(k)$	input voltage
$V_{out}(k)$	output voltage
V_{50}	total voltage required for 50% successful defibrillation
$\dot{V}_A$	alveolar ventilation
$\dot{V}_D$	dead space ventilation
$\dot{V}_E$	minute volume
$\dot{V}_{O_2}$	oxygen consumption
$\dot{V}_A/\dot{Q}$	ventilation-perfusion ratio
$V_T(t)$	thoracic voltage
ω_n	natural frequency

x	variable optical attenuation absorbance subscript
$\xi(t)$	performance function
y	nonspecific optical attenuation source subscript
$y(k)$	output signal, primary signal
Z_b	body tissue impedance
$Z_c(t)$	time-varying (changing) impedance
Z_0	constant impedance
Z_{sk}	skin impedance
$Z_T(t)$	thoracic impedance
ζ	damping ratio
$[\ ]$	concentration
$\cdot$	derivative superscript
$\hat{\ }$	estimate superscript
$-$	mean superscript, normalized subscript
$\sim$	measured superscript
$*$	optimal superscript
T	transpose superscript

I

FILTERS

The hospital environment is an infinite source of signal distortion. In older monitors, 60 Hz interference from power lines may distort the signal of interest. During surgery, electromagnetic interference is generated by the electrosurgical unit used for cautery. In unanesthesized patients, patient motion is a significant source of distortion. Even respiration and blood pressure may obscure the signal of interest. These and other noise sources make filtering a necessary requirement before further digital signal processing and/or control can be applied to the signal of interest.

If the noise source is restricted to a particular frequency band, frequency-selective filters may be used to minimize the noise. Under certain constraints of a periodic signal, a pseudorandom binary sequence filter may be used, which essentially functions as a bandpass filter that does not require frequency specifications. These strategies are discussed in Chapters 1 and 2.

Alternatively, if the noise and signal frequency bands overlap, but a reference source for the noise is present, then an adaptive filter may be used to minimize the noise. An adaptive filter possesses a structure that is adjustable in such a way that its performance improves through contact with its environment. This strategy is discussed in Chapters 3 and 4.

Finally, if the noise and signal frequency bands overlap, but the noise is minimal when transformed to another domain, then time-frequency or time-scale analysis may be used to minimize the noise. After transformation, important signal characteristics may be recovered. These strategies are discussed in Chapters 5 and 6.

1

SYSTEM THEORY AND FREQUENCY-SELECTIVE FILTERS

System theory is the study of dynamic systems from experimental data. A complete treatment of system theory, including frequency-selective filters, could easily fill several textbooks. Given the constraint of a single chapter, only introductory system theory concepts that lay the foundation for other chapters are reviewed in Chapter 1. Similarly, the discussion of frequency-selective filters is abbreviated out of necessity. It is assumed that the reader has a working knowledge of linear system theory, Fourier transforms, z-transforms, Laplace transforms, and the basic concepts of stochastic processes. For a review of this material, the reader is referred to [1].

1.1 INPUT–OUTPUT DESCRIPTION

Suppose you are acquiring respiration data in the hospital operating room while 60 Hz interference noise is present. Your data acquisition system includes an analog lowpass filter with a cutoff frequency of 1 MHz. Based on **discrete time,** k, the **input sequence,** $u(k)$, is the respiration data, and the **disturbance** (or noise) **sequence,** $v(k)$, is the 60 Hz noise. The system is the analog-to-digital converter and lowpass filter; the **output sequence,** $y(k)$, is the acquired data.

In general, a dynamic system can be represented by the diagram in Figure 1.1. Here, the system is driven by $u(k)$ and $v(k)$. The user can control $u(k)$ but not $v(k)$. The output provides useful information about the system. For a dynamic system, the control action at time k will influence the output at times greater than k. For the purposes of this discussion, $u(k)$, $v(k)$, and $y(k)$ are restricted to scalars. The system is therefore a **single input, single output (SISO) model.**

For the moment, let us assume that there is no disturbance, more than one input may be present (not SISO), and the system is represented by $\mathbf{H}(k)$. Here, $\mathbf{H}(k)$ represents the **system operator** by which the input is mapped into the output:

4 SYSTEM THEORY AND FREQUENCY-SELECTIVE FILTERS

$$y(k) = \mathbf{H}[u(k)] \tag{1.1}$$

This operator may map more than one input sequence into one output sequence, thus constituting a many-to-one mapping. For example, a square law device can be characterized by the mapping

$$y(k) = u^2(k) \tag{1.2}$$

We can restrict the system to one-to-one mapping and **linearity** by constraining the system to follow **superposition:**

$$\mathbf{H}[au_1(k) + bu_2(k)] = a\mathbf{H}[u_1(k)] + b\mathbf{H}[u_2(k)] \tag{1.3}$$

for arbitrary constants a and b. We can further restrict the system to be **time-invariant**, such that a delay of the input sequence causes a corresponding shift in the output sequence. Specifically, for all delays, k_0, an input sequence with values $u_1(k) = u(k - k_0)$ must produce the output sequence with values $y_1(k) = y(k - k_0)$.

1.1.1 Linear Time-Invariant Systems

Given the properties of linearity and time-invariance, the input–output description is immensely simplified. Now the **linear time-invariant (LTI) system** is completely characterized by its **impulse response**, $h(k)$, where

$$y(k) = u(k)*h(k) = \sum_{n=-\infty}^{\infty} u(n)h(k-n) \tag{1.4}$$

where * is the convolution operator. Eq. (1.4) is commonly called the **convolution sum**, and is commutative, associative, and distributive. Given an **impulse**, $\delta(k)$ (where $\delta(k) = 0$ for all k except $\delta(1) = 1$) as the input function, the impulse response is the response of a system to an impulse such that

$$h(k) = \delta(k)*h(k) = \sum_{n=-\infty}^{\infty} \delta(n)h(k-n) \tag{1.5}$$

The LTI system may be **causal** if the output sequence value at index $k = k_0$ depends only on the input sequence values for $k \leq k_0$. This implies that if $u_1(k) = u_2(k)$ for all $k \leq k_0$, then $y_1(k) = y_2(k)$ for $k \leq k_0$.

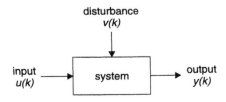

Figure 1.1 Discrete time system.

The LTI system may be **stable** in the bounded-input bounded-output sense if and only if every bounded input sequence produces a bounded output sequence. That is, the impulse response must be **absolutely summable:**

$$\sum_{n=-\infty}^{\infty} |h(n)| < \infty \qquad (1.6)$$

Many times, it may be efficient to compute the impulse response through initial calculations in the frequency domain. In the frequency domain, the Fourier transform of the system impulse response, otherwise known as the **transfer function,** is easily calculated, and followed by conversion back to the time domain. Conversion to and from the frequency domain is accomplished using the Fourier transform pair:

$$U(e^{j\omega}) = \Im\{u(k)\} = \sum_{k=-\infty}^{\infty} u(k)e^{-j\omega k} \qquad (1.7)$$

$$u(k) = \Im^{-1}\{U(e^{j\omega})\} = \frac{1}{2\pi} \int_{-\pi}^{\pi} U(e^{j\omega})e^{j\omega k} \, d\omega \qquad (1.8)$$

where $\Im$ is the Fourier operator and ω equals the discrete frequency in units of radians/sample. Discrete frequency is calculated as

$$\omega = \frac{2\pi(\text{frequency in Hertz})}{(\text{sampling frequency in Hertz})} \qquad (1.9)$$

Although the interval $-\pi < \omega < \pi$ has been used in Eq. (1.8), any interval of length 2π can be used. The Fourier transform representation exists for any finite length sequence, since it converges to a finite sum, and is absolutely summable. For a sequence of infinite length, the Fourier transform representation may or may not exist.

Using the **convolution theorem** (also known as the **dual domain theory of duality**), which states that convolution in the time domain translates to multiplication in the frequency domain, the transfer function, $H(e^{j\omega})$, can be calculated from $U(e^{j\omega}) = \Im\{u(k)\}$ and $Y(e^{j\omega}) = \Im\{y(k)\}$ as

$$H(e^{j\omega}) = \frac{Y(e^{j\omega})}{U(e^{j\omega})} \qquad (1.10)$$

1.1.2 Random Signals

In the previous section, we assumed that the signals are **deterministic,** that is, each value of a sequence is uniquely determined by a mathematical expression, table of data, or rule of some type. However, a system may be so complex that it is extremely difficult to describe. In such cases, modeling the system as a **stochastic process** is analytically useful. Again, through calculations in the frequency domain, the transfer function can be estimated.

A **stochastic signal** is assumed to be a member of an ensemble of signals that is characterized by a set of **probability density functions.** For a specific signal at a specific time, the signal's amplitude is assumed to be determined by an underlying scheme of probabilities. Let us assume that our system is stable, linear, and time-invariant, and that our input, $u(k)$, is real-valued and **wide-sense stationary**. A wide-sense stationary signal is assumed to possess a constant mean and to not vary with time. Assuming a system impulse response, $h(k)$, the system output, $y(k)$, is then a sample function of an output random process related to the input process by the linear transformation

$$y(k) = \sum_{n=-\infty}^{\infty} u(n)h(k-n) \tag{1.11}$$

which is the same result as Eq. (1.4). Stochastic signals are not absolutely summable, and consequently do not directly have Fourier transforms. However, many of the signal properties can be summarized in terms of **autocorrelation** sequences, for which the Fourier transforms do exist. Autocorrelation sequences are finite-energy sequences that tend to die out over time. Let us illustrate this by defining the autocorrelation and **power density spectrum**.

The autocorrelation function of a signal, $u(k)$, is defined as

$$\phi_{uu}(k) = \sum_{n=-\infty}^{\infty} u(n)u(k+n) \tag{1.12}$$

The autocorrelation for a noise sequence uniformly distributed in the range $\{-1, 1\}$ is illustrated in Figure 1.2. Similarly, the **cross-correlation** function of two signals, $u(k)$ and $y(k)$, is defined as

$$\phi_{uy}(k) = \sum_{n=-\infty}^{\infty} u(n)y(k+n) \tag{1.13}$$

Transforming these functions to the frequency domain leads to a calculation of the transfer function. The power density spectrum (or power spectrum) is the Fourier transform of the autocorrelation function:

$$\Phi_{uu}(e^{j\omega}) = \Im\{\phi_{uu}(k)\} \tag{1.14}$$

Similarly, the **cross power density spectrum** is the Fourier transform of the cross-correlation function:

$$\Phi_{uy}(e^{j\omega}) = \Im\{\phi_{uy}(k)\} \tag{1.15}$$

Using these functions, the transfer function, $H(e^{j\omega})$, can be calculated as

$$H(e^{j\omega}) = \frac{\phi_{uy}(e^{j\omega})}{\phi_{uu}(e^{j\omega})} \tag{1.16}$$

1.2 LINEAR CONSTANT COEFFICIENT DIFFERENCE EQUATIONS

A general model structure for the LTI system is the **autoregressive moving average, exogeneous input (ARMAX) model,** which consists of the linear constant coefficient difference equation:

1.2 LINEAR CONSTANT COEFFICIENT DIFFERENCE EQUATIONS

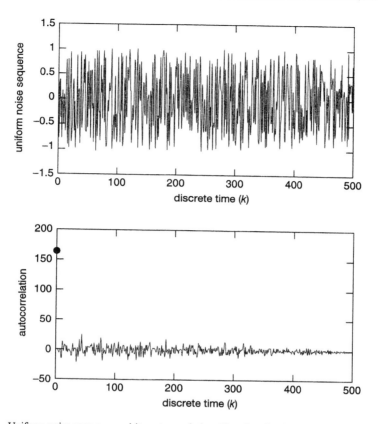

Figure 1.2 Uniform noise sequence and its autocorrelation. Note that the first autocorrelation sample has a magnitude of 165.5.

$$\sum_{n=0}^{N} a_n y(k-n) = \sum_{n=0}^{M} b_n u(k-n) + \sum_{n=0}^{P} c_n e(k-n) \tag{1.17}$$

where $e(k)$ is a sequence of independent and identically distributed (iid) random variables with zero mean, otherwise known as **white noise,** and $a_0 = c_0 = 1$. The model is autoregressive because the output, $y(k)$, looks back on past values of itself:

$$y(k) = -\sum_{n=1}^{N} a_n y(k-n) + \sum_{n=0}^{M} b_n u(k-n) + \sum_{n=0}^{P} c_n e(k-n) \tag{1.18}$$

The model also possesses a moving average, $\sum_{n=0}^{P} c_n e(k-n)$, and an exogeneous, or external, input, $u(k)$.

A subset of the ARMAX model is the controlled, **autoregressive (ARX) model,** assuming white noise is not present:

$$\sum_{n=0}^{N} a_n y(k-n) = \sum_{n=0}^{M} b_n u(k-n) \tag{1.19}$$

8   SYSTEM THEORY AND FREQUENCY-SELECTIVE FILTERS

For the ARX model, the output, $y(k)$, for a given input, $u(k)$, is not uniquely specified, as auxiliary information or conditions are required. [Recall that homogeneous equations such as Eq. (1.19) do not possess forcing functions, as do nonhomogeneous equations such as Eq. (1.18).] If it can be assumed that the system is initially at rest, then the system is linear, time-invariant, and causal.

1.2.1 z-transform Representation

A linear, time-invariant, discrete time system is implemented using adders, multipliers, and memory for storing delayed sequence inputs. In analog discrete time implementations such as switched-capacitor filters, the delays are implemented by charge storage devices. The unit delay system is represented by the system function z^{-1}. The ARX model in Eq. (1.19), a general Nth order difference equation, is represented by the block diagram of Figure 1.3. This representation is known as the **Direct-form realization.**

Using unit delays and the z-transform **time-shifting property**, $Z\{u(k-n)\} = z^{-n}U(z)$, the ARX model in Eq. (1.19) can be rewritten by taking the z-transform on both sides as

$$\sum_{n=0}^{N} a_n z^{-n} Y(z) = \sum_{n=0}^{M} b_n z^{-n} U(z) \tag{1.20}$$

The corresponding transfer function is

$$H(z) = \frac{Y(z)}{U(z)} = \frac{\sum_{n=0}^{M} b_n z^{-n}}{\sum_{n=0}^{N} a_n z^{-n}} = \frac{b_0 \left(1 + \frac{b_1}{b_0} z^{-1} + \ldots + \frac{b_M}{b_0} z^{-M}\right)}{1 + a_1 z^{-1} + \ldots + a_N z^{-N}} \tag{1.21}$$

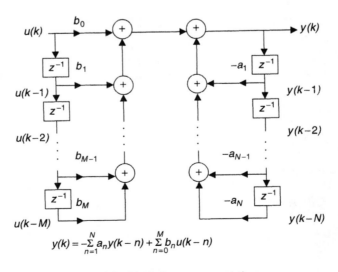

Figure 1.3   Block diagram representation.

recalling that $a_0 = 1$. The transfer function can also be expressed in its factored form as

$$H(z) = \frac{b_0 \prod_{n=1}^{M}(1 - f_n z^{-1})}{\prod_{n=1}^{N}(1 - g_n z^{-1})} \quad (1.22)$$

The roots of the numerator, f_n, are the **zeros** of the transfer function. The roots of the denominator or characteristic equation, g_n, are the **poles** of the transfer function. The transfer function is stable if and only if it has no poles outside the unit circle. The unit circle is described by $|z| = 1$.

Conversion from the z-domain to the frequency domain involves the substitution $z = e^{j\omega}$:

$$H(e^{j\omega}) = H(z)|_{z=e^{j\omega}} \quad (1.23)$$

$$H(e^{j\omega}) = \frac{\sum_{n=0}^{M} b_n e^{-j\omega n}}{\sum_{n=0}^{N} a_n e^{-j\omega n}} \quad (1.24)$$

$$H(e^{j\omega}) = \frac{b_0 \prod_{n=1}^{M}(1 - f_n e^{-j\omega})}{\prod_{n=1}^{N}(1 - g_n e^{-j\omega})} \quad (1.25)$$

1.3 BASIC FREQUENCY-SELECTIVE FILTER CONCEPTS

Keeping all these representations of the transfer function in mind, we are now ready to design a frequency-selective filter. Ideally, a frequency-selective filter passes only frequencies of interest, based on a cutoff frequency, ω_c. Each filter form is named for the frequency range that is passed: **lowpass** filter, **highpass** filter, **bandpass** filter, **bandstop** (all frequencies except a certain band or range) filter. For example, a 1 Hz (or 2π radians/sec) sine wave with 60 Hz (or 120π radians/sec) interference may be lowpass filtered to remove its noise artifact (Figure 1.4). For all these filters, if the ARMAX system is initially at rest, then the system is linear, time-invariant, and causal.

1.3.1 Filter Applications

A frequency-selective filter can be used to remove noise if, after taking the Fourier transform of a noisy signal, the signal and noise components appear in separate frequency bands. More importantly, a frequency-selective filter can be used as a **prefilter** to avoid **aliasing**.

Aliasing is a distortion that occurs when a continuous signal that is **bandlimited**, or restricted to frequency components below the **Nyquist frequency**, Ω_N, is then sampled at a frequency less than the **Nyquist rate**, $2\Omega_N$. Ω represents continuous time frequency in radians/sec. The underlying foundation of this result is Shannon's **sampling theorem**. According to this theorem, samples of a continuous bandlimited signal that have been ac-

10 SYSTEM THEORY AND FREQUENCY-SELECTIVE FILTERS

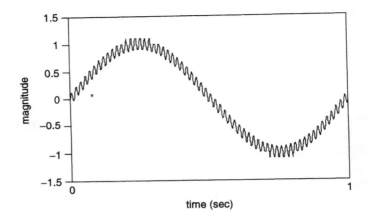

Figure 1.4 A 1 Hz sine wave with 60 Hz interference.

quired frequently enough are sufficient to represent the signal exactly. For example, note the distortion in a $1/\pi$ radian/sec sine wave (Nyquist rate = $2/\pi$ radians/sec) when it is sampled at $4/3\pi$ radians/sample (Figure 1.5). The resulting data are samples of a sine wave of $1/3\pi$ radians/sample.

Typically, a continuous time signal is sampled at a very high sampling rate, far above the Nyquist rate of the signal. The utilized data acquisition system passes the signal through an analog lowpass filter with a high cutoff frequency before analog/digital conversion. The user then prefilters the digital signal using a sharp cutoff frequency filter to avoid aliasing, and **downsamples** the data to the desired sampling rate. The choice of filter depends on phase linearity and efficiency (related to model order) constraints that are discussed in later sections. Downsampling to a rate of (original sampling rate/G) can be accomplished by inputting every Gth sample or averaging every G samples. This process is illustrated in Figure 1.6.

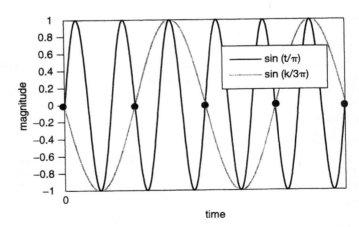

Figure 1.5 Aliasing example. The function $\sin(t/\pi)$ is sampled at frequency $4/3\pi$ radians/sec, resulting in samples from the function $\sin(k/3\pi)$.

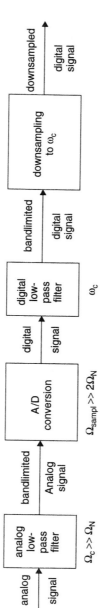

Figure 1.6 Typical data acquisition process.

1.3.2 Filter Specifications

Before a frequency-selective filter can be designed, it must be specified. Specifications are based on the frequency response magnitude and phase.

For the frequency response in Eq. (1.25), the **magnitude**, $|H(e^{j\omega})|$, is

$$|H(e^{j\omega})| = \frac{|b_0|\prod_{n=1}^{M}|1 - f_n e^{-j\omega}|}{\prod_{n=1}^{N}|1 - g_n e^{-j\omega}|} \quad (1.26)$$

It is common practice to transform these products into a corresponding sum of terms by considering $20 \log_{10}|H(e^{j\omega})|$, the **gain in decibels** (dB):

$$\text{gain in dB} = 20 \log_{10}|b_0| + \sum_{n=1}^{M} 20 \log_{10}|1 - f_n e^{-j\omega}| - \sum_{n=1}^{N} 20 \log_{10}|1 - g_n e^{-j\omega}| \quad (1.27)$$

Attenuation of the frequency response occurs when the magnitude is less than 1, or the gain is negative.

From Eq. (1.25), the **phase response**, $\angle H(e^{j\omega})$, is

$$\angle H(e^{j\omega}) = \angle b_0 + \sum_{n=1}^{M} \angle[1 - f_n e^{-j\omega}] - \sum_{n=1}^{N} \angle[1 - g_n e^{-j\omega}] \quad (1.28)$$

The **group delay** is defined as the negative derivative of the phase response:

$$\text{grd}[H(e^{j\omega})] = \frac{-d}{d\omega} \angle H(e^{j\omega}) \quad (1.29)$$

The two main filter specifications are the **passband** and **stopband**, as shown in Figure 1.7. The passband is the region within which the magnitude of the frequency response must approximate unity with an error of δ_1, such that

$$\left(1 - \frac{\delta_1}{2}\right) \leq |H(e^{j\omega})| \leq \left(1 + \frac{\delta_1}{2}\right), |\omega| \leq \omega_p \quad (1.30)$$

where ω_p is the passband frequency. The stopband is the region in which the magnitude response must approximate zero with an error less than δ_2:

$$|H(e^{j\omega})| \leq \delta_2, \omega_s \leq |\omega| \leq \pi \quad (1.31)$$

where ω_s is the stopband frequency. The transition is then just the region between the passband and stopband frequencies.

1.3.3 Filter Types

Once the filter is specified, the designer must choose whether to implement the filter as an **infinite** or **finite impulse response** (IIR or FIR, respectively) **filter.** An IIR filter is described by the ARX model of Eq. (1.19), with the constraint that $M \leq N$. Its name is based

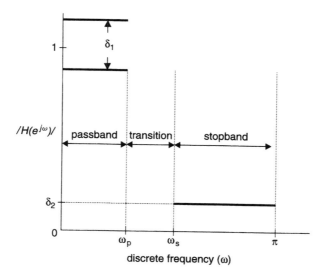

Figure 1.7 Specifications for a digital lowpass filter.

on the recognition that long division of the transfer function numerator by its denominator generates an infinite number of terms. Conversely, an FIR filter is described by an exogenous input, without white noise:

$$y(k) = \sum_{n=0}^{M} b_n u(k-n) \qquad (1.32)$$

As is evident from the equation, the FIR transfer function contains a finite number of terms.

The choice between an FIR or IIR filter depends on the constraints of the design problem. FIR systems can be easily designed to have exactly linear phase or generally linear phase. On the other hand, IIR systems can be described by closed form equations, which lead to a more efficient design.

1.4 DESIGN OF IIR DIGITAL FILTERS FROM ANALOG FILTERS

The widely accepted approach for designing IIR filters involves six steps:

1. Choose the analog filter model.
2. Calculate the model order and cutoff frequency.
3. Find the s-plane pole locations.
4. Factor the continuous time transfer function.
5. Obtain a lowpass filter prototype.
6. Transform from lowpass filter prototype to desired lowpass, highpass, bandpass, or bandstop filter.

These steps are summarized in this section.

1.4.1 Common Analog Filter Models

Two common analog filters are the **Butterworth** and **Chebyshev filters.** The choice of filter depends on the desired frequency characteristics.

The continuous time Butterworth filter is defined by the magnitude-squared function

$$|H_c(j\Omega)|^2 = \frac{1}{1 + \left(\dfrac{j\Omega}{j\Omega_c}\right)^{2N}} \qquad (1.33)$$

where Ω_c is the continuous cutoff frequency, and N is the model order (Figure 1.8a). Butterworth filters are defined by the property that the magnitude response is maximally flat in the passband. As a consequence, the first $(2N-1)$ derivatives of the magnitude-squared function equal zero at $\Omega = 0$ for an Nth order filter. This filter also possesses a magnitude response that is monotonic in the passband and stopband. As N increases, the filter characteristics become sharper (Figure 1.8b).

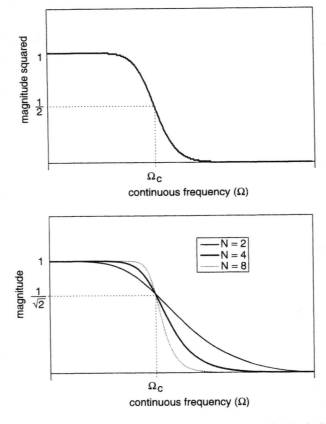

Figure 1.8 Lowpass Butterworth filter. (a) Magnitude-squared response. (b) Magnitude responses for $N = 2$, 4, and 8.

1.4 DESIGN OF IIR DIGITAL FILTERS FROM ANALOG FILTERS

By substituting the **Laplace frequency,** $s = j\Omega$, we obtain the Laplace function

$$H_c(s)H_c(-s) = \frac{1}{1 + \left(\dfrac{s}{j\Omega_c}\right)^{2N}} \tag{1.34}$$

The roots of the denominator polynomial, or poles, are located at

$$s_n = (-1)^{1/2N}(j\Omega_c) = \Omega_c e^{(j\pi/2N)(2n+N-1)}, \qquad n = 0, 1, \ldots, 2N-1 \tag{1.35}$$

Plotting these poles in the s plane reveals that the $2N$ poles are equally spaced in intervals of π/N radians. (Figure 1.9a). To obtain a stable and causal filter, we should choose a pole for each pair of roots in the left half plane of the s-plane.

Similarly, the Chebyshev filter is defined by the magnitude-squared function

$$|H_c(j\Omega)|^2 = \frac{1}{1 + \varepsilon^2 V_N^2\left(\dfrac{\Omega}{\Omega_c}\right)} \tag{1.36}$$

where $V_N(x)$ is the Nth order Chebyshev polynomial and ε is the allowable passband **ripple** (Figure 1.10). The Chebyshev polynomial is defined as

$$V_N(x) = \cos(N \cos^{-1} x) \tag{1.37}$$

Type I Chebyshev filters are defined by the property that the magnitude response is **equiripple** in the passband and monotonic in the stopband. The magnitude-squared function ripples between 1 and $1/(1 + \varepsilon^2)$ for $0 \leq \Omega/\Omega_c \leq 1$ and decreases monotonically for $\Omega/\Omega_c > 1$. The Chebyshev poles lie on an ellipse in the s-plane (Figure 1.9b). Again, to obtain a stable and causal filter, we should choose a pole for each pair of roots in the left half plane of the s-plane.

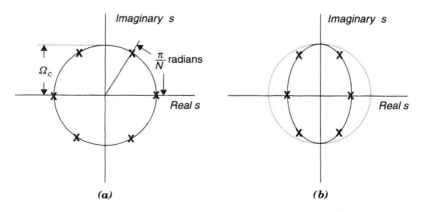

Figure 1.9 Pole Locations for a third-order (a) Butterworth filter, and (b) Type I Chebyshev filter. N = number of pole pairs = model order.

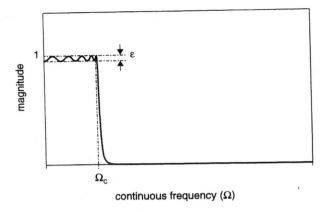

Figure 1.10 Chebyshev lowpass filter magnitude response.

With the Chebyshev filter, the accuracy of the filter approximation is uniformly distributed over the passband, stopband, or both. Compared to the Butterworth filter, this is a more efficient distribution of the accuracy, leading to a relatively lower order Chebyshev filter.

1.4.2 Model Order and Cutoff Frequency Calculations

Once the user has chosen the filter type, the model order and cutoff frequency are calculated from specific frequency specifications for the magnitude function. Since original specifications are given in discrete frequency, they must first be converted to continuous frequency using the relation

$$\Omega = \frac{2}{T_d}\tan\left(\frac{\omega}{2}\right) \quad (1.38)$$

where T_d is the sampling interval. Since the effect of T_d is cancelled during a later design step, we choose $T_d = 1$, which is referred to as a normalized sampling interval. For example, suppose we want to design a discrete time Butterworth filter with magnitude, $|H(e^{j\omega})|$, for which

$$0.9000 \leq |H(e^{j\omega})| \leq 1, \quad 0 \leq |\omega| \leq 0.609 \quad (1.39)$$

$$|H(e^{j\omega})| \leq 0.200, \quad 0.881 \leq |\omega| \leq 2.008 \quad (1.40)$$

Using Eq. (1.38), the continuous specifications are calculated as

$$0.90 \leq |H_c(j\Omega)| \leq 1, \quad 0 \leq |\Omega| \leq 0.2\pi \quad (1.41)$$

$$|H_c(j\Omega)| \leq 0.2, \quad 0.3\pi \leq |\Omega| \leq \pi \quad (1.42)$$

Since the magnitude response of an analog Butterworth filter is a **monotonically** decreasing, or smoothly decreasing, function of frequency, these specifications will be satisfied if

1.4 DESIGN OF IIR DIGITAL FILTERS FROM ANALOG FILTERS

$$|H_c(j0.2\pi)| \geq 0.9 \tag{1.43}$$

$$|H_c(j0.3\pi)| \leq 0.2 \tag{1.44}$$

Using Eqs. (1.43) and (1.44) with equality and substituting into Eq. (1.33) yields

$$1 + \left(\frac{0.2\pi}{\Omega_c}\right)^{2N} = \frac{1}{0.9^2} \tag{1.45}$$

$$1 + \left(\frac{0.3\pi}{\Omega_c}\right)^{2N} = \frac{1}{0.2^2} \tag{1.46}$$

Taking the natural logarithms of both sides of Eqs. (1.45) and (1.46) and solving for $\ln(\Omega_c)$ yields

$$\ln(\Omega_c) = -0.5 \ln\left(\frac{1}{0.9^2} - 1\right) + N \ln 0.2\pi = -0.5 \ln\left(\frac{1}{0.2^2} - 1\right) + N \ln 0.3\pi \tag{1.47}$$

Solving for N results in

$$N = \frac{\ln\left(\frac{1}{0.9^2} - 1\right) - \ln\left(\frac{1}{0.2^2} - 1\right)}{2(\ln 0.2\pi - \ln 0.3\pi)} = 05.707 \tag{1.48}$$

Substituting N into Eq. (1.45) yields $\Omega_c = 0.7134$. Since model order, N, must be an integer, we round N up to the nearest integer $N = 6$, in order to meet or exceed the specification. Substituting $N = 6$ into Eq. (1.45) yields $\Omega_c = 0.7090$. With this value of Ω_c, the passband specification of the continuous filter will be met exactly, and the stopband specification will be exceeded.

1.4.3 Pole Locations and Transfer Function

The calculated values of N and Ω_c are used to find the poles of the magnitude function. Continuing with our Butterworth filter example, we require the substitution

$$e^{j\theta} = \cos\theta + j\sin\theta \tag{1.49}$$

to find the poles. Using Eqs. (1.35) and (1.49), we can calculate the poles in the left-half plane for $N = 6$ and $\Omega_c = 0.7090$ (Table 1.1). Only the left-half plane poles are used to describe the magnitude (not magnitude-*squared*) function.

Once the left-hand poles are known, the Laplace transfer function is easily determined. In our Butterworth filter example, each complex pole pair, $(-\alpha \pm j\beta)$, is used to find the corresponding **quadratic equation**

$$s^2 + 2\alpha s + \alpha^2 + \beta^2 = 0 \tag{1.50}$$

The quadratic equations are factors of the transfer function denominator:

$$H_c(s) = \frac{1}{(s^2 + 0.368s + 0.503)(s^2 + 1.002s + 0.502)(s^2 + 1.370s + 0.503)} \tag{1.51}$$

Table 1.1 Left Hand-plane poles in Butterworth filter example

s_n	Pole
s_1	$-0.184 + j\,0.685$
s_2	$-0.501 + j\,0.501$
s_3	$-0.685 + j\,0.184$
s_4	$-0.685 - j\,0.184$
s_5	$-0.501 - j\,0.501$
s_6	$-0.184 - j\,0.685$

1.4.4 Transformations

To obtain the lowpass filter prototype for our design, the bilinear transformation is used to convert the analog transfer function to its discrete form. The bilinear transform is an algebraic transformation between the variables s and z that maps the entire $j\Omega$ axis in the s-plane to one revolution of the unit circle in the z-plane. The bilinear transformation corresponds to replacing s by

$$s = \frac{2}{T_d}\left(\frac{1-z^{-1}}{1+z^{-1}}\right) \quad (1.52)$$

As before, we use $T_d = 1$, since the sampling interval has no influence in our design. Since our discrete time specifications are mapped to continuous time specifications, and the continuous time filter is mapped back to discrete time, the effect of T_d is cancelled. The discrete cutoff frequency, ω_c, is calculated from

$$\omega_c = 2\arctan\left(\frac{\Omega_c T_d}{2}\right) \quad (1.53)$$

Continuing with our example, substitution of Eq. (1.52) into (1.51) results in the lowpass filter prototype

$$H(z) = \frac{(1+z^{-1})^6}{(5.239 - 6.994z^{-1} + 3.767z^{-2})(6.506 - 6.996z^{-1} + 2.498z^{-2})(7.243 - 6.994z^{-1} + 1.763z^{-2})} \quad (1.54)$$

Obviously, this transfer function could be further factored so that the first term of each quadratic equals 1. Using Eq. (1.53), the discrete cutoff frequency is $\omega_c = 0.6814$ radians/sample.

For efficient implementation, the lowpass prototype is used to construct other filters. If an alternate filter type such as a highpass filter or a lowpass filter with another cutoff frequency is desired, a final frequency transformation is necessary. First, let us redefine $H_{lp}(Z)$, not $H(z)$ [as shown in Eq. (1.54)], as the transfer function of the lowpass filter prototype. Also, let us define ω_p as the lowpass filter prototype cutoff frequency. To obtain the final filter, $H(z)$, we replace every occurrence of Z^{-1} in $H_{lp}(Z)$ with the transformation given in Table 1.2 [2].

1.4 DESIGN OF IIR DIGITAL FILTERS FROM ANALOG FILTERS

Table 1.2 Transformations from a Lowpass digital filter prototype with cutoff frequency ω_p [2]

Filter type	Transformation	Parameter formulas
Lowpass	$Z^{-1} = \dfrac{z^{-1} - \alpha}{1 - \alpha z^{-1}}$	$\alpha = \dfrac{\sin\left(\dfrac{\omega_p - \omega_c}{2}\right)}{\sin\left(\dfrac{\omega_p + \omega_c}{2}\right)}$ ω_c = desired cutoff frequency
Highpass	$Z^{-1} = \dfrac{z^{-1} + \alpha}{1 + \alpha z^{-1}}$	$\alpha = -\dfrac{\cos\left(\dfrac{\omega_p + \omega_c}{2}\right)}{\cos\left(\dfrac{\omega_p - \omega_c}{2}\right)}$ ω_c = desired cutoff frequency
Bandpass	$Z^{-1} = -\dfrac{z^{-2} - \dfrac{2\alpha k}{k+1} z^{-1} + \dfrac{k-1}{k+1}}{\dfrac{k-1}{k+1} z^{-2} - \dfrac{2\alpha k}{k+1} z^{-1} + 1}$	$\alpha = \dfrac{\cos\left(\dfrac{\omega_{c2} + \omega_{c1}}{2}\right)}{\cos\left(\dfrac{\omega_{c2} - \omega_{c1}}{2}\right)}$ $k = \cot\left(\dfrac{\omega_{c2} - \omega_{c1}}{2}\right)\tan\left(\dfrac{\omega_p}{2}\right)$ ω_{c1} = desired lower cutoff frequency ω_{c2} = desired upper cutoff frequency
Bandstop	$Z^{-1} = \dfrac{z^{-2} - \dfrac{2\alpha k}{1+k} z^{-1} + \dfrac{1-k}{1+k}}{\dfrac{1-k}{1+k} z^{-2} - \dfrac{2\alpha k}{1+k} z^{-1} + 1}$	$\alpha = \dfrac{\cos\left(\dfrac{\omega_{c2} + \omega_{c1}}{2}\right)}{\cos\left(\dfrac{\omega_{c2} - \omega_{c1}}{2}\right)}$ $k = \tan\left(\dfrac{\omega_{c2} - \omega_{c1}}{2}\right)\tan\left(\dfrac{\omega_p}{2}\right)$ ω_{c1} = desired lower cutoff frequency ω_{c2} = desired upper cutoff frequency

To finish our Butterworth example, suppose we wish to transform our lowpass filter prototype with prototype cutoff frequency $\omega_p = 0.6814$ to a highpass filter with cutoff frequency $\omega_c = 2$. From Table 1.2, we calculate that $\alpha = -0.2886$. Using the highpass transformation in Table 1.2 and Eq. (1.54), the highpass filter transfer function is

$$H(z) = \frac{0.006 - 0.039z^{-1} + 0.098z^{-2} - 0.130z^{-3} + 0.098z^{-4} - 0.039z^{-5} + 0.006z^{-6}}{1 + 1.623z^{-1} + 1.765z^{-2} - 1.058z^{-3} + 0.415z^{-4} + 0.089z^{-5} + 0.009z^{-6}} \quad (1.55)$$

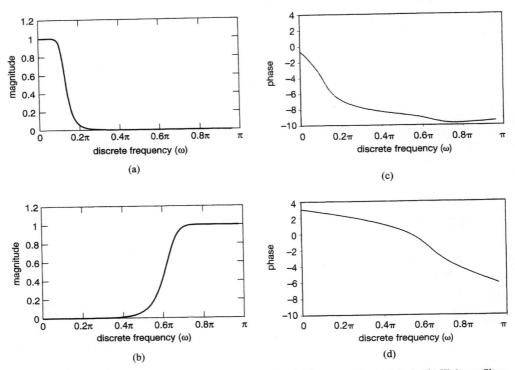

Figure 1.11 Magnitude and phase plots for filter examples. (a) Lowpass filter magnitude. (b) Highpass filter magnitude. (c) Lowpass filter phase. (d) Highpass filter phase.

The magnitude and phase plots for the lowpass filter prototype and highpass filter are given in Figure 1.11.

1.4.5 Other IIR Filter Methods

Other methods besides this impulse-invariant approach may be used to design IIR filters. Rather than defining an analog filter impulse response, an analog filter step response may be defined and used as the basis of a filter design. Moreover, equiripple IIR filters may be designed using **linear programming** (an optimization methodology), or by iteratively using the **Remez exchange algorithm** on the numerator and denominator of the transfer function [3]. These methods are beyond the range of our discussion.

1.5 DESIGN OF FIR FILTERS BY WINDOWING

In contrast to IIR filter design, which relies on a continuous to discrete time transformation, FIR filter design is usually restricted to implementation in discrete time. The simplest method of FIR filter design is the **window method,** which involves truncating, or "windowing," an ideal frequency response. Only basic windowing concepts are presented in this section, since most medical instrumentation applications employ IIR, rather than FIR, filters. IIR filters are typically chosen because of their closed form and lower implementation order.

1.5.1 Windowing

Given a desired impulse response, $h_d(k)$, we can define a new system with impulse response, $h(k)$, such that

$$h(k) = \begin{cases} h_d(k), & 0 \leq k \leq M \\ 0, & \text{otherwise} \end{cases} \qquad (1.56)$$

Note that M is the order of the system polynomial in Eq. (1.32). As shown in Eq. (1.56), $(M+1)$ is the length or duration of the impulse response.

Eq. (1.56) can be represented as the product of the desired impulse response and a finite duration window, $w(k)$:

$$h(k) = h_d(k)w(k) \qquad (1.57)$$

where $w(k)$ is often described by one of the equations in Table 1.3. The simplest window is the **rectangular window**, which involves truncation at sample M:

$$w(k) = \begin{cases} 1, & 0 \leq k \leq M \\ 0, & \text{otherwise} \end{cases} \qquad (1.58)$$

Table 1.3 Common windows

Window type	Shape	Equation
Rectangular		$w(k) = \begin{cases} 1, & 0 \leq k \leq M \\ 0, & \text{otherwise} \end{cases}$
Bartlett (triangular)		$w(k) = \begin{cases} 2k/M, & 0 \leq k \leq M \\ 2 - 2k/M, & M/2 \leq k \leq M \\ 0, & \text{otherwise} \end{cases}$
Hanning		$w(k) = \begin{cases} 0.5 - 0.5\cos(2\pi k/M), & 0 \leq k \leq M \\ 0, & \text{otherwise} \end{cases}$
Hamming		$w(k) = \begin{cases} 0.54 - 0.46\cos(2\pi k/M), & 0 \leq k \leq M \\ 0, & \text{otherwise} \end{cases}$
Blackman		$w(k) = \begin{cases} 0.42 - 0.5\cos(2\pi k/M) + 0.08\cos(4\pi k/M), & 0 \leq k \leq M \\ 0, & \text{otherwise} \end{cases}$

The rectangular, Hanning, Hamming, and Blackman windows are all special cases of the generalized cosine window. Each of these windows is composed of combinations of sinusoidal sequences with frequencies 0, $2\pi/(M-1)$, and $4\pi/(M-1)$. By summing the individual terms to form the window, the low-frequency peaks in the frequency domain combine in such a way as to minimize undesired frequency characteristics.

Assuming that the Fourier transforms of $h(k)$ and $w(k)$ exist, the **Windowing theorem** may be used to calculate the transform function of Eq. (1.57) as

$$H(e^{j\omega}) = \frac{1}{2\pi} \int_{-\pi}^{\pi} H_d(e^{j\theta}) W(e^{j(\omega-\theta)}) d\theta \qquad (1.59)$$

Looking at Eq. (1.59), $H(e^{j\omega})$ is the period convolution of the desired ideal frequency response with the Fourier transform of the window. Thus, this frequency response is a "smeared" version of the desired response $H_d(e^{j\omega})$. For a **unit step function** [$w(k) = 1$ for all k], $W(e^{j\omega})$ is a period impulse train with period 2π, resulting in $H(e^{j\omega}) = H_d(e^{j\omega})$ to mimic $H_d(e^{j\omega})$. $H(e^{j\omega})$ will only deviate from $H_d(e^{j\omega})$ where $H_d(e^{j\omega})$ changes abruptly.

All the windows in Table 1.3 have been defined with **generalized linear phase**. Linear phase results from **even symmetry** [$f(k) = f(-k)$] about the sample $M/2$:

$$w(k) = \begin{cases} w(M-k), & 0 \le k \le M \\ 0, & \text{otherwise} \end{cases} \qquad (1.60)$$

1.5.2 Filter Forms

The desired impulse responses for a lowpass, highpass, and bandpass filter are given below:

$$h_{\text{lowpass}}(k) = \begin{cases} \dfrac{\sin[\omega_c(k-M/2)]}{\pi(k-M/2)}, & -\infty < k < \infty \\ \dfrac{\omega_c}{\pi}, & k = M/2 \end{cases} \qquad (1.61)$$

$$h_{\text{highpass}}(k) = \begin{cases} \dfrac{\sin[\pi(k-M/2)] - \sin[\omega_c(k-M/2)]}{\pi(k-M/2)}, & -\infty < k < \infty \\ 1 - \dfrac{\omega_c}{\pi}, & k = M/2 \end{cases} \qquad (1.62)$$

$$h_{\text{bandpass}}(k) = \begin{cases} \dfrac{\sin[\omega_{c2}(k-M/2)] - \sin[\omega_{c1}(k-M/2)]}{\pi(k-M/2)}, & -\infty < k < \infty \\ \dfrac{1}{\pi}(\omega_{c2} - \omega_{c1}), & k = M/2 \end{cases} \qquad (1.63)$$

In designing an FIR filter, a delay, d, must be included in the desired impulse response to obtain a causal filter. In each desired impulse response, the delay is set to $M/2$ to guarantee symmetry and therefore linear phase.

1.5 DESIGN OF FIR FILTERS BY WINDOWING

Design of an FIR filter by windowing involves three steps:

1. Specify the desired order, M, and cutoff frequency, ω_c (or frequencies, ω_{c1} and ω_{c2}, for a bandpass filter).
2. Select a window in Table 1.3 and multiply it by the desired filter form.
3. Convolve the resulting filter function with the input to obtain a filtered output.

For example, suppose we want to design a 4th order Hamming lowpass filter with $\omega_c = \pi/4$. Using the formula in Table 1.3, the Hamming coefficients are $w(0) = w(4) = 0.08$, $w(1) = w(3) = 0.54$, and $w(2) = 1$ (Figure 1.12a). We multiply this window by the ideal impulse response of Eq. (1.61) to obtain the impulse response. The calculated lowpass coefficients are $h_{lp}(0) = h_{lp}(4) = 0.159155$, $h_{lp}(1) = h_{lp}(3) = 0.225079$, and $h_{lp}(2) = 0.25$. The calculated impulse response is composed of $h(0) = h(4) = 0.012732$, $h(1) = h(3) = 0.121543$, and $h(2) = 0.25$. The impulse response can be normalized by dividing each coefficient by the sum of all five coefficients, such that the center of the passband possesses a magnitude of exactly 1. With $\Sigma_{k=0}^{4} h(k) = 0.51855$, the scaled impulse response becomes $h_s(0) = h_s(4) = 0.0246$, $h_s(1) = h_s(3) = 0.2344$, and $h_s(2) = 0.4821$ (Figure 1.12b). Taking the Fourier transform of this impulse response results in the magnitude and phase response of Figures 1.12c and d, respectively.

1.5.3 Other FIR Filter Methods

Other methods besides windowing may be used to design FIR filters. For example, Parks and McClellan reformulated the filter design problem as a problem in polynomial approx-

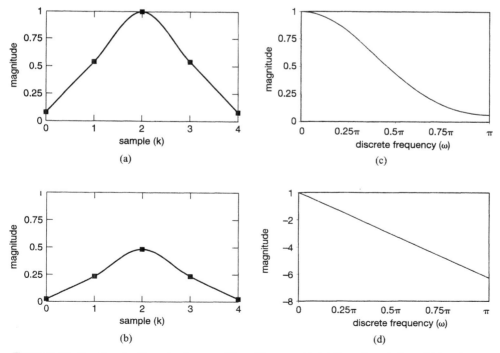

Figure 1.12 Fourth-order Hamming lowpass filter with $\omega_c = \pi/4$. (a) Hamming window, (b) impulse response, (c) magnitude response, (d) phase response.

24 SYSTEM THEORY AND FREQUENCY-SELECTIVE FILTERS

imation [4]. Further, FIR filters may be designed using the Remez exchange algorithm [3]. Both methods are beyond the range of our discussion.

1.6 PSEUDORANDOM BINARY SEQUENCE FILTER

Up to this point, the filters we have designed have passed a specified frequency range that contains the signal of interest. If the majority of system noise resides outside this range, then the system **signal-to-noise ratio** (SNR) is increased. Another method for preserving the signal of interest involves filtering a periodic signal through a **pseudorandom binary sequence** (PRBS) filter. A pseudorandom binary sequence is composed of two values that appear to be introduced randomly but are reproducible by deterministic means. Due to correlation properties of this filter, the periodic signal is amplified, whereas the noise amplitude remains constant. With this alternate method, the signal-to-noise ratio is also increased. In essence, the PRBS filter functions as a bandpass filter that does not require frequency specifications.

The PRBS filter is one of many **spread-spectrum** techniques originally developed for military communications but now finding applications in the commercial arena, particularly in cellular telephony. Spread spectrum refers to a **modulation** technique that spreads the spectrum of a signal by using a very wideband spreading signal. The spreading signal is chosen to have properties that facilitate **demodulation** of the transmitted signal by the intended receiver, and which make demodulation by an unintended receiver as difficult as possible.

1.6.1 PRBS Properties

The PRB sequence consists of a predetermined binary sequence of ones and zeros of **maximal length,** $M = 2^N - 1$, where N is the filter order. These maximal length sequences, or **M-sequences,** possess several useful properties:

1. An M-sequence contains one more one than zero. The number of ones in the sequence is $\frac{1}{2}(N+1)$.
2. The **modulo-2,** or binary, sum of an M-sequence and any phase shift of the same sequence is another phase of the same M-sequence.
3. The **periodic autocorrelation function,** $\phi_{mm}(k)$, has two values and is given by

$$\phi_{mm}(k) = \begin{cases} 1.0, & k = iM \\ \dfrac{-1}{M}, & \text{otherwise} \end{cases} \quad (1.64)$$

where i is any integer and M is the sequence period [see Eq. (1.12)].

The M-sequences for $N \leq 5$ are listed in Table 1.4. Alternative sequence representations include octal format and the **primitive polynomial,** $g(D)$. For primitive polynomial representation, the bit number of the original sequence is translated into a power of D. Therefore, 1110100, or 164_o, can also be represented as $g(D) = D^2 + D^4 + D^5 + D^6$.

1.6 PSEUDORANDOM BINARY SEQUENCE FILTER

Table 1.4 *M*-sequences

N	M	M-sequence
2	3	110
3	7	1110100
4	15	111101011001000
5	31	1111100110100100001010111011000

1.6.2 Filter Construction

The PRBS filter is based on the *M*-sequence, and consists of two parts: the **encoder** and **decoder**. The encoder, $e(k)$, is the original *M*-sequence. The decoder, $d(k)$, is a similar code in which each zero has been replaced by minus one. Using Eq. (1.13), cross-correlation of the encoder and decoder gives an unusual result of

$$\phi_{ed}(k) = \begin{cases} \dfrac{M+1}{2}, & k=0 \\ 0, & \text{otherwise} \end{cases} \tag{1.65}$$

This cross-correlation can be used to filter a periodic signal. Each incoming period, τ, of the input signal is encoded, that is, multiplied by the next value of the encoder sequence. The encoded signal is then passed through its transmission medium, which may add system noise. The received signal is decoded by periodic convolution with the decoder sequence, such that

$$\text{decoder output}(k) = \sum_{i=0}^{M-1} d(i) \cdot \text{encoder output}(i\tau + k) \tag{1.66}$$

The PRBS filtering process is illustrated in Figure 1.13.

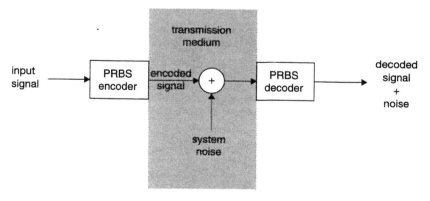

Figure 1.13 PRBS filter process.

1.6.3 Advantages

Why would an engineer bother using such a complicated filter strategy? In practice, systems are often **perturbed** with a signal as a means of determining the system impulse response. Unfortunately, system noise may be acquired during this perturbation. Recall that our received signal is amplified $(M + 1)/2$ times. Even if system noise is received, it is not amplified. Therefore, the PRBS filter increases the received signal-to-noise ratio. The cost of this filter, however, is that the output can only be calculated after M periods have occurred.

Another advantage of this filter is the reduction of signal **stackup.** Stackup occurs when the **time decay** of a perturbed signal response back to its original value is longer than its period (Figure 1.14). In other words, the onset of a periodic signal, which results from a periodic perturbation, occurs faster than its time decay. The PRBS filter eliminates stackup because only one perturbation response is preserved.

As an example, let us calculate the filter response of a periodic sequence {5 3 2 0} ($\tau = 4$) to a second-order filter, with $M = 3$. From Table 1.4, the encoder sequence is $e(k) = \{1\ 1\ 0\}$; the corresponding decoder sequence is $d(k) = \{1\ 1\ -1\}$. Multiplying each period by the encoder sequence results in the encoder output of {5 3 2 0 5 3 2 0 0 0 0 0 5 3 2 0 5 3 2 0 0 0 0 0}. Using Eq. (1.66), convolving this sequence with the decoder output results in the decoder output {10 6 4 0 0 0 0 0 0 0 0 0}.

Now let us add uniform noise in the range {0 1} to the encoder output of this sequence in a third-order filter, with $M = 7$ (Figure 1.15). This noise significantly degrades the encoder output sequence. However, if we decode the output and divide the result by $(M + 1)/2$, the final result is now closer to the original sequence for the first period. Although the normalized decoded sequence still does not equal the original, the noise offset is now more uniform.

1.7 SUMMARY

In this introductory chapter, we have reviewed basic concepts of the input–output description. Within the constraint of a linear time-invariant system, we have considered deter-

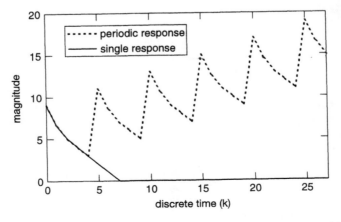

Figure 1.14 Stackup phenomenon occurs when time decay of original response to perturbation is greater than response period.

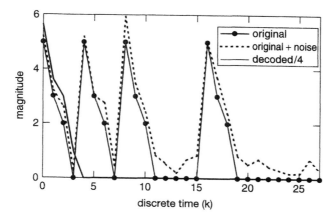

Figure 1.15 PRBS filter results for $M = 7$.

ministic and random inputs and representations of the transfer function in the Fourier and z-domains. We have also discussed methods for filtering signals in a specified frequency range. Each filter form is named for the frequency range that is passed: lowpass filter, highpass filter, bandpass filter, and bandstop filter. The typical design of an infinite impulse response filter involves conversion of an analog filter to its discrete counterpart. The typical design of a finite impulse response filter involves windowing, or truncation, of an ideal frequency response. We have also introduced the concept of a pseudorandom binary sequency filter, based on signal encoding and decoding using a maximal length sequence. In effect, the PRBS filter functions as a bandpass filter without the need for frequency specifications.

If signal noise is generally limited to a certain frequency range, then IIR filters can be used to increase the signal-to-noise ratio. IIR filters are typically chosen over FIR filters because of their closed form and higher efficiency. If signal noise is contained within the signal frequency range, a PRBS filter can be used to minimize the effect of this noise.

1.8 REFERENCES

[1]. Oppenheim, A. V. and Schafer, R. W. *Discrete-Time Signal Processing*. Prentice Hall: Englewood Cliffs, NJ, 1989.

[2]. Constantinides, A. G. Spectral transformations for digital filters. *Proc IEE, 117,* 1585–1590, 1970.

[3]. Parks, T. W. and Burrus, C. S. *Digital Filter Design*. Wiley: New York, 1987.

[4]. Parks, T. W. and McClellan, J. H. A program for the design of linear phase finite impulse response filters. *IEEE Trans Audio Electroacoustics, AU-20,* 195–199, 1972.

Further Reading

Linear System Theory and Frequency-selective Filters: Ref. [1], Chapters 2, 5, and 6.

IIR and FIR Filters: Ref. [1], Chapter 7.

Pseudorandom Binary Sequence Filter: Ziemer, R. E. and Peterson, R. L. *Digital Communications and Spread Spectrum Systems*. Macmillan: New York, 1985, Chapter 8.

Stochastic Processes: Papoulis, A. *Probability, Random Variables, and Stochasic Processes*, 3rd ed., McGraw-Hill: New York, 1991.

1.9 RECOMMENDED EXERCISES

IIR Filters: see Ref. [1], Problems 7.3, 7.5, 7.6, 7.7b, 7.8b, 7.10b, 7.12, 7.14, 7.17, 7.18, and 7.20.

FIR Filters: see Akay, M. *Biomedical Signal Processing*. Academic Press: San Diego, 1994. Exercises in Section 3: 4, 5, 6, and 7 (excluding Kaiser window).

PRBS Filter: see Ref. [3], Exercise 8-20, Problems 8-11, 8-12, and 8-20.

2

LOW FLOW RATE OCCLUSION DETECTION USING RESISTANCE MONITORING

In this chapter, we discuss an industrial application of the PRBS filter: low flow rate occlusion detection using resistance monitoring. During **resistance** monitoring, **pressure** waveforms are processed to obtain estimates of the resistance to flow. Unfortunately, **noise artifact** in the pressure data may bias these estimates. Noise artifact may be induced by the patient, clinician, or medical instrumentation. The PRBS filter increases the signal-to-noise ratio sufficiently to minimize the effect of noise artifact during resistance calculations.

2.1 PHYSIOLOGY OF INTRAVENOUS DRUG ADMINISTRATION

A **drug,** or therapeutic agent, may be administered by many routes. **Oral ingestion** is the most common because it is the safest, most convenient, and most economic method of drug administration. However, the **parenteral administration,** or administration by methods other than oral ingestion, of drugs has certain advantages over oral administration. Parenteral administration may be essential for a drug to be absorbed in active form. **Bioavailability,** the extent to which a drug reaches its site of action, is usually more rapid and more predictable when a drug is given parenterally than by mouth. Therefore, the effective dose can be more accurately selected. In emergency situations, parenteral administration is efficient and may be a necessity.

The major routes of parenteral administration are **intravenous** (IV, within a vein), **subcutaneous** (implanted under the skin), and **intramuscular** (within a muscle). **Absorption** is the rate at which a drug leaves its site of administration. Absorption from subcutaneous and intramuscular sites occurs by simple **diffusion** along the gradient from drug **depot,** or site of administration, to **plasma,** the fluid portion of blood and **lymph** in which cells are suspended.

Drugs administered intravenously are subject to possible first-pass **elimination** in the lung prior to **distribution** to the rest of the body. The lungs serve as a temporary clearing site for a number of agents, as a filter for particulate matter that may be given intravenously, and as a route of elimination for volatile substances. The desired concentration of a drug in blood is obtained with an accuracy and immediacy not possible by any other procedure. This occurs because of the rapid speed by which the entire blood supply is circulated (within 60 seconds in a healthy adult).

Intravenous administration can be subdivided into **injections,** or transient administrations, and **infusions,** or sustained administrations using an infusion device. An injection is modeled as an impulse; an ideal infusion is modeled as a unit step function. Intravenous administration typically occurs in the **peripheral** and **central** veins. In addition to drugs, fluids, electrolytes, nutrition, and blood are administered intravenously. Before an infusion occurs, a **catheter,** or needle, must be inserted and connected to an infusion device.

2.1.1 Catheter Insertion and Connection to an Infusion Device

During **venipuncture,** or puncture of the vein, a nurse selects the area for catheter insertion. The vein is distended using a tourniquet a short distance above the intended puncture site. A cleansing agent such as iodine is externally administered to provide an antibacterial barrier around the penetration site. The skin over the vein is initially pierced with the catheter at about a 30 degree angle; the angle is immediately lowered to avoid puncturing the back wall of the vein. Once the catheter is in the vessel, it is carefully advanced further into the vein; the IV **administration set,** or tubing, from an infusion device is then attached. After the infusion begins successfully, the catheter and tubing are secured to the skin using tape.

2.1.2 Fluid Flow

Flow through the vein can be can be roughly approximated using **Hagen–Poiseuille's equation** for **laminar,** or nonturbulent, flow of a **Newtonian** fluid through a rigid tube:

$$Q = \frac{\pi r^4 (P_1 - P_2)}{8 \eta L} \tag{2.1}$$

where Q is the flow; P_1 and P_2 are the pressure at the tube inlet and outlet, respectively; r is the tube radius; L is the tube length; and η is the **fluid viscosity.** Newtonian refers to a fluid whose stress–strain relationship is linear, following Newton's law. The viscosity of water at 20°C is 0.001 Pa-s; the viscosity of blood at all temperatures is approximately 0.0004 Pa-s. Rearranging Eq. (2.1) results in a fluid analog of the classic **Ohm's law:**

$$(P_1 - P_2) = Q \frac{8 \eta L}{\pi r^4} \tag{2.2}$$

$$(P_1 - P_2) = Q \cdot R \tag{2.3}$$

where R is the fluid resistance. The conventional units for pressure and flow are mmHg and ml/hr, respectively. Therefore, a derived unit for resistance is mmHg-hr/ml, where 1

Table 2.1 Resistance values for catheter components (sample size = 10). Reprinted from [1] with kind permission from CRC Press, Boca Raton, Florida.

Component	Length (cm)	Fluid resistance (mmHg-hr/ml)
Standard Administration Set	91–213	0.0043–0.0053
Extension tube for central venous pressure monitoring	15	0.0155
19-gauge epidural catheter	91	0.2904–0.4971
18-gauge needle	6–9	0.0141–0.0179
23-gauge needle	2.5–9	0.1652–0.3440
25-gauge needle	1.5–4.0	0.5251–1.4120
Vicra-Quick-Cath Catheter, 18 gauge	5	0.0129
Extension set with 0.22 micron air-eliminating filter		0.6230
0.2 micron filter		0.5550

mmHg-hr/ml equals 4.8×10^{-14} Pa-s/m^3. Resistance values for several catheter components using distilled water and flow rates of 100, 200, and 300 ml/hr are given in Table 2.1.

2.1.3 Potential Complications

Frequent monitoring of the IV site and clear, concise documentation assist the health care team in delivering safe and effective therapy. However, even with the utmost care, complications in peripheral IV therapy still occur. Flow obstruction complications include **upstream** (above the infusion device) and **downstream** (below the infusion device) **occlusions** and **positional catheters.** An upstream occlusion is a blockage that occurs between the infusion device and the administration fluid; a downstream occlusion occurs between the infusion device and patient. The occlusion is often caused by a forgotten clamp on the IV tubing, accidental kinking of the tubing, or a clot in the catheter. A positional catheter occurs when a catheter tip wedges against the internal lining of the vein wall, restricting fluid flow. The positional state is caused by patient movement, such as wrist flexation.

More serious complications occur at the venipucture site. **Infiltration** and **extravasation** occur when the IV fluid is infused into the **extravascular tissue,** or tissue surrounding the blood vessel. Infiltration involves the infusion of an IV solution or medication; extravasation involves the infusion of a **vesicant,** an agent capable of causing a blister or tissue destruction. Depending on the potential for the solution to irritate the vein and the length of infusion, an infiltration or extravasation may cause severe tissue damage. **Phlebitis** occurs when the vein becomes inflamed from the infusion fluid.

Other complications include **hematoma, catheter embolism,** and **local infection.** Hematoma is the leaking of blood into the surrounding tissue, and may occur during IV insertion if the catheter pierces the back of the vein. A catheter embolism occurs when a piece of the IV catheter breaks off and floats freely in the vessel, causing a decrease in blood pressure and pain along the vein. A local infection is contamination, usually bacterial, at the IV site.

Infrequent, systemic complications include sepsis, an infection that has entered the patient's circulation; air embolism, the entry of air into the patient's circulatory system; and circulatory overload, the infusion of fluids at a rate greater than the patient's system can accommodate. Other system complications include speed shock—a reaction to a substance, unfamiliar to the body, that is rapidly infused into the circulation—and an allergic reaction.

2.2 INTRAVENOUS INFUSION DEVICES

An intravenous delivery system generally consists of three major components: the fluid or drug reservoir, the catheter system for transferring the fluid or drug from the reservoir into the vasculature through a venipuncture, and a device for regulating and/or generating flow. The catheter system typically includes the catheter and administration set. The administration set minimally consists of a spike, a hard plastic tube with a sharp point for penetrating the reservoir; a drip chamber, a plastic tube for visually verifying flow; tubing; and a roller clamp for regulating fluid flow (Figure 2.1). For some infusion pumps, a special tubing segment in the administration set interfaces with the infusion device pumping mechanism. An optional filter may remove particulate matter suspended in the infusion solution; optional extension tubing is added when extra tubing length is required. Infusion devices include gravity flow regulation, controllers, syringe pumps, and volumetric infusion pumps.

2.2.1 MECHANICAL FLOW REGULATORS

The first infusion devices were merely fluid reservoirs positioned above the patient, with a mechanical clamp around the attached tubing. Mechanical flow regulators still comprise the largest segment of IV infusion systems, and are commonly used to administer fluids and electrolytes. They are most useful when the patient is not fluid restricted and the acceptable therapeutic rate range of the drug is relatively wide, with minimal risk of serious adverse reactions.

Flow in these systems is driven by gravity and is controlled by the roller clamp, which provides a controlled resistance. By placing the fluid reservoir 60–100 cm above the patient's **right atrium,** a hydrostatic pressure gradient, P_h, equal to 1.34 mmHg per cm of elevation is provided. The pressure gradient is minimally reduced by the physiologic mean pressure in the veins, P_v. From Eq. (2.3), flow is then calculated as

$$Q = \frac{P_h - P_v}{R_{mfr} + R_n} \quad (2.4)$$

where R_{mfr} is the resistance to flow through the mechanical flow regulator and R_n is the resistance in the remainder of the delivery system. The clinician adjusts the roller clamp

Figure 2.1 Baxter 1C8109 Administration Set with slide and roller clamps. Courtesy of Baxter Healthcare Corporation, Round Lake, Illinois.

while estimating flow rate by counting the frequency of drops falling through the drip chamber.

This simple flow regulator has many disadvantages. First, it can not be used for arterial infusions since the higher vascular pressure (> 100 mmHg) exceeds available hydrostatic pressure. Second, the compliance, or flexibility, of the tubing may cause large fluctuations in flow rate (> 10%). Further, administration sets with preset flow rates (i.e., 10–20 drops/ml) are only approximations. These preset flow rates are corrupted by drip chamber tolerances and differences in fluid specific gravity and surface tension.

2.2.2 Controllers

Drug infusion accuracy was improved in the early 1970s by the invention of the infusion controller [2]. This device controlled the rate of a gravity-fed infusion by counting drops. A drop sensor was placed around the drip chamber, and provided feedback for automatic rate adjustment. Though flow rate accuracy was improved, the accuracy was still limited by the rate and viscosity dependence of the drop size. Rate accuracy was further corrupted by administration set motion associated with movement and improper positioning of the drip chamber. As with the mechanical flow regulator, maximum flow rate was also limited by available hydrostatic pressure.

2.2.3 Volumetric Infusion Pumps

The innovation in the late 1970s of a **volumetric infusion pump** overcame the hydrostatic pressure limitation of the infusion controller [3]. Rather than relying on hydrostatic pressure to produce flow, a volumetric infusion pump displaces constant flow. In the first volumetric infusion pump, reservoir fluid was displaced from a small, disposable **cassette** using a single **piston** at a predetermined flow rate to the patient. Fluid flow from early cassette pumps was not uniform, as fluid flow was discontinued as the cassette was refilled (Figure 2.2b). In contrast, current pumps generate pressure with high flow rate accuracy and uniformity (Figure 2.2c) for a wide range of infusion rates (0.1 to 1000.0 ml/hr) using **linear peristaltic** mechanisms (Figure 2.3). In these mechanisms, a portion of the administration set tubing is positioned in a linear channel against a rigid backing plate. An array of **cam-driven actuators** sequentially occlude the tubing starting with the section nearest the reservoir, forcing fluid toward the patient with a sinusoidal wave, or peristaltic, action.

2.2.4 Syringe Pumps

Mechanical flow regulators, controllers, and volumetric infusion pumps are capable of large patient volume delivery, in excess of 100 ml. When a small volume is desired, **syringe pumps** are used (Figure 2.4). In the United States, syringe pumps are used primarily in **neonatal**, or premature infant, hospital units, where the patients are fluid restricted. In Europe, hospitals prefer to administer small fluid volumes to the majority of their patients.

For this pump, the syringe acts as both reservoir and volumetric pumping chamber. A precision **leadscrew** in the pump produces constant linear advancement of the syringe **plunger.** Typically, pumps accept syringes ranging in size from 5–100 ml. Flow rate ac-

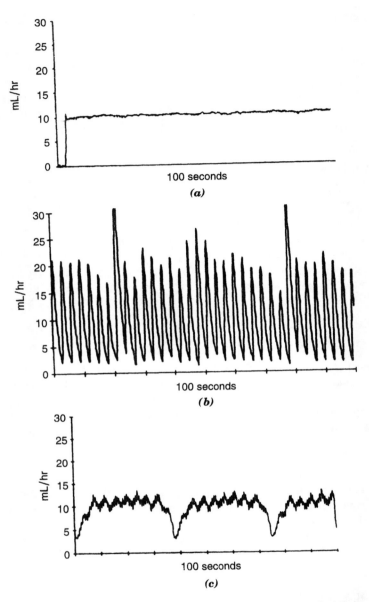

Figure 2.2 Continuous flow pattern for representative (a) syringe, (b) cassette, and (c) linear peristaltic pump at 10 ml/hr. Reprinted from [1] with kind permission from CRC Press, Boca Raton, Florida.

curacy and uniformity are determined by the pumps' displacement characteristics and syringe characteristics such as tolerance on the internal syringe diameter. When the leadscrew is correctly positioned against the plunger, the resulting flow may be quite uniform (Figure 2.2a).

For a clinician, the choice of infusion delivery system is based on desired fluid volume, flow uniformity, and flow accuracy.

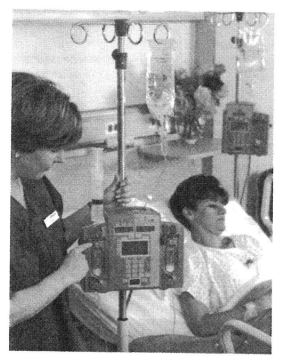

Figure 2.3 The ALARIS Medical Systems Signature Edition® Infusion pump currently employs technologies that have evolved from the research described in Section 2.4. Courtesy of ALARIS Medical Systems, San Diego, CA.

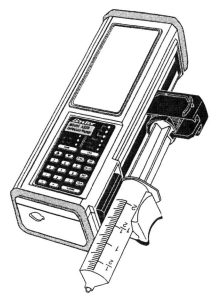

Figure 2.4 Baxter Auto Syringe AS50 Infusion Pump. Courtesy of Baxter Healthcare Corporation, Round Lake, Illinois.

2.3 PROBLEM SIGNIFICANCE

80% of hospitalized patients in the United States receive IV therapy annually [1]. A subset of these patients is the infants and small children who require infusion flow rates under 1.1 ml/hr. In this section, the significance and difficulty of detecting downstream occlusions at these low flow rates is discussed. A solution for faster downstream occlusion detection, implemented with a PRBS filter, is described.

2.3.1 Clinical Significance

Flow obstruction, or occlusion, can have significant consequences. The effect of a drug with a short **half-life**, or time for the amount of drug in the body to decrease to 50%, may be compromised if flow to the patient is occluded. When an upstream occlusion occurs, no fluid is pumped into the administration set. However, until a downstream occlusion is detected, an infusion device continues to infuse, storing fluid in the administration set. When the downstream occlusion is eliminated, the stored volume is delivered to the patient in a **bolus.** With concentrated drugs, this bolus can produce a large perturbation in the patient's status.

Neonates and small children are particularly sensitive to undetected downstream occlusions. Many of the patients in this population are receiving **vasoactive** drugs such as **dopamine,** which enhance heart muscle contractility, thus increasing blood pressure. Such drugs have short distribution half-lives on the order of 2 minutes [4]. Typically, drug infusions for this population are administered at low flow rates less than 1.1 ml/hr. Since traditional downstream occlusion detection times are inversely proportional to flow rate, the time to detection is prolonged. Moreover, these patients are obviously fluid-restricted. For example, a 23 week gestation neonate may weigh less than 1000 grams [5]. If not detected, the prolonged time until downstream occlusion detection may result in a significant fluid bolus to the patient.

2.3.2 Traditional Pressure-Based Downstream Occlusion Detection

When complete occlusion occurs between the infusion pump and patient, the resistance to flow approaches infinity. From Eq. (2.3), the generated pressure also rises over time. By this principle, the traditional method of downstream occlusion detection utilizes a pressure transducer positioned immediately below the pumping mechanism. Based on a predetermined pressure threshold, P_{alarm}, often set by the clinician, an alarm is generated when the pressure exceeds this threshold. Until the threshold is reached, the pump continues to propel fluid into the section of tubing between the pump and occlusion. The time by which the pressure rises in this tubing section, or **time to alarm** (TTA), is proportional to the tubing compliance, C_{tubing}, and inversely proportional to the flow rate:

$$TTA = \frac{P_{\text{alarm}} \cdot C_{\text{tubing}}}{\text{flow rate}} \qquad (2.5)$$

Compliance is the volume increase in a closed tube per applied pressure. Using a fixed alarm threshold of 500 mmHg, a flow rate of 1.0 ml/hr, and a representative tube compliance of 1 μl/mmHg, the time to alarm is 30 minutes. Under these conditions, the IVAC 560 and imed Gemini PC infusion pumps alarm in 12 minutes using their respective administration sets, which each possess a compliance of 0.4 μl/mmHg [6]. Although the pressure threshold in both these pumps may be lowered to decrease the time to alarm, this may result in an unacceptable increase in false alarms due to transient pressure increase during patient movement.

2.3.3 Market Forces

Occlusion detection has been used as part of a market strategy to sell infusion pumps. In 1997 alone, hospital sales of volumetric infusion pumps and their associated disposables (administration sets and options) generated $162,276,192 and $342,962,311, respectively, in the United States. These sales estimates exclude sales to federal hospitals and nursing homes. During this timeframe, the dominant manufacturer of infusion disposables was Baxter Healthcare [7]. In the current hospital maintenance organization (HMO) climate of cost containment, Baxter has been able to bundle its sales of disposables and fluid reservoirs, thus offering hospitals significant savings. Historically, the competing marketing strategy of IVAC Corporation, now ALARIS Medical Systems, has been to offer technically superior infusion pumps, since it manufactures infusion disposables but not fluid reservoirs. The IVAC Signature Edition® infusion pump is the first pump that uses resistance, rather than pressure, for faster detection of downstream occlusions [8]. As described above, pressure increases gradually over time during a complete downstream occlusion. In contrast, resistance increases almost immediately.

2.3.4 Past Resistance Studies

In 1996, James Philip and his group published research conducted in 1987 describing a method for resistance calculation [9]. In 118 adult patients scheduled for elective surgery, Philip used a modified IVAC 560 pump for infusion therapy into a peripheral vein. The pump was interfaced to a microcomputer that controlled the flow rate. Within a 2 minute interval, fluid with flow rates of 0, 50, 100, 200, and 300 ml/hr was administered; the corresponding pressures generated were recorded. 10 to 15 seconds were allowed to elapse after each change in flow rate to ensure that a stable pressure had been reached. Total resistance, R_{total}, was calculated as the slope of pressure versus flow

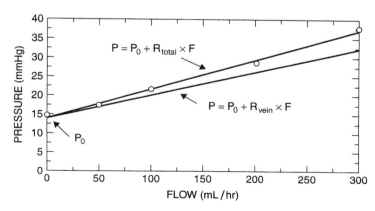

Figure 2.5 Pressure-flow relationship for a single patient. F = flow. Reprinted from [9] with kind permission from Kluwer Academic Publishers.

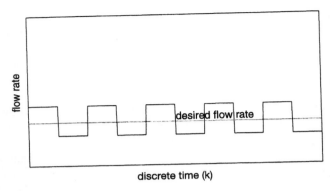

Figure 2.6 Alternating flow rates for resistance calculation. From [10].

(Figure 2.5). The pressure in the peripheral vein without flow, P_0, was the measured value without flow referenced to the right atrium. The resistance for all tubing and catheter combinations used in this study, R_{sys}, was determined, and ranged from 0.011 to 0.034 mmHg-hr/ml. The median resistance of the patient's vein, R_{vein}, the difference between total and system resistance, was calculated as 0.022 mmHg-hr/ml. Although resistance can be calculated with this method in an experimental setting, it is an impractical method for general hospital use since the calculation significantly affects fluid flow uniformity.

Philip streamlined this method by proposing that an infusion pump continuously deliver fluid at two similar flow rates, with the desired flow rate as the mean flow rate [10]. As shown in Figure 2.6, each adjacent pumping cycle of a linear peristalic pump delivers a different flow rate. The delivered flows and resulting pressures can be used to continuously calculate resistance as

$$R = \frac{(P_2 - P_1)}{Q_2 - Q_1} \quad (2.6)$$

This method is suitable for calculating resistance at high flow rates, such as greater than 50 ml/hr [8], since a rapid flow rate change results in a rapid pressure change. However, although this method is certainly an improvement over Philip's earlier design, it is not suitable for low flow rates because it affects flow uniformity. For a mean flow rate of 1.0 ml/hr, the smallest possible increase of 0.1 ml/hr translates into a flow change of 10%. Further, a long time interval is required to detect the resulting pressure change.

2.4 RESISTANCE MONITORING IN THE IVAC SIGNATURE EDITION® PUMP

An alternate method for resistance calculation is determination of the ratio of the mean value of a pressure waveform generated by a flow waveform to the mean value of a flow waveform. Because the pressure waveform may possess noise artifact, the flow and pressure waveforms are preprocessed using a pseudorandom binary sequence filter (see Section 1.6) before the resistance is calculated.

2.4.1 Data Acquisition and Processing

In 1994, pressure data were acquired from 5 adult subjects who had agreed to saline infusions in a **metacarpal** vein at the back of the hand, using an experimental infusion pump prototype [6]. A pressure waveform was generated in response to each PRBS-encoded flow waveform (Figure 2.7). A different sequence length was used for each flow rate: $M = 7$ for 1.1 ml/hr, $M = 15$ for 10 ml/hr, and $M = 31$ for 50 ml/hr [8]. The sequence lengths were chosen so that the first decoded pressure waveform would be available within 2 minutes of initial infusion. The infusions were conducted at these flow rates during 17 potential noise sources. These noise sources included eating with a catheterized arm, blood pressure cuff inflation, and coughing. From previous studies, it was known that the pa-

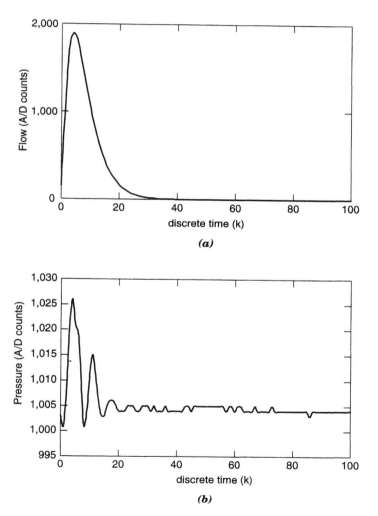

Figure 2.7 Lowpass filtered (a) flow and (b) pressure waveforms (period = 1900 samples, f_s = 1000 Hz) at 1.1 ml/hr. Note the pressure offset of approximately 1000 A/D counts. The resistance between the two signals is 0.020 mmHg-hr/ml. Courtesy of ALARIS Medical Systems, San Diego, CA.

tient's blood pressure also acted as a noise source. No occlusions were generated or occurred spontaneously during the infusions.

The data acquisition and processing systems used in this study are shown in Figure 2.8. The pump prototype contained a pressure transducer, interfaced to the administration set, for acquiring catheter pressure data. Pressure data from this transducer were analog filtered using a second-order Butterworth analog filter with a cutoff frequency of 100 Hz. The analog data were transmitted to a data acquisition board (National Instruments DAQPad 1200, Austin, TX), where they were digitized with a sampling frequency of 1000 Hz and stored in binary format. After the data were recorded, they were digitally filtered using a second-order Butterworth filter with a cutoff frequency of 100 Hz, and downsampled, or **decimated,** to 100 Hz within the software program LabVIEW (National Instruments, Austin, TX). The pressure and corresponding flow waveforms were then PRBS decoded. The flow waveforms were known because they were programmed in the pump prototype.

2.4.2 Data Analysis

Within LabVIEW, the raw and filtered flow/pressure waveform pairs were processed to calculate resistance, using a method that will be described in Chapter 7. The resulting resistance values were similar to values obtained from dividing the mean pressure by the mean flow [Eq. (2.3)] in a system free of noise artifact. A resistance of 1.5 mmHg-hr/ml was chosen as the threshold for a downstream occlusion alarm [8], based on extensive bench testing. The number of false occlusion alarms was determined for raw and filtered waveforms during each noise source. These alarms were used as the response for testing the effects of flow rate, PRBS filtering, and noise sources within a **Design of experiments (DOE) factorial design** [11]. Factorial design is a statistical technique that enables the user to determine the statistically significant factors that affect a response such as false occlusion alarm.

Based on this statistical analysis, flow rate did not significantly impact alarm frequency. However, PRBS filtering significantly reduced the number of false alarms. Moreover, with PRBS filtered waveforms, patient blood pressure artifact was no longer observed. Wheelchair self-ambulation was shown to be the noise source that generated the most false alarms. Wheelchair self-ambulation is a particularly severe noise source because the patient's catheter may vibrate at a different frequency during wheelchair travel than the vibration frequency of the infusion pump, which is pushed on a separate IV pole. In Figure 2.9, one sequence length ($M = 31$) of raw pressure waveforms during wheelchair ambula-

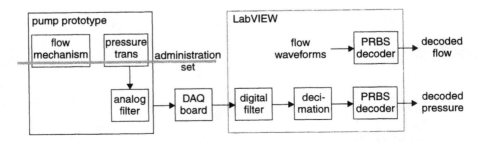

Figure 2.8 Data acquisition and processing systems.

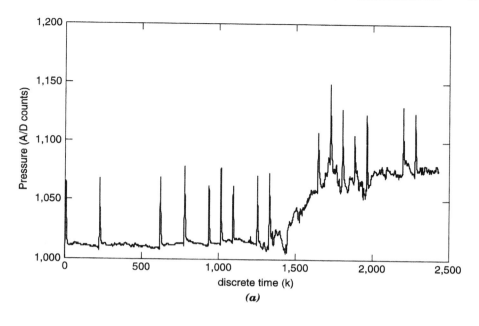

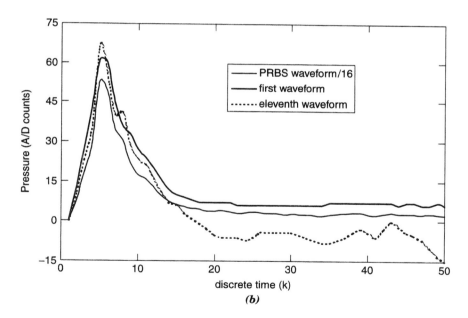

Figure 2.9 (a) Typical raw pressure during wheelchair self-ambulation, at a flow rate of 50 ml/hr. (b) The first raw waveform, the eleventh waveform, and the PRBS filter output scaled by 1/16 are shown. For comparison purposes, the baseline value of each waveform has been subtracted. Courtesy of ALARIS Medical Systems, San Diego, CA.

tion and the subsequent PRBS filtered waveform are illustrated. In this example, the flow rate is 50 ml/hr [6]. Unfortunately, more data from this study are not available because results were not publicly disclosed.

2.4.3 Implementation

Based on these results, the PRBS filter was included in the design of the IVAC Signature Edition® pump (Figure 2.3). Within the flow rate range 0.6–1.4 ml/hr, a sequence length $M = 7$ was used. The sequence length for each flow rate range was chosen so that the duration of one sequence occurred in no less than approximately 52 seconds, with a mechanism revolution volume of 184 µl [12]. The resistance occlusion threshold was chosen to be 2 mmHg-hr/ml [6]. A **median filter** for the seven most current resistance values was added to further decrease the probability of false alarms from noise artifact [12]. During median filtering, an odd number of sample values is sorted, and the middle or median value is used as the filter output [13].

Resistance monitoring, using a PRBS filter, significantly reduces the time to alarm for a downstream occlusion at low flow rates. The Signature Edition® standard administration set possesses a compliance of 0.4 µl/mmHg. For a typical pressure threshold of 500 mmHg and flow rate setting of 1 ml/hr, the typical time to alarm for a full occlusion would be 12 minutes. This time to alarm would be increased if motion artifact were present. With implementation of the PRBS and median filters, a full occlusion at 1 ml/hr is typically determined in only 8 minutes. In addition to minimizing the time during which critical medication is not administered, this significant reduction in the time to alarm occurs at a lower pressure, thus minimizing the risk of a bolus infusion. The time to alarm would be decreased further if less than seven resistance samples were used in the median filter.

2.5 SUMMARY

As will be demonstrated throughout this textbook, novel instrumentation features, directly resulting from digital signal processing, are being used to increase the market share of medical instrumentation companies. In this chapter, we discussed the history of infusion therapy, which has advanced from the use of mechanical flow regulators and infusion controllers to the use of volumetric infusion and syringe pumps for delivery of fluid with greater flow accuracy and uniformity. At low flow rates, such as those less than 1.1 ml/hr, downstream occlusions can not be detected quickly using pressure-based monitoring systems. Neonates and small children may only receive infusions with these low flows since they are fluid restricted and often receive drugs with short half-lives. Thus, they are severely compromised by slow detection of downstream occlusions. Within the IVAC Signature Edition® infusion pump, resistance-based detection of downstream occlusions enables much faster time to alarm (8 versus 12 minutes at 1.1 ml/hr). The resistance calculations are minimally affected by noise artifact, due to implementation of a pseudorandom binary sequence filter.

2.6 REFERENCES

[1]. Voss, G. I. and Butterfield, R. D. Parenteral infusion devices. In *The Biomedical Engineering Handbook,* edited by J. D. Bronzino. CRC Press: Boca Raton, FL, 1995, pp. 1311–1321.

[2]. Georgi, H. W. *Parenteral administration fluid flow control system.* U.S. Patent 3,736,930. Jun. 5, 1973.

[3]. Jenkins, J. A., Flatten, O. H., and Hyman, O. E. *IV pump.* U.S. patent 3,985,133. Oct. 12, 1976.

[4]. Steinberg, C. and Notterman, D. A. Pharmacokinetics of cardiovascular drugs in children. *Clin Pharmacokinet, 27,* 345–367, 1994.

[5]. Bregman, J. Developmental outcome in very low birthweight infants: current status and future trends. *Pediatr Clin North Am, 45,* 673–690, 1998.

[6]. ALARIS Medical Systems, San Diego, CA. 2000.

[7]. IMS Health, *Hospital Supply Index.* IMS America: Plymouth Meeting, PA, 1997.

[8]. Voss, G. I., Butterfield, R. D., Baura, G. D., and Barnes, C. W. *Fluid flow impedance monitoring system.* U.S. Patent 5,609,576. Mar. 11, 1997.

[9]. Scott, D. A., Fox, J. A., Cnaan, A., Philip, B. K., Lind, L. J., Palleiko, M. A., Stelling, J. M., and Philip, J. H. Resistance to fluid flow in veins. *J Clin Mon, 12,* 331–337, 1996.

[10]. Philip, J. H. *Intravenous fluid flow monitor.* U.S. patent 4,898,576. Feb. 6, 1990.

[11]. Box, G. E. P., Hunter, W. G., and Hunter, J. S. *Statistics for Experimenters: An Introduction to Design, Data Analysis, and Model Building.* Wiley: New York, 1978.

[12]. Butterfield, R. D. and Farquhar, A. *Fluid flow resistance monitoring system.* U.S. Patent 5,803,917. Sept. 8, 1998.

[13]. Tukey, J. W. Nonlinear (Nonsuperposable) methods for smoothing data. *Conf Rec EASCON'74,* 673, 1974.

[14]. Sugimachi, M. National Cardiovascular Center Research Institute, Osaka, Japan. Research data, 1992.

Further Reading

Infusion Therapy: Booker, M. F. and Ignatavicius, D. D. *Infusion Therapy: Techniques and Medications.* Saunders: Philadelphia, 1996.

Infusion Pump Flow Accuracy: Auty, B. Equipment for intravenous infusion: some aspects of performance. *Aggressologie, 29,* 824–828, 1988.

2.7 MATLAB EXERCISES

As discussed in the preface, each even-numbered application chapter contains exercises based on a monitoring technology. The exercises may be analyzed directly, using the techniques described in the accompanying theoretical chapter. However, it is more efficient to utilize the software program Matlab and its Toolboxes. For documentation of a particular Matlab function while working in the Matlab Command Window, type "help name," where name is the function name.

2.8 INTRAARTERIAL BLOOD PRESSURE EXERCISES

Engineer A at Startup Company has been asked to determine the digital signal processing algorithm for the **intraarterial blood pressure** module of a new instrument. Blood pressure monitoring will be a secondary feature of this instrument, and will be obtained from **aortic** (main artery delivering oxygenated blood from the left side of the heart) data. Engineer A is aware that continuous intraarterial, rather than intermittent noninvasive, monitoring is required during situations when cardiac function is rapidly changing, such as acute hypertensive (short-lived high blood pressure) crises, use of potent vasoactive drugs, or **hypotensive** (low blood pressure) anesthesia. She also knows that the anesthesiologist inserting the blood pressure catheter typically uses the **radial artery** (wrist) site, and that the arterial waveform originating in the aorta becomes distorted as it is transmitted through the cardiovascular system. These changes result from the summation of the forward pulse wave and reflected waves from the peripheral vessels. The impulse response between the radial and aortic artery needs to be determined before **systolic** (peak), **diastolic** (baseline), and **mean** values can be predicted. We recreate her analysis below.

2.8.1 Clinical Data

The aortic and radial data files in Figure 2.10 were simultaneously recorded from a patient with **atrial fibrillation** (erratic heart rhythm due to continuous, rapid discharges from areas of the atria) due to **rheumatic heart disease.** Ten seconds of data were recorded with a sampling frequency of 200 Hz and cutoff of 1 MHz, and saved in units of mmHg. These data were generously provided by Dr. Masaru Sugimachi [14].

Please move the *aortic.dat* and *radial.dat* files from *ftp://ftp.ieee.org/uploads/press/baura* to your *Matlab\bin* directory. Input the data using "load file.dat -ascii."

2.8.2 Useful Matlab Functions

The following functions from the Signal Processing Toolbox are useful for completing this exercise: abs, butter, csd, fft, filter, and psd.

2.8.3 Exercises

1. Plot the magnitude responses of the aortic and radial waveforms. What is the bandwidth of each signal?
2. Lowpass filter the data using a fourth-order Butterworth filter with a cutoff frequency of 20 Hz. (a) Plot the original and filtered data. How do the original and filtered waveforms differ in terms of (b) general shape (explanation should include

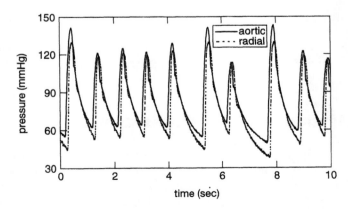

Figure 2.10 Simultaneously recorded aortic and radial blood pressure waveforms from a subject with atrial fibrillation due to rheumatic heart disease [14].

bandwidth discussion); (c) average values of systolic, diastolic, and mean pressures; (d) initial samples at beginning of waveform; (e) phase?

3. Using only samples 255 to 500 of each filtered waveform, (a) calculate and plot the power spectral density of the aortic waveform and the cross-power spectral density of the aortic and radial waveforms; (b) calculate and plot the ratio to determine the transfer function.

4. In Chapter 7, we will discuss a method for estimating the order of a linear system's impulse response. Assume that the impulse response between the radial and aortic waveforms is a ninth-order linear system. Draw an electrical circuit that represents a ninth-order system.

5. What is the fundamental limitation of predicting the radial waveform from each patient's aortic waveform?

3

ADAPTIVE FILTERS

In Chapter 1, we observed that noise restricted to a specific frequency band may be reduced through a frequency-selective filter, thus increasing the signal-to-noise ratio. Often, however, the noise and signal bands overlap. In this case, a filtering strategy that may be useful, given certain constraints, is the **adaptive filter**. An adaptive filter possesses a structure that is adjustable in such a way that its performance improves through contact with its environment. Such behavior is the result of several constraints on the **signal,** $u(k)$; the **noise associated with the signal,** $n_0(k)$; the **reference noise source,** $n_1(k)$; and the **filter output,** $\hat{y}(k)$:

1. $u(k)$, $n_0(k)$, $n_1(k)$, and $\hat{y}(k)$ are statistically **stationary** (not variable with time).
2. $u(k)$, $n_0(k)$, $n_1(k)$, and $\hat{y}(k)$ have **zero means.**
3. $u(k)$ is uncorrelated with $n_0(k)$ and $n_1(k)$.
4. $n_0(k)$ is correlated with $n_1(k)$.

When an adaptive filter is used as an **adaptive noise canceller,** the reference noise source is filtered and subtracted from a **primary input** containing the signal and noise to eliminate the noise by cancellation (Figure 3.1).

In this chapter, we discuss some of the fundamental theories underlying the adaptive filter and adaptive noise cancellation. Adaptive noise cancellation research was first conducted in the late 1950s [1–4] and remains an active research topic. In order to restrict our discussion of adaptive filters to a single chapter, other adaptive filter applications such as **system identification, equalization,** and **control** will not be considered. However, references for these applications are provided at the end of the chapter.

3.1 ADAPTIVE NOISE CANCELLATION PROOF

The basic structure of the system we are considering is shown in Figure 3.1. A signal, $u(k)$, is transmitted to a sensor that receives the signal plus an uncorrelated noise, $n_0(k)$. The combined signal and noise, $[u(k) + n_0(k)]$, form the primary input to the canceller. In later sec-

3.1 ADAPTIVE NOISE CANCELLATION PROOF

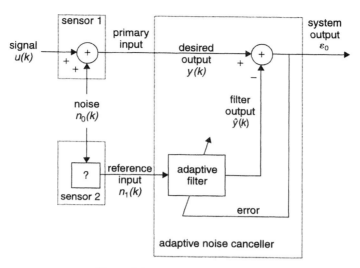

Figure 3.1 Adaptive noise canceller.

tions, the primary input is also referred to as the desired output, $y(k)$. A second sensor receives a noise, $n_1(k)$, which is uncorrelated with the signal but correlated in an unknown way with noise, $n_0(k)$. This sensor provides the reference input to the canceller. The noise, $n_1(k)$, is filtered to produce an output, $\hat{y}(k)$, that is a close replica of $n_0(k)$. This output is subtracted from the primary input, $[u(k) + n_0(k)]$, to produce the system output, or output error,

$$\varepsilon_o(k) = u(k) + n_0(k) - \hat{y}(k) \tag{3.1}$$

By feeding the system output back to an adaptive filter and adjusting the filter to minimize **system output power** (i.e., the power of the system error), the output becomes a best approximation (in the **least mean squares sense**) to the signal, $u(k)$. "Least mean squares" refers to minimization of the mean squared error between an approximation and the true signal. Surprisingly, the four constraints listed above are all that are required for the adaptive filter, as shown in the following proof.

We begin by squaring the filter output in Eq. (3.1) to obtain

$$\varepsilon_o^2(k) = u^2(k) + [n_0(k) - \hat{y}(k)]^2 + 2u(k)[n_0(k) - \hat{y}(k)] \tag{3.2}$$

The **expected value**, $E\{\ \}$, of a function is its mean. Taking the **expectations** on both sides of Eq. (3.2) results in

$$E\{\varepsilon_o^2(k)\} = E\{u^2(k)\} + E\{[n_0(k) - \hat{y}(k)]^2\} + 2E\{u(k)[n_0(k) - \hat{y}(k)]\} \tag{3.3}$$

Note that $u(k)$ is uncorrelated with $n_0(k)$ and with $\hat{y}(k)$. The cross-correlation of two functions, which was defined in Eq. (1.13), is the expected value of the product of these functions. Therefore, the last term in Eq. (3.3) equals zero, and Eq. (3.3) simplifies to

$$E\{\varepsilon_o^2(k)\} = E\{u^2(k)\} + E\{[n_0(k) - \hat{y}(k)]^2\} \tag{3.4}$$

As the adaptive filter is adjusted to minimize $E\{\varepsilon_o^2(k)\}$, the signal power, $E\{u^2(k)\}$ is unaffected. The resulting minimum output power is therefore

$$E_{\min}\{\varepsilon_o^2(k)\} = E\{u^2(k)\} + E_{\min}\{[n_0(k) - \hat{y}(k)]^2\} \tag{3.5}$$

When the minimum of $E\{\varepsilon_o^2(k)\}$ is reached, $E\{[n_0(k) - \hat{y}(k)]^2\}$ is also minimized. The filter output, $\hat{y}(k)$, is then a best least mean squares estimate of the primary noise, $n_0(k)$. Since the signal power in the output remains constant, minimizing the total output power maximizes the output signal-to-noise ratio.

Note that this proof does not assume a particular structure for the adaptive filter. To simplify our discussion, we constrain the filter structure to that of a linear filter such as the FIR or IIR filter of Eq. (1.32) or (1.19), respectively. Calculation of the coefficients in each filter is determined through **optimization** techniques.

3.2 OPTIMIZATION CONCEPTS

Let us review some of the concepts behind optimization in preparation for calculating the coefficients of our linear adaptive filter. Given a model, coefficients may be calculated based on a **performance function**, $\xi[\mathbf{w}(k)]$, where $\mathbf{w}(k)$ is a weight vector containing the model coefficients, $w_i(k)$:

$$\mathbf{w}(k) = [w_0(k) w_1(k) \cdots w_M(k)]^T \tag{3.6}$$

The performance function is a criterion that enables the optimum set of operating conditions to be identified. In many engineering applications, an economic criterion is selected. Within the context of **system identification,** a measure of the error between the estimated and true system is used.

Once the performance function is known, an **optimization method** is selected for estimating the optimum solution of the performance function. Two fundamental classes of methods for finding the minimum of unconstrained functions of several variables are the **direct search methods** and **gradient methods.** Direct search methods such as the **simplex search** require only calculation of current values of the performance function. The **gradient**, $\nabla \mathbf{u}[\mathbf{w}(k)]$, is a vector consisting of partial derivatives of any function, $u(k)$, with respect to each coefficient, $w_i(k)$:

$$\nabla \mathbf{u}[\mathbf{w}(k)] \equiv \frac{\partial u(k)}{\partial \mathbf{w}(k)} = \left[\frac{\partial u(k)}{\partial w_0(k)} \quad \frac{\partial u(k)}{\partial w_1(k)} \cdots \frac{\partial u(k)}{\partial w_M(k)} \right]^T \tag{3.7}$$

Gradient methods, as their classification implies, require calculation of accurate values of at least the first, and sometimes the second, derivative of the performance function. As an example, let us define a function $u(k)$ such that

$$u(k) = 8w_0^2(k) + 4w_0(k)w_1(k) + 5w_1^2(k) \tag{3.8}$$

Using Eq. (3.7), the gradient is calculated as

$$\nabla \mathbf{u}[\mathbf{w}(k)] \equiv \frac{\partial u(k)}{\partial \mathbf{w}(k)} = [16w_0(k) + 4w_1(k) \quad 4w_0(k) + 10w_1(k)]^T \tag{3.9}$$

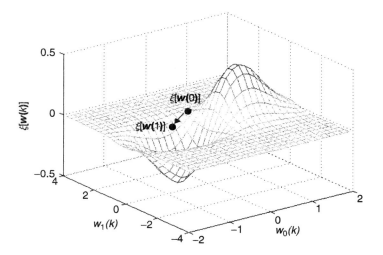

Figure 3.2 Optimization iteration for $\xi[\mathbf{w}(k)] = w_0(k)e^{[-w_0^2(k)-w_1^2(k)]}$ with $\mathbf{w}(0) = [0\ 1]^T$ and $\mu = 0.5$.

One commonly used gradient method is the **method of steepest descent, or Cauchy's method.** As its name implies, iterations of this method are based on the steepest changes in the value of the performance function. The simple algorithm is expressed as:

$$\mathbf{w}(k+1) = \mathbf{w}(k) - \mu\{\nabla \mathbf{u}[\mathbf{w}(k)]\} \tag{3.10}$$

where μ is a constant that regulates the step size and has dimensions of reciprocal signal power. One advantage of Cauchy's method is its guaranteed move toward a minimum when searching within a **convex,** or outward curving, region of a function. Note from Eq. (3.10) that, given a convex region, we are assured that $\mathbf{w}(k+1)$ is less than $\mathbf{w}(k)$.

For example, suppose our performance function is

$$\xi[\mathbf{w}(k)] = w_0(k)e^{[-w_0^2(k)-w_1^2(k)]} \tag{3.11}$$

Let $\mathbf{w}(0) = [0\ 1]^T$; therefore $\xi[\mathbf{w}(0)] = 0$ and $\nabla[\mathbf{w}(0)] = [0.368\ 0]^T$. Using Eqs. (3.9) and (3.10) with a step size of 0.5, $\mathbf{w}(1) = [-0.184\ 1]^T$ and $\xi[\mathbf{w}(1)] = -0.070$. As shown in Figure 3.2, with a single iteration, the value of the performance function has decreased toward the global minimum since the search occurs in the convex region of the function.

3.3 LEAST MEAN SQUARES ALGORITHM FOR FINITE IMPULSE RESPONSE FILTERS

We are now ready to discuss the classic least mean squares (LMS) algorithm, first proposed by Widrow and Hoff [2, 5]. Widrow used an FIR filter which he called an adaptive linear combiner.

3.3.1 Performance Function

The form of the adaptive linear combiner is

$$\hat{y}(k) = \sum_{n=0}^{M} w_n(k) u(k-n) \qquad (3.12)$$

where $\hat{y}(k)$ is the filter output that approximates a **desired signal**, $y(k)$. The weights $w_n(k)$ vary with discrete time, unlike the coefficients b_n in the FIR filter of Eq. (1.32). Eq. (3.12) can be rewritten as

$$\hat{y}(k) = \mathbf{w}^T(k)\mathbf{u}(k) \qquad (3.13)$$

where the input vector is

$$\mathbf{u}(k) = [u(k)\ u(k-1) \cdots u(k-M)]^T \qquad (3.14)$$

As the name of the algorithm implies, the performance function is the mean instantaneous squared error between the desired and filtered output. Calculating the mean from the expected value of the squared error results in

$$\xi[\mathbf{w}(k)] = E\{\varepsilon_o^2(k)\} = E\{y(k) - \hat{y}(k)\}^2 \qquad (3.15)$$

Substituting Eq. (3.13) and noting that the expected value of any sum is the sum of expected values leads to

$$\xi[\mathbf{w}(k)] = E\{y^2(k)\} + E\{\mathbf{w}^T(k)\mathbf{u}(k)\mathbf{u}^T(k)\mathbf{w}(k)\} - E\{2y(k)\mathbf{u}^T(k)\mathbf{w}(k)\} \qquad (3.16)$$

Assume that $\varepsilon_o(k)$, $y(k)$ and $\mathbf{u}(k)$ are statistically stationary, and that the weights are no longer adjusted (independent of time, having reached steady state). Because the expected value of a product is the product of expected values when variables are statistically stationary, the performance function simplifies to

$$\xi(\mathbf{w}) = E\{y^2(k)\} + \mathbf{w}^T E\{\mathbf{u}(k)\mathbf{u}^T(k)\}\mathbf{w} - 2E\{y(k)\mathbf{u}^T(k)\}\mathbf{w} \qquad (3.17)$$

Let us further simplify Eq. (3.16) by defining $\underline{\mathbf{R}}(k)$ as the square matrix:

$$\underline{\mathbf{R}}(k) = E\{\mathbf{u}(k)\mathbf{u}^T(k)\} \qquad (3.18)$$

$$= E\left\{\begin{array}{cccc} u^2(k) & u(k)u(k-1) & \cdots & u(k)u(k-M) \\ u(k-1)u(k) & u^2(k-1) & \cdots & u(k-1)u(k-M) \\ \vdots & \vdots & \vdots & \vdots \\ u(k-M)u(k) & u(k-M)u(k-1) & \cdots & u^2(k-M) \end{array}\right\} \qquad (3.19)$$

This matrix can be thought of as the input correlation matrix because the cross terms are the cross-correlations among the input components. Similarly, let us define $\mathbf{p}(k)$ as the column vector:

$$\mathbf{p}(k) = E\{y(k)\mathbf{u}(k)\} = E\{y(k)u(k)\ y(k)u(k-1) \cdots y(k)u(k-M)\}^T \qquad (3.20)$$

When **u**(k) and y(k) are stationary, the elements of **R**(k) and **p**(k) are all constant second-order statistics. The performance function is then

$$\xi(\mathbf{w}) = E\{y^2(k)\} + \mathbf{w}^T\mathbf{R}(k)\,\mathbf{w} - 2\mathbf{p}(k)^T\mathbf{w} \tag{3.21}$$

Looking at Eq. (3.21), the performance function is quadratic, with a bowl-shaped performance surface that is concave upward (for example, see global minimum in Figure 3.2). Therefore, only a single global minimum exists; no local minima are present.

3.3.2 Optimization

The LMS algorithm uses the method of steepest descent for optimization, with an estimate of the gradient. From Eqs. (3.7) and (3.21), the gradient of the adaptive linear combiner is

$$\nabla\xi[\mathbf{w}(k)] = \frac{\partial\xi[\mathbf{w}(k)]}{\partial\mathbf{w}(k)} = 2\mathbf{R}(k)\,\mathbf{w}(k) - 2\mathbf{p}(k) \tag{3.22}$$

However, we assumed in our calculation of **R**(k) and **p**(k) in the last section that **u**(k) and y(k) are statistically independent. Since this is generally not true, Widrow and Hoff developed the LMS algorithm by using $\varepsilon_o^2(k)$ as a estimate of $\xi(\mathbf{w}(k)) = E\{\varepsilon_o^2(k)\}$. In this way, complicated calculations of **R**(k) and **p**(k) are avoided. The estimated gradient, $\hat{\nabla}\xi[\mathbf{w}(k)]$, is then just

$$\hat{\nabla}\xi[\mathbf{w}(k)] = \frac{\partial\hat{\xi}[\mathbf{w}(k)]}{\partial\mathbf{w}(k)} = \frac{\partial}{\partial\mathbf{w}(k)}\varepsilon_o^2(k) \tag{3.23}$$

Substituting Eq. (3.13) and knowing that the error is the difference between the desired and filtered output yields

$$\hat{\nabla}\xi[\mathbf{w}(k)] = \frac{\partial\hat{\xi}[\mathbf{w}(k)]}{\partial\mathbf{w}(k)} = \frac{\partial}{\partial\mathbf{w}(k)}[y(k) - \mathbf{w}^T(k)\mathbf{u}(k)]^2 \tag{3.24}$$

$$\hat{\nabla}\xi[\mathbf{w}(k)] = -2\varepsilon_o(k)\mathbf{u}(k) \tag{3.25}$$

With this simple estimate of the gradient, we use the method of steepest descent in Eq. (3.10) to specify the LMS algorithm as

$$\mathbf{w}(k+1) = \mathbf{w}(k) + 2\mu\varepsilon_o(k)\mathbf{u}(k) \tag{3.26}$$

Again, μ is the gain constant that regulates the speed and stability of adaptation. The LMS algorithm is guaranteed to converge to the optimal solution only if the inverse of the maximum **eigenvalue**, or root, λ_{max}, of **R**(k) is greater than the gain constant, which in turn is greater than zero:

$$\frac{1}{\lambda_{max}} > \mu > 0 \tag{3.27}$$

The rate of convergence is slow. Additionally, the LMS algorithm is sensitive to the eigenvalue spread [ratio of largest to smallest eigenvalue of **R**(k)].

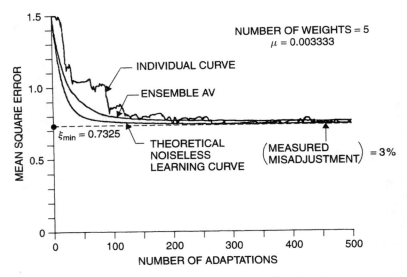

Figure 3.3 Learning curves for FIR adaptive filter. From [6].

3.3.3 Example

As an illustration, let us simulate a primary input, $[u(k) + n_0(k)]$, with a bandpass filtered white Gaussian signal, $n_0(k)$, which has been added to another independent white Gaussian signal, $u(k)$. Because a true reference input is unavailable, we use the primary input, delayed by 1 sample, as an approximate reference input, $n_1(k)$. This delay causes the broadband signal components in the reference input to become decorrelated from those in the primary input.

We assume our adaptive filter has five weights, and initially set them to zero. The gain constant is set to 0.003333. As the filter adapts, the mean squared error is plotted as a function of number of adaptations, or iterations. The results are shown in Figure 3.3. Here, the mean squared error between estimated and observed data is plotted for an individual curve, ensemble average, and theoretical noiseless learning curve. The individual curve refers to results for one set of inputs; the ensemble average refers to the mean of 200 curves. The theoretical curve shows how the process would evolve if noise were not present when the gradient was calculated at each step. Note that the ensemble average learning curve falls to within 2% of its converged value at around iteration 200. The residual difference between the ensemble learning curve and $E_{\min}\{\varepsilon_o^2(k)\}$ ($\xi_{\min}$ in the Figure) is 3% [6].

3.4 INFINITE IMPULSE RESPONSE FILTERS

As stated in Chapter 1, infinite impulse response filters are much more efficient than finite impulse response filters. Recalling Eq. (1.19), the basic form of the IIR filter is

$$\hat{y}(k) = \sum_{n=1}^{N} a_n(k)\hat{y}(k-n) + \sum_{n=0}^{M} b_n(k)u(k-n) \qquad (3.28)$$

Because of the fewer computations required, an adaptive IIR filter is potentially preferable to an IIR filter if several issues can be resolved:

1. Unlike FIR coefficients, not all choices of IIR coefficients are stable.
2. Feedback of the output makes calculations of the gradient substantially more complex.
3. **High resonance** poles may have long time constants, which complicate IIR, but not FIR, convergence analysis.

Unfortunately, a robust adaptive IIR algorithm does not exist, but remains a current research goal. One obvious approach to optimization of an adaptive IIR filter would be to extend the LMS algorithm for IIR coefficients, thus simplifying gradient calculations. However, in 1977, Widrow and McCool showed that this extended algorithm may converge to false minima, as the mean squared error is not quadratic and is sometimes multimodal [7]. We present an alternative approach developed by Larimore & Johnson in this section—the simplified hyperstable adaptive recursive filter (SHARF) [8]. Although far from perfect, SHARF and its convergence and stability properties have been extensively studied.

3.4.1 Hyperstability

Before discussing the SHARF algorithm, let us introduce the concept of hyperstability. As before, assume that $H(z)$ is a rational scalar transfer function for a linear time-invariant system driven by input, $u(k)$, and responding with output $y(k)$. This system is hyperstable if and only if its transfer function is **strictly positive real** (SPR), that is,

$$Re[H(z)] > 0 \tag{3.29}$$

The SPR restriction guarantees that the output will remain bounded. As a consequence, the zeros of a hyperstable transfer function all reside within the unit circle. The transfer function contributes less than ±90 degrees of phase at all frequencies. For example, let us investigate a simple second-order system. For the transfer function

$$H(z) = \frac{1}{1 - a_1 z^{-1} - a_2 z^{-2}} \tag{3.30}$$

only certain conjugate pole pairs are SPR. The SPR region is shown in relation to the unit circle as an unshaded oval in Figure 3.4. Note that the SPR region does not include the vicinity near $z = 1$, where poles of an oversampled continous process would cluster.

Now let us introduce a single zero to our transfer function such that

$$H(z) = \frac{1 + c_1 z^{-1}}{1 - a_1 z^{-1} - a_2 z^{-2}} \tag{3.31}$$

The addition of a zero purposely deforms the SPR region within which unknown poles may be found. As shown in Figure 3.5, $c_1 = 0$ produces the original oval region of Figure 3.4. As c_1 becomes negative and $H(z)$ possesses a zero on the positive real axis, the SPR

54 ADAPTIVE FILTERS

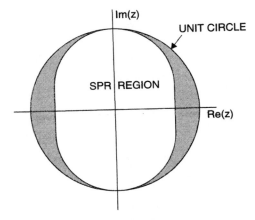

Figure 3.4 Strictly positive real region for a second-order system. From [8].

region is deformed toward $z = 1$. For $c_1 = -1$, the SPR region becomes circular tangential with the unit circle at $z = 1$, and encompasses the low frequency pole locations.

3.4.2 The Hyperstable Adaptive Recursive Filter

Before we can understand SHARF, we must understand its foundation—the hyperstable adaptive recursive filter (HARF). Conceptually, using the HARF method, we multiply the system transfer function by specially selected zeros, in order to ensure that the poles are strictly positive real. To develop the theory behind the HARF algorithm, let us model our desired output, $y(k)$, as an ARMA process, where

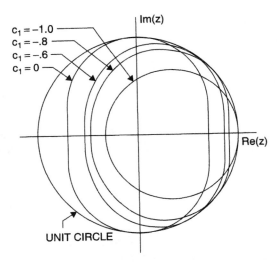

Figure 3.5 Strictly positive real region with error smoothing for a second-order system. From [8].

3.4 INFINITE IMPULSE RESPONSE FILTERS

$$y(k) = \sum_{n=1}^{N_y} \tilde{a}_n \hat{y}(k-n) + \sum_{n=0}^{M_y} \tilde{b}_n u(k-n) \tag{3.32}$$

Note that $\tilde{a}_n$ and $\tilde{b}_n$ are constant. We assume that sufficient variables are provided in the adaptive filter to span the parameter space generating $y(k)$, in other words, that $M \geq M_y$ and $N \geq N_y$. Without loss of generality, we can assume that $M = M_y$ and $N = N_y$, with any excess coefficients equaling zero.

Let us define the IIR weight vector, $\mathbf{w}(k)$, as

$$\mathbf{w}(k) = [b_0(k) \cdots b_M(k) \vdots a_1(k) \cdots a_N(k)]^T \tag{3.33}$$

$$\mathbf{w}(k) = [w_0(k) w_1(k) \cdots w_{M+N}(k)]^T \tag{3.34}$$

where $a_n(k)$ and $b_n(k)$ are the coefficients of Eq. (3.28). As before, the performance function is

$$\xi[\mathbf{w}(k)] = E\{\varepsilon_o^2(k)\} = E\{y(k) - \hat{y}(k)\}^2 \tag{3.35}$$

Based on Eqs. (3.28) and (3.32), the error becomes

$$\varepsilon_o(k) = \sum_{n=1}^{N} \{\tilde{a}_n y(k-n) - a_n \hat{y}(k-n)\} + \sum_{n=0}^{M} \{\tilde{b}_n - b_n(k)\} u(k-n) \tag{3.36}$$

It is sufficient to choose $b_n(k) \equiv \tilde{b}_n$ and $a_n(k) \equiv \tilde{a}_n$ to minimize the performance function. Now the second term of Eq. (3.36) disappears, leaving

$$\varepsilon_o(k) = \sum_{n=1}^{N} \{\tilde{a}_n y(k-n) - \tilde{a}_n \hat{y}(k-n)\} \tag{3.37}$$

$$\varepsilon_o(k) = \sum_{n=1}^{N} \tilde{a}_n \varepsilon_o(k-n) \tag{3.38}$$

Let us assume that Eq. (3.32) is hyperstable; therefore,

$$\lim_{k \to \infty} \varepsilon_o(k) = 0 \tag{3.39}$$

Once steady state has occurred, this choice of coefficients yields a minimal squared error.

The implementation of the HARF algorithm is shown in Figure 3.6. Note that $n_1(k)$ is filtered twice—first to obtain the filter output, $\hat{y}(k)$, and second to obtain an auxiliary process, $f(k)$. The filter output is calculated as

$$\hat{y}(k) = \sum_{n=1}^{N} a_n(k) f(k-n) + \sum_{n=0}^{M} b_n(k) n_1(k-n) \tag{3.40}$$

The auxiliary process is generated using the original IIR coefficients, shifted forward in time by one sample:

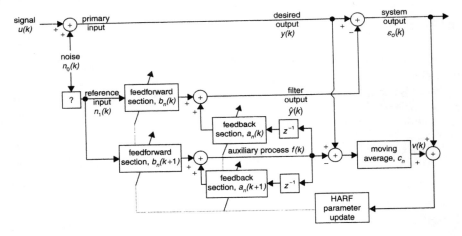

Figure 3.6 HARF Algorithm.

$$f(k) = \sum_{n=1}^{N} a_n(k+1)f(k-n) + \sum_{n=0}^{M} b_n(k+1)n_1(k-n) \qquad (3.41)$$

With convergence of the coefficients, $a_n(k+1) = a_n(k)$ and $b_n(k+1) = b_n(k)$. Thus, $\hat{y}(k)$ assymptotically converges to $f(k)$.

In order to guarantee stability, we update the filter coefficients as

$$a_n(k) = a_n(k-1) + \frac{\mu_n}{q(k)} f(k-n-1) \cdot$$

$$\{[y(k-1) - \hat{y}(k-1)] + \sum_{n=1}^{P} c_n[y(k-n-1) - f(k-n-1)]\}, \quad 1 \le n \le N \quad (3.42)$$

$$b_n(k) = b_n(k-1) + \frac{\rho_n}{q(k)} n_1(k-n-1) \cdot$$

$$\{[y(k-1) - \hat{y}(k-1)] + \sum_{n=1}^{P} c_n[y(k-n-1) - f(k-n-1)]\}, \quad 1 \le n \le M \quad (3.43)$$

where $q(k)$ is a normalizing factor greater than unity:

$$q(k) = 1 + \sum_{n=1}^{N} \mu_n f^2(k-n-1) + \sum_{n=0}^{M} \rho_n n_1^2(k-n-1) \qquad (3.44)$$

and μ_n and ρ_n are arbitrary positive constants. From Eqs. (3.42) and (3.43), the update of each coefficient is basically a product of two components. The first component is the value of the signal corresponding to the given weight in output Eq. (3.40). The second component in brackets depends on the instantaneous performance of the filter, disguised by a moving average expression. As such, for a given quality of performance, the largest adjustment will be made to the coefficients contributing the most to the output via Eq. (3.40).

The P c_n coefficients are chosen by the algorithm designer so that the transfer function

$$H(z) = \frac{1 + \sum_{n=1}^{P} c_n z^{-n}}{1 - \sum_{n=1}^{N} a_n z^{-n}} \quad (3.45)$$

is strictly positive real. Under these conditions, it can be proven that the moving average quantity

$$v(k) = [y(k) - f(k)] + \sum_{n=1}^{P} c_n [y(k-n) - f(k-n)] \quad (3.46)$$

converges to zero [9]. As a result, $\hat{y}(k)$ converges to $f(k)$, which in turn converges to $y(k)$ [8]. Note that a_n are usually not exactly known.

3.4.3 The Simplified Hyperstable Adaptive Recursive Filter

The HARF algorithm described above is computationally complex. An auxiliary process, $f(k)$, and normalized scale factor, $q(k)$, are computed at each iteration. The algorithm can be simplified substantially by specifying that the rate constants μ and ρ are sufficiently small, as in successful gradient approximation procedures. Therefore, the coefficients change very little at each iteration, and

$$a_n(k+1) \cong a_n(k) \quad (3.47)$$

$$b_n(k+1) \cong b_n(k) \quad (3.48)$$

Substitution of Eqs. (3.47) and (3.48) into Eqs. (3.40) and (3.41) leads to

$$f(k) \cong \hat{y}(k) \quad (3.49)$$

Thus, the Eq. (3.40) becomes

$$\hat{y}(k) = \sum_{n=1}^{N} a_n(k) \hat{y}(k-n) + \sum_{n=0}^{M} b_n(k) n_1(k-n) \quad (3.50)$$

and the moving average process of Eq. (3.46) becomes

$$v(k) \cong [y(k) - \hat{y}(k)] + \sum_{n=1}^{P} c_n [y(k-n) - \hat{y}(k-n)] \quad (3.51)$$

$$v(k) \cong \varepsilon_o(k) + \sum_{n=1}^{P} c_n \varepsilon_o(k-n) \quad (3.52)$$

Eq. (3.52) is a simple moving average of the output error.

To further simplify the algorithm, note that $q(k)$ in Eq. (3.44) is a simple normalizing factor that controls the instantaneous adaptation rate, reducing the effective step size for large values of filter input and output. Since we have already assumed that constants μ and ρ are sufficiently small, $q(k)$ is approximately equal to 1.

With our approximations of $f(k)$, $v(k)$, and $q(k)$, Eqs. (3.42) and (3.43) simplify to

$$a_n(k) = a_n(k-1) + \mu_n \hat{y}(k-n-1)v(k-1), \quad 1 \leq n \leq N \quad (3.53)$$

$$b_n(k) = b_n(k-1) + \rho_n n_1(k-n-1)v(k-1), \quad 0 \leq n \leq M \quad (3.54)$$

Eqs. (3.50) to (3.54) represent the simplified hyperstable adaptive recursive filter, SHARF (Figure 3.7). Although substantial computational savings have been realized, the simplified algorithm no longer rigorously satisfies the hyperstability condition. Convergence is no longer guaranteed for arbitrary positive μ and ρ. However, slow adaptation maintains close approximation to a hyperstable structure [8].

3.4.4 Comparison of the SHARF and LMS Algorithms

Let us compare the SHARF and LMS algorithms with a simple simulation. Assume that the signal, $u(k)$, is a simple periodic waveform represented by Figure 3.8b, and that the signal noise, $n_0(k)$, is represented by a unity power white Gaussian noise process, $G(z)$, which is transmitted through a path represented by two poles such that

$$n_0(k) = \frac{0.1630 G(z)}{1 - 1.739 z^{-1} + 0.81 z^{-2}} \quad (3.55)$$

The signal noise is a narrowband process with a center frequency equal to the sampling frequency divided by 24. 1000 samples of the primary input are illustrated in Figure 3.8a. Also assume that the reference noise, $n_1(k)$, has encountered a transmission path represented by two zeros that cancel the original two poles such that $n_1(k) = G(z)$. This implies that the optimal adaptive filter is

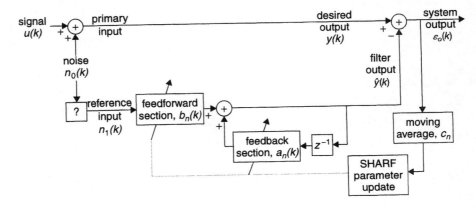

Figure 3.7 SHARF Algorithm.

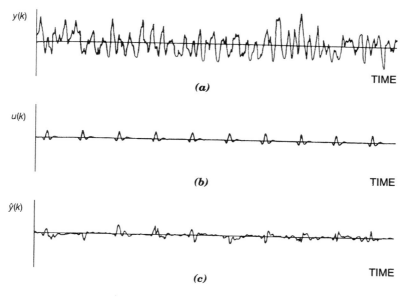

Figure 3.8 SHARF performance. (a) Noise-corrupted primary input. (b) Signal component. (c) Estimated signal from the noise canceller output. From [8].

$$H^*(k) = \frac{0.1630}{1 - 1.739z^{-1} + 0.81z^{-2}} \quad (3.56)$$

In this way, the filter output, $\hat{y}(k)$, would cancel out the signal noise to isolate the signal. In the time domain, the optimal impulse response is given by

$$h^*(k) = 0.1630(0.9)^k \cdot [\cos(0.2618k) + 3.732 \sin(0.2618k)] \quad (3.57)$$

This impulse response possesses an envelope that decays to under 10% only after 21 terms.

Using the SHARF algorithm with $M = 1$, $N = 2$, $\mu = 0.1$, $\rho = 0.005$, $P = 1$, and $c_1 = -1$ results in convergence to approximately 13 dB improvement in about 2000 adaptations. As shown in Figure 3.8c, the signal spikes are easily detectable in the residual noise after SHARF processing. As we demonstrated in Section 3.4.1, a second-order system of the form shown in Eq. (3.31) is guaranteed to be strictly positive real with $c_1 = -1$, and is therefore hyperstable [9].

In contrast, the LMS algorithm is implemented with $M = 6$ and an adaptive rate chosen to provide convergence in about 1500 iterations, requiring approximately a comparable total computational effort to the SHARF implementation. The resulting filter provides only about 5 dB noise suppression. To obtain noise suppression comparable to SHARF, more FIR coefficients are required for the same number of iterations, which severely increases the computation (Table 3.1). Thus, the SHARF IIR implementation is much more efficient and effective than the LMS FIR implementation. However, convergence is no longer guaranteed for arbitary positive μ and ρ, and the choice of c_n may be difficult for higher-order systems [8].

Table 3.1 Noise suppression obtained from increasing LMS FIR coefficients, with approximately 1500 iterations. From [8].

Coefficient number	Noise suppression (dB)
6	5
9	10
12	11
15	12

3.4.5 Other IIR Algorithms

Although we have examined the SHARF algorithm in depth, by no means is it the definitive IIR adaptive filter approach. Other stability approaches include fixing the pole locations [10] and designing adaptive notch filters [11]. In spite of their stability limitations, IIR algorithms continue to be investigated because of the desire for greater computational efficiency [11, 12].

3.5 ADAPTIVE NOISE CANCELLATION

In the previous sections, we discussed two types of adaptive filters. We now discuss the application of adaptive filtering to adaptive noise cancellation. Three common implementations of adaptive noise cancelling are presented in this section. The specific examples are based on the work of Bernard Widrow [13].

3.5.1 Adaptive Noise Canceller

The adaptive noise canceller has been discussed throughout this chapter. Here, the primary input is the combined signal and noise, $[u(k) + n_0(k)]$, and the reference input is a second noise, $n_1(k)$, which is uncorrelated with the signal but correlated in an unknown way with noise, $n_0(k)$. A classic example of this approach is the cancellation of the maternal heartbeat in fetal electrocardiography. Although this example uses more than one input, which is beyond the scope of our general discussion, the results are easily understood.

During recording of the fetal electrocardiogram (ECG), interference results from the maternal heartbeat, which has an amplitude 2 to 10 times greater than that of the fetal heartbeat. Interference also results from the background noise of muscle activity and fetal motion, with an amplitude greater than or equal to that of the fetal heartbeat. To minimize this interference, Widrow's group used four chest leads to record the maternal heartbeat and other multiple reference inputs for the canceller. A single abdominal lead recorded the combined maternal and fetal heartbeats that served as a primary input. The data were prefiltered with a bandwidth of 0.3 to 75 Hz, digitized with a sampling rate of 512 Hz, and recorded on tape. The recorded data were input to a multichannel LMS adaptive filter. Each of the reference inputs was processed with 32 coefficients with nonuniform (log periodic) spacing and a total delay of 129 ms.

As shown in Figure 3.9b, baseline drift and 60 Hz interference are clearly observable in the primary input obtained from the abdominal lead. Further, the maternal heartbeat (Figure 3.9a) dominates in the primary input recording. The maternal heartbeat and

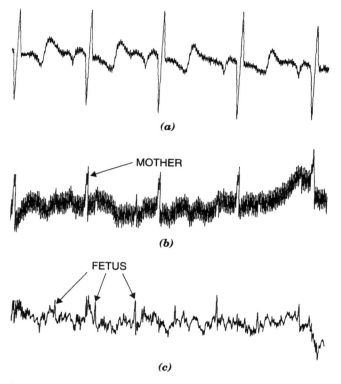

Figure 3.9 Result of adaptive noise cancellation within fetal ECG. (a) Reference input (chest lead). (b) Primary input (abdominal lead). (c) Noise canceller output. From [13].

three other chest recordings containing the 60 Hz interference served as a reference to reduce these interferences. In the filter output (Figure 3.9c), the fetal heartbeat is clearly discernable [13].

3.5.2 Adaptive Predictor

Up to this point, it has been assumed that an external reference source is readily available. If a reference source is not present, adaptive noise cancelling may still be viable under a special condition. In circumstances where a broadband signal is corrupted by periodic interference, a fixed delay may be applied to the primary input to create the reference input (Figure 3.10). The delay must be chosen to cause the broadband signal components in the reference input to become decorrelated from those in the primary input. Because of their periodic nature, the interference components will remain correlated with each other. This implementation is called an **adaptive predictor,** which is a type of **adaptive line enhancer.** We have already analyzed the outputs of an adaptive predictor in the example in Section 3.3.3.

3.5.3 Adaptive Self-Tuning Filter

Another application of adaptive noise cancelling is the **adaptive self-tuning filter,** which is a second type of adaptive line enhancer. Once again, we assume a primary input with

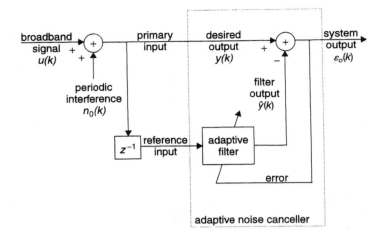

Figure 3.10 Adaptive predictor.

mixed broadband and periodic components and a reference input created from the delayed primary input. However, instead of isolating the broadband signal, we now wish to isolate the periodic signal. This is easily accomplished by taking the system output from the adaptive filter (Figure 3.11).

As an example, let us combine colored Gaussian noise and a sine wave to simulate a primary input (Figure 3.12a). We delay the primary input, and input this reference input to an LMS adaptive filter. As shown in Figure 3.12b, the adaptive filter output closely approximates the actual sine wave. If we look at the filter impulse response after convergence (Figure 3.13a), it bears a close resemblance to a sine wave. If the broadband input had been white noise, the impulse response would have been sinusoidal. The corresponding transfer function (Figure 3.13b) possesses a magnitude of nearly one at the frequency

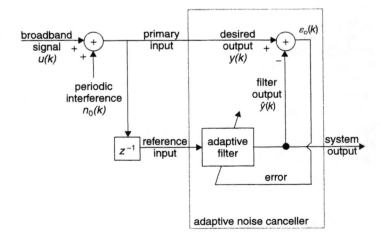

Figure 3.11 Adaptive self-tuning filter.

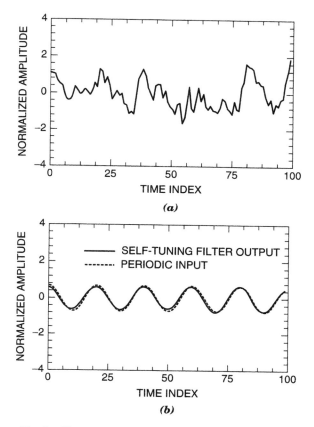

Figure 3.12 Adaptive self-tuning filter experiment. (a) Input signal (colored Gaussian noise plus sine wave). (b) Self-tuning filter output plotted with sine wave input. From [13].

of the interference. Although the phase shift at this frequency is not zero, the sum of this phase shift and the chosen delay forms an integral multiple of 360 degrees [13].

3.6 SUMMARY

In this chapter, we have reviewed the basic concepts underlying adaptive filters. An adaptive filter possesses a structure that is adjustable in such a way that its performance improves through contact with its environment. Such behavior results from the constraints that the signal is uncorrelated with the signal noise and reference noise, and that the signal noise and reference noise are correlated. The filter may be implemented with a linear or nonlinear model.

As discussed in Chapter 1, a linear time-invariant filter can be implemented as a finite or infinite response filter. The classic adaptive FIR filter is the adaptive linear combiner proposed by Widrow and Hoff. The coefficients of this filter are determined using the least mean squares algorithm. The theory underlying another algorithm, the recursive least squares algorithm, is discussed in Chapter 7. Research on adaptive IIR filters re-

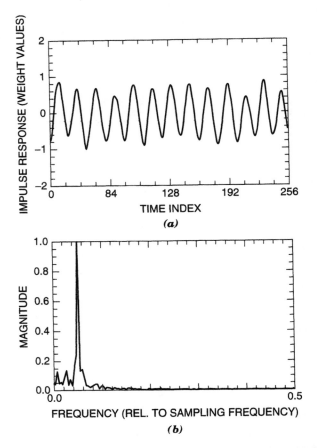

Figure 3.13 Adaptive self-tuning filter characteristics. (a) Impulse response after convergence. (b) Transfer function magnitude after convergence. From [13].

mains a current topic of investigation, as they are more efficient than FIR filters but are also subject to instability because of feedback coefficients. Two adaptive IIR filters, the hyperstable and simplified hyperstable adaptive recursive filters, are discussed in this chapter to illustrate the considerations involved in developing a stable adaptive IIR filter algorithm.

When an adaptive filter is used as an adaptive noise canceller, the reference noise source is filtered and subtracted from a primary input containing the signal and noise to eliminate the noise by cancellation. If an external reference noise source is not readily available, adaptive noise cancelling may still be viable. In circumstances in which a primary input consists of mixed broadband and periodic components, it may be delayed to create a reference noise source. The delay is chosen to cause the broadband signal components in the reference input to become decorrelated from those in the primary input. Because of their periodic nature, the interference components will remain correlated with each other. This filter is called an adaptive line enhancer. When the signal consists of the broadband components, the implementation is called an adaptive predictor. When the signal consists of the periodic components, the implementation is called a self-tuning filter.

3.7 REFERENCES

[1]. Howells, P. *Intermediate frequency side-lobe canceller.* U.S. Patent 3,202,990. Aug. 24, 1965.

[2]. Widrow, B. and Hoff, M., Jr. Adaptive switching circuits. *IRE WESCON Conv Rec,* Part 4: 96–104, 1960.

[3]. Rosenblatt, F. The perceptron: a perceiving and recognizing automaton, Project PARA. *Cornell Aeronaut Lab Rep* 85-460-1, Jan, 1957.

[4]. Gabor, D., Wilby, W. P. L., and Woodcock, R. A universal nonlinear filter predictor and simulator which optimizes itself by a learning process. *Proc Inst Elect Eng, 108B,* 1960.

[5]. Widrow, B. and Stearns, S. D. *Adaptive Signal Processing.* Prentice Hall: Englewood Cliffs, NJ, 1985.

[6]. Widrow, B., McCool, J. M., Larimore, M. G., and Johnson, C. R., Jr. Stationary and nonstationary learning characteristics of the LMS adaptive filter. *Proc IEEE, 64,* 1151–1162, 1976.

[7]. Widrow, B. and McCool, J. M. Comments on "an adaptive recursive LMS filter." *Proc IEEE, 65,* 1402–1404, 1977.

[8]. Larimore, M.G., Treichler, J. R., and Johnson, C. R., Jr. SHARF: an algorithm for adapting IIR digital filters. *IEEE Trans ASSP,* ASSP-28, 428–440, 1980.

[9]. Johnson, C. R., Jr. A convergence proof for a hyperstable adaptive recursive filter. *IEEE Trans Inform Theory,* IT-25, 745–749, 1979.

[10]. Williamson, G. A. and Zimmermann, S. Globally convergent adaptive IIR filters based on fixed pole locations. *IEEE Trans Sig Proc, 44,* 1418–1427, 1996.

[11]. Regalia, P. A. *Adaptive IIR filtering in signal processing and control.* Marcel Dekker: New York, 1995.

[12]. Shynk, J. J. Adaptive IIR filtering. *IEEE ASSP Mag, 6,* 4–21, 1989.

[13]. Widrow, B., Glover, J. R., Jr., McCool, J. M., Kaunitz, J., Williams, C. S., Hearn, R. H., Zeidler, J. R., Dong, E., Jr., and Goodlin, R. C. Adaptive noise cancelling: principles and applications. *Proc IEEE, 63,* 1692–1716, 1975.

Further Reading

System Identification, and Equalization: Haykin, S. *Adaptive Filter Theory,* 2nd ed., Prentice Hall: Englewood Cliffs, NJ, 1991.

Transform Domain Filters: Shynk, J. J. Frequency-domain and multirate adaptive filtering. *IEEE SP Mag, 9,* 14–37, 1992.

Nonlinear Adaptive Filters: Haykin, S., *Neural Networks: a Comprehensive Foundation.* 2nd ed., Prentice Hall: Englewood Cliffs, NJ, 1999.

Kalluri, S. and Arce, G. R., Adaptive weighted myriad filter algorithms for robust signal processing in α-stable noise environments. *IEEE Trans Sig Proc, 46,* 322–344, 1998.

3.8 RECOMMENDED EXERCISES

FIR Filters. See Ref. [5], Exercises 6.1, 6.8, 6.9, 6.10.
IIR Filters. See Ref. [5], Exercises 8.10 and 8.11.
Adaptive Noise Cancellation. See Akay, M. *Biomedical Signal Processing.* Academic Press: San Diego, 1994. Computer Experiments 7.2, 7.3, 7.5, 8.1, and 8.3.i.

4

IMPROVED PULSE OXIMETRY

In this chapter, we discuss an industrial application of the adaptive noise canceller: elimination of noise artifact from pulse oximetry monitoring. During **pulse oximetry, tissue oxygenation** is estimated by monitoring oxygen **saturation** in arterial blood. Unfortunately, motion and other artifacts in the saturation data may bias these estimates. The adaptive filter increases the signal-to-noise ratio by canceling the effects of these noise sources.

4.1 PHYSIOLOGY OF OXYGEN TRANSPORT

Metabolism, the totality of chemical reactions that occur in animal cells, can be subdivided into two major categories. During **catabolism,** the processes related to degradation of complex substances with subsequent generation of energy, oxygen, O_2, is required for the **oxidation** of organic substances. During **anabolism,** the processes concerned primarily with the assembly of complex organic molecules, carbon dioxide, CO_2, is released. The process during which these gases are exchanged between the cells and their surroundings is known as **respiration.**

As shown in Figure 4.1, gas transport occurs as four subprocesses. During the **ventilation** portion of pulmonary respiration, oxygen is transported from the outside air by **convective,** or bulk flow, transport to the **alveoli** of the lung. Each alveolus is a tiny, thin-walled sac in the lungs filled with **capillaries,** or small blood vessels. During the alveolar **diffusion** portion of pulmonary respiration, oxygen is moved along a decreasing concentration gradient from the alveoli to the blood in the lung capillaries. During blood gas transport, the circulating blood convectively transports oxygen to tissue capillaries. Finally, during tissue respiration, oxygen is diffused from the tissue capillaries into the surrounding cells. The removal of the carbon dioxide formed as a byproduct of anabolism occurs by the same four subprocesses, in reverse order.

Blood vessels that transport oxygenated blood are called arteries. Blood vessels that transport deoxygenated blood are called veins.

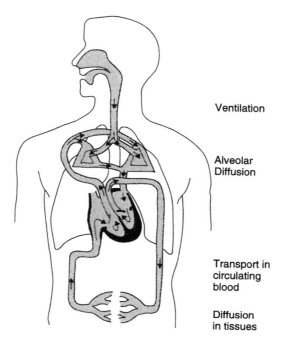

Figure 4.1 Gas transport.

4.1.1 Pulmonary Respiration

Ventilation of the alveoli necessary for gas exchange is the result of rhythmic alternation of **inspiration,** or breathing in, and **expiration,** or breathing out. Expansion of the **thoracic,** or chest, cavity is due to elevation of the ribs and flattening of the diaphragm. During lung expansion, pressure is reduced in the lungs; this pressure differential causes air to flow into the body. Ventilation depends on the **tidal volume,** or volume of each breath (V), and **respiration frequency,** or number of breaths per unit time (f_r).

During inspiration, the inspired air is partly cleaned as it passes through the nose. It also begins to be moistened and warmed as it passes through the nose and throat, eventually becoming 100% saturated with water vapor and 37°C in the lower airways. The **glottis,** or opening to the **trachea,** or windpipe, dilates, causing the inspired air to flow at the highest velocity. The air flows through the trachea, through its branches that are called the **bronchi,** and through the tubular extensions called **bronchioles.** Because the bronchi are also dilated due to stimulation by the **sympathetic nerves,** airflow is facilitated to the alveoli (Figure 4.2). The volume of the conducting airways, from the trachea to the brochioles, is called the **anatomical dead space** because gas is not exchanged.

Calculations of ventilation are based on the expiratory phase. The **expiratory tidal volume,** V_E, is composed of the **dead space tidal volume,** V_D, and **alveolar space tidal volume,** V_A, such that

$$V_E = V_D + V_A \tag{4.1}$$

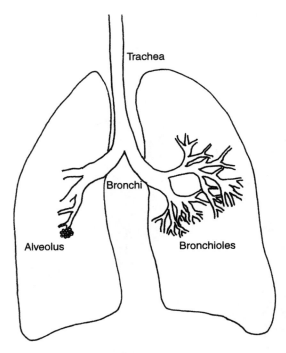

Figure 4.2 Air flow path through the lungs. Right lobe shows detail of bronchioles. Left lobe shows branching up to alveolus.

The minute volume, the gas volume breathed in or out per minute ($\dot{V}_E$), is by definition the product of the tidal volume and respiratory frequency:

$$\dot{V}_E = V_E \cdot f_r \tag{4.2}$$

The respiratory frequency of a typical adult is 14 breaths per minute. With a typical tidal volume of 0.5 liters/min, the typical adult minute volume is 7 liters/min. Taking the derivative of Eq. (4.1) provides an alternate calculation for the minute volume, where

$$\dot{V}_E = \dot{V}_D + \dot{V}_A \tag{4.3}$$

$\dot{V}_D$ is the dead space ventilation and $\dot{V}_A$ is the alveolar ventilation. Typically, the alveolar volume is 70% of the tidal volume.

Within the alveoli, the gas concentrations can be determined using **Dalton's law.** By Dalton's law, each gas in a mixture exerts a **partial pressure**, P_{gas}, proportional to its share of the total volume, F_{gas}. The alveolar gas mixture contains oxygen, carbon dioxide, nitrogen, a small amount of noble gases, and water vapor. The alveolar oxygen and carbon dioxide partial pressures, $P_{A\,O2}$ and $P_{A\,O2}$, respectively, depend on the ratio of alveolar ventilation to **lung perfusion,** $\dot{Q}$, the lung blood supply. In a person at rest with healthy lungs, the ventilation–perfusion ratio, $\dot{V}_A/\dot{Q}$, equals 0.9 to 1.0.

The alveolar partial pressure of oxygen is kept high ($P_{A\,O2}$ = 100 mmHg), whereas the oxygen partial pressure of venous blood entering the lung capillaries is considerably lower ($P_{v\,O_2}$ = 40 mmHg). In contrast, the alveolar partial pressure of carbon dioxide is kept low ($P_{A\,CO_2}$ = 40 mmHg), whereas the carbon dioxide partial pressure of venous blood is higher ($P_{v\,CO_2}$ = 46 mmHg). These partial pressure differences drive the diffusion of oxygen and carbon dioxide between adjacent capillary and alveolus cells, a distance of only 0.5 μm. The net effect of gas transport is reflected in the partial pressures in the systemic arteries. In a healthy young adult, the arterial oxygen partial pressure, P_{aO_2}, is 95 mmHg; the arterial carbon dioxide partial pressure, $P_{a\,CO_2}$, is 40 mmHg.

Respiration is controlled by respiratory neurons in the brain stem.

4.1.2 Blood Gas Transport

Since oxygen and carbon dioxide, like other gases, are not very soluble in blood, they can not be readily transported by blood. Instead, oxygen from the lungs and carbon dioxide from the tissues bind to molecules of **hemoglobin** within **erythrocytes,** or red blood cells. Hemoglobin consists of four **protein** chains. A **heme** molecule within each chain contains the iron ion, Fe^{++}. This ion may reversibly bind to an O_2 or CO_2 molecule without a change in **valence,** or charge, resulting in **oxyhemoglobin** (HbO_2) or **deoxyhemoglobin** (Hb), respectively.

The red color of a hemoglobin solution such as blood results from the relatively strong absorption of short-wavelength light by this substance. Thus, a considerable amount of the light in the blue part of the spectrum is absorbed, whereas most of the red, long-wavelength light is transmitted. Deoxygenated blood absorbs light more strongly than oxygenated blood at long wavelengths and less strongly at short wavelengths. Therefore, venous blood appears darker, with a bluish-red color, than arterial blood.

Functional arterial oxygen saturation ($S_a O_2$) is defined as the ratio of oxyhemoglobin concentration to the total concentration of arterial hemoglobin available for reversible oxygen binding:

$$S_a O_2 = \frac{[HbO_2]}{[Hb] + [HbO_2]} \tag{4.4}$$

where [ ] denotes concentration and it is assumed that other species of hemoglobin are not present. According to the **law of mass action,** the oxygen saturation of hemoglobin depends on the prevailing oxygen partial pressure. This relationship is graphically represented by the oxygen **dissociation curve** (Figure 4.3). As shown by the sigmoidal nature of the dissociation curve, oxygen transport is very efficient. Hemoglobin is nearly fully saturated in the lungs, yet is efficiently released in the capillaries. At low oxygen pressures, hemoglobin acts as if it were binding oxygen very weakly, yet as more oxygen is bound the affinity becomes greater. This sigmoidal curve is most simply characterized by the half-saturation pressure, P_{50}, the oxygen partial pressure corresponding to 50% oxygen saturation. For arterial blood under normal conditions (pH = 7.4, temperature = 37°C), P_{50} is approximately 27 mmHg. The dissociation curve is affected by temperature, pH, $P_{a\,CO_2}$, and disease states.

Like oxygen, carbon dioxide bound hemoglobin is dependent on the prevailing carbon dioxide partial pressure. However, the CO_2 dissociation curve differs from that for oxygen in that CO_2 binding exhibits no saturation. The chemical binding of CO_2 is also more complicated because the acid–base balance in the blood must be maintained.

Oxygenated blood travels through the **left atrium** to the **left ventricle** of the heart,

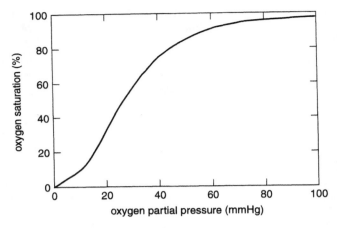

Figure 4.3 Oxygen dissociation curve of hemoglobin for pH = 7.4 and temperature = 37°C

where it is pumped through the aorta to arteries. The arteries branch into smaller **arterioles** and even smaller capillaries. Deoxygenated blood in the capillaries travels through **venules,** larger veins, and the **right atrium** to the **right ventricle,** where it is pumped to the lungs (Figure 4.1). The circulatory system is described in greater detail in Chapter 6.

4.1.3 Tissue Respiration

Like gas exchange in the lungs, the exchange of respiratory gases between the capillary blood and tissue cells occurs by diffusion. However, the particular partial pressures of oxygen and carbon dioxide in different parts of the circulatory system differ from those of the lung. The rate of respiration is determined by the partial-pressure gradient, the size of the exchange area, the length of the diffusion path, and the magnitude of the diffusion resistance of various structures through which the molecules pass.

4.1.4 Hypoxia

When the supply of oxygen to a tissue is inadequate to meet its needs, **tissue hypoxia** occurs. This condition may be caused by a disturbance in respiratory gas exchange in the lung or gas transport in the blood. During **arterial hypoxia,** the ventilation–perfusion ratio, $\dot{V}_A/\dot{Q}$, decreases. Consequently, the arterial partial pressure and concentration of oxygen decrease. During severe arterial hypoxia, the oxygen supply to the tissues is restricted and only moderate physical exertion is possible. Under these conditions, the oxygen partial pressure in the capillary blood can fall to very low levels, so that **venous hypoxia** occurs. Once the oxygen partial pressure gradient between the blood and tissue no longer suffices to enable an adequate release of oxygen, energy metabolism is restricted.

During **anemic hypoxia,** the oxygen capacity of the blood is diminished. Anemic hypoxia is caused by blood loss, inadequate hemoglobin synthesis (**anemia**), formation of **methemoglobin** (MetHb), or carbon monoxide poisoning. Methemoglobin is a hemoglobin mutation that does not bind oxygen. Carbon monoxide binds with greater **affinity** to hemoglobin than does oxygen. Even in low concentrations, carbon monoxide can displace oxygen from hemoglobin to form **carboxyhemoglobin** (COHb), and make it unavailable for oxygen transport.

During **ischemic hypoxia,** organ perfusion is restricted. First, the normal amount of oxygen is extracted from the blood as it flows through the capillaries, so a greater **arteriovenous** difference in oxygen concentration exists. As a direct consequence, the oxygen partial pressure drops substantially along the capillary. Because the oxygen partial pressure gradient between blood and tissue is lowered concurrently, the oxygen supply to the cells may become inadequate.

Continuous monitoring of arterial oxygen saturation does not identify hypoxia. Rather, it identifies **hypoxemia,** insufficient oxygen in the blood. However, of currently monitored variables, arterial oxygen saturation is the best indication that any oxygenation problem exists or is about to occur. Arterial oxygen saturation is continuously monitored during surgery, surgical recovery, and postoperative and critical care.

4.2 IN VITRO OXYGEN MEASUREMENTS

The most common method for measuring arterial oxygenation is the **Clark P_{O_2} electrode.** Using this electrode, the current produced by a negative electrode in a blood sample is directly proportional to the availability of oxygen molecules at the electrode tip. Although this electrode may be used to make **in vivo,** or within the organism, measurements of the arterial partial pressure of oxygen when used as a **catheter** electrode, it is more commonly used within **blood gas analyzers** (Figure 4.4). Because blood gas analyzer **in**

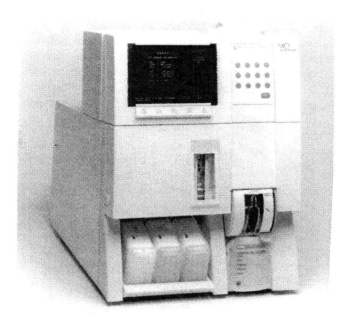

Figure 4.4 IL1640 Blood Gas Analyzer. Courtesy of Instrumentation Laboratory, Lexington, MA.

vitro (or outside of the organism) measurements are time consuming and require that blood be drawn, in vitro continuous **oximetry** methods have been investigated as an alternative. Oximetry refers to the optical measurement of oxyhemoglobin saturation in the blood. Recall that the relationship between oxygen saturation and oxygen partial pressure is the oxygen dissociation curve given in Figure 4.3.

4.2.1 Pulse Oximetry Calculations

As described by the **Beer–Lambert law, transmitted light,** I, resulting from **incident light,** I_0, traveling through a uniform medium of **path length,** l, containing an absorbing substance with **concentration,** c, decreases as

$$I = I_0 e^{-\varepsilon(\lambda)cl} \tag{4.5}$$

Here, ε is the **extinction coefficient** of the absorbing substance at a specific **wavelength,** λ. The **unscattered absorbance,** or **optical density,** $A(\lambda)$, of this process is the negative natural log of the ratio of transmitted light to incident light:

$$A(\lambda) = -\ln\frac{I}{I_0} = \varepsilon(\lambda)cl \tag{4.6}$$

When more than one substance absorbs light in a medium, each absorber contributes its part to the **total absorbance,** $A_t(\lambda)$, as

$$A_t(\lambda) = \sum_{i=1}^{n} \varepsilon_i(\lambda)c_i l_i \tag{4.7}$$

where n equals the number of absorbers.

A basic demonstration of oximetry involves transmitting two discrete wavelengths of light, one at a time, through a blood sample of known length containing only oxyhemoglobin and deoxyhemoglobin, and detecting the transmitted light with one **photodetector** (Figure 4.5). Because the extinction coefficients of oxyhemoglobin and deoxyhemoglobin differ at each wavelength (Figure 4.6), their respective concentrations can be

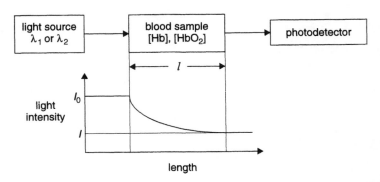

Figure 4.5 Oximetry experimental apparatus.

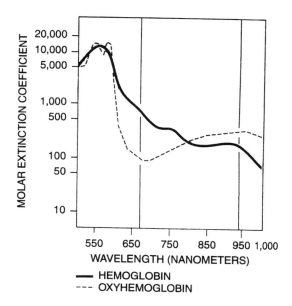

Figure 4.6 The extinction coefficients of deoxyhemoglobin (or hemoglobin) and oxyhemoglobin as functions of wavelength. Wavelengths employed by the pulse oximeter (660 and 940 nm) are indicated. Reprinted from [1] with kind permission from Kluwer Academic Publishers.

determined from Eq. (4.7) by solving for two unknown concentrations using two wavelength equations. These concentrations are then substituted into Eq. (4.4) to determine the oxygen saturation. The **co-oximeter** uses four wavelengths to determine deoxyhemoglobin, oxyhemoglobin, carboxyhemoglobin, and methemoglobin from a single blood sample.

Unfortunately, light traveling through biologic tissue such as the finger or earlobe is absorbed by more than two substances. Primary absorbers are the skin pigmentation, bones, and the arterial and venous blood. The light is also attenuated by light scattering, geometric factors, and characteristics of the light emitter and detector elements. Using two wavelengths, the total absorbance can be written as

$$A_t(\lambda_1) = \varepsilon_o(\lambda_1)c_o l_o + \varepsilon_d(\lambda_1)c_d l_d + \varepsilon_x(\lambda_1)c_x l_x + A_y(\lambda_1) \qquad (4.8)$$

$$A_t(\lambda_2) = \varepsilon_o(\lambda_2)c_o l_o + \varepsilon_d(\lambda_2)c_d l_d + \varepsilon_x(\lambda_2)c_x l_x + A_y(\lambda_2) \qquad (4.9)$$

where the subscript o refers to arterial oxyhemoglobin, d refers to arterial deoxyhemoglobin, x refers to variable absorbances not from arterial blood, and y refers to nonspecific sources of optical attenuation. It is assumed that dysfunctional forms of hemoglobin (i.e., MetHb and COHb) are not present.

To isolate the contributions of arterial blood, only the pulsating absorbances are analyzed, thus leading to "pulse" oximetry. By taking the time derivative of the absorbances, the last two constant terms in Eqs. (4.8) and (4.9) go to zero. Additionally, an assumption is made that the blood path length changes, dl_o/dt and dl_d/dt, are equivalent. Thus, the ratio, R, of the two absorbance derivatives remains constant:

74 IMPROVED PULSE OXIMETRY

$$R = \frac{\frac{dA_t(\lambda_1)}{dt}}{\frac{dA_t(\lambda_2)}{dt}} = \frac{\varepsilon_o(\lambda_1)c_o\frac{dl_o}{dt} + \varepsilon_d(\lambda_1)c_d\frac{dl_d}{dt}}{\varepsilon_o(\lambda_2)c_o\frac{dl_o}{dt} + \varepsilon_d(\lambda_2)c_d\frac{dl_d}{dt}} \qquad (4.10)$$

$$R = \frac{\varepsilon_o(\lambda_1)c_o + \varepsilon_d(\lambda_1)c_d}{\varepsilon_o(\lambda_2)c_o + \varepsilon_d(\lambda_2)c_d} \qquad (4.11)$$

Recalling from Eq. (4.4) that functional arterial oxygen saturation is calculated from $c_o = [HbO_2]$ and $c_d = [Hb]$ leads to

$$S_pO_2 = \frac{\varepsilon_d(\lambda_1) - \varepsilon_d(\lambda_2)R}{[\varepsilon_d(\lambda_1) - \varepsilon_o(\lambda_1)] - [\varepsilon_d(\lambda_2) - \varepsilon_o(\lambda_2)]R} \qquad (4.12)$$

where S_pO_2 denotes S_aO_2 measurements made by pulse oximetry.

In actual implementation, common **light emitter diodes** (LEDs) are used as light sources and **photodiodes** are used as photodetectors. The wavelengths chosen are red near 660 nm and near-infrared in the range of 890–950 nm. LEDs are not monochromatic light sources, but have bandwidths between 20–50 nm. Further, the assumption that the two path length changes are equal is not necessarily true, and second-order scattering effects have been neglected. To account for these deviations from theory, an empirical relationship between R and S_pO_2 is determined for each unique pulse oximetry sensor design. The relationship is modeled as a function of form

$$S_pO_2 = \frac{a - bR}{c - dR} \qquad (4.13)$$

A typical calibration curve is shown in Figure 4.7. In preparation for monitoring, an oximeter probe containing two LEDs and one photodiode is placed on an appendage such

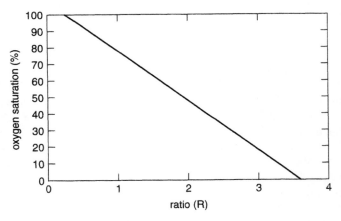

Figure 4.7 Typical pulse oximetry calibration curve.

as a finger or ear lobe. During each heartbeat, the ratio, R, is measured, and a corresponding S_pO_2 value from a lookup table in the pulse oximeter memory is displayed.

4.2.2 Ear Oximeters

The first oximeter was developed by Ludwig Nicolai in 1931. His apparatus consisted of the blue-green bands of a mercury vapor light, a rotating wheel filter, a vacuum photoelectric cell, and a vacuum tube amplifier. By occluding the circulation, Nicolai obtained exponential curves for the decay of oxyhemoglobin. Nicolai's associate Kurt Kramer made the first precise measurements of oxygen saturation using an incandescent light and a red filter. He confirmed that the Beer–Lambert law was precisely correct in hemoglobin solution.

During World War II, Glenn Millikan, the son of famed physicist Robert Millikan, developed a lightweight ear oxygen meter for which he coined the term "oximeter." Funded by the United States military, the device consisted of a servo system in which the oximeter reading controlled the oxygen supply to an aviator's mask, which was built with an attached oximeter. Infrared light was transmitted through a green filter to the ear lobe, and was detected by a photocell. A few years later, in 1948, Earl Wood modified Millikan's earpiece, improving the infrared filter and adding an inflatable balloon with which the ear could be made bloodless for initial zero setting. Around 1964, Robert Shaw constructed a self-calibrating, eight-wavelength ear oximeter. Hewlett-Packard marketed this device in 1970. Because of its large earpiece and high cost, it was rarely used for clinical monitoring but was widely used in research laboratories [2].

4.2.3 Pulse Oximeters

Conventional pulse oximetry was invented by Takuo Aoyagi at Nihon Kohden in 1972. Using the ratio of red to infrared light absorption of the pulsating components within the measuring site, which were assumed to only be arterial blood, oxygen saturation was calculated without calibration. Although considering only pulsating components eliminated the need for calibration, this oximeter, and later the pulse oximeter from Minolta, used the Beer–Lambert law, and grossly overestimated S_pO_2 readings under 90% [3].

To compensate for this overestimation, more recent pulse oximeters base their calculations on **calibration curves** derived from studies in healthy volunteers. The Biox Company of Boulder, Colorado, was the first to file patents for a calibration curve method in 1981. However, Biox, now Ohmeda, only marketed oximeters to pulmonary function laboratories. In contrast, Willliam New, a physician who founded Nellcor with Jack Lloyd in 1981, recognized the potential importance of pulse oximeters. Nellcor marketed their oximeters to all hospital units with patients that were limited in their ability to regulate their own oxygen supply. When the market for pulse oximetry expanded, Ohmeda sued Nellcor for patent infringement; the case was later settled out of court [4]. By 1989, when pulse oximeters were evaluated in the June issue of *Health Devices,* 14 manufacturers were marketing this technology as a standalone device or monitor module [5].

4.3 PROBLEM SIGNIFICANCE

Continuous monitoring of arterial hemoglobin oxygen saturation using pulse oximetry has been a standard of care in the operating room since 1990 and in the recovery room since 1992 [6]. As stated by Severinghaus and Astrup in their historical review of oximetry, "pulse oximetry is arguably the most significant technological advance ever made in monitoring the well-being and safety of patients during anesthesia, recovery, and critical care" [2].

4.3.1 Clinical Significance

Despite universal agreement on its importance, pulse oximetry measurements are not always accurate. Oxygen saturation measurements using pulse oximetry are accuracy-limited by the **signal-to-noise ratio,** SNR. Motion artifact may induce spurious pulses similar to arterial pulses, usually yielding a red-to-infrared absorbance ratio near 1. A ratio of 1 is associated with different S_pO_2 values for different instruments. For example, the early Ohmeda Biox 3700 monitor displayed an S_pO_2 reading of 85% under these circumstances [7]. In a prospective study of 9578 recovery room patients, patient motion was responsible for 56% of the 106 cases in which pulse oximetry was completely discontinued and in 29% of the 123 cases in which it was temporarily discontinued [8]. Similarly, **electrocautery** induces spurious pulses. Electrocautery devices are used during surgery to cut and stop bleeding by applying a radio frequency spark between a probe and tissue. Electrocautery artifact is caused by the wide spectrum radio frequency emissions detected by the photodiode in the pulse oximetry sensor.

The SNR is also decreased during **poor peripheral perfusion.** When low-amplitude pulses are present as a result of **hypovolumia** (low volume), **hypotension** (low blood pressure), **hypothermia** (low temperature), or **cardiac bypass,** pulse oximetry readings may be intermittent or unavailable. In the same prospective study of 9578 recovery room patients referred to previously, poor peripheral perfusion was responsible for 20% of the 106 cases in which pulse oximetry was completely discontinued and in 46% of the 123 cases in which it was temporarily discontinued [8]. Unfortunately, accurate readings are most desirable during these situations.

Pulse oximetry monitoring is also limited by high concentrations of **dysfunctional** hemoglobins (i.e., methemoglobin, carboxyhemoglobin) or interference from physiologic **dyes** such as **methylene blue.** Because calibration curves are based on data from healthy volunteers, the calibration data is based on S_pO_2 levels in the range of 70 to 100%, which can be easily induced. S_pO_2 levels below 70% are extrapolated from these data points, and may not be accurate [3].

4.3.2 Traditional Empirical Methods

The first pulse oximeters did not compensate for low SNR due to motion artifact or low perfusion. Although second-generation products attempted compensation, the implemented "signal processing" consisted of **empirical** processing of the input signals. For example, the Nellcor N-200 pulse oximeter implemented an algorithm called C-LOCK. **Electrocardiogram** (ECG) as well as pulse oximetry data were input. According to Nellcor's patent, the occurrence of the R wave portion of the ECG signal was detected, and the time delay by which an arterial pulse followed the R wave was determined. With this information, a time window was derived in which arterial waveform features for the ratio calculation were to be expected. Signals within the time window were evaluated for correlation with past arterial waveforms based on shape criteria (Figure 4.8) [9]. In a clinical study of 4 Hz motion artifact, the C-LOCK algorithm, which can be powered off in the Nellcor

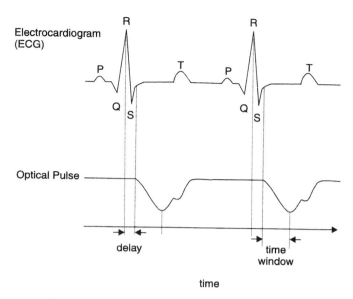

Figure 4.8 Nellcor C-LOCK patent algorithm. Based on [9].

N-200, was shown to double the total error rate, compared to the Nellcor N-200 without the C-LOCK algorithm, from 3.44 to 6.88%. In this study, total error rate was defined as the sum of the average number of times the difference between the instrument saturation and the reference saturation was more than 3% in absolute value, plus the average number of times the instrument displayed saturation zeros [10].

Similarly, Criticare pulse oximeters implemented an algorithm in which the R wave portion of the ECG signal was used to determine a time window for valid pulse oximetry arterial waveforms, which were then ensemble averaged to minimize noise [11]. In general, empirical methods for noise compensation perform poorly, because they can not compensate for noise sources not evaluated during bench top simulations. Because of the complexity of the hospital environment, all noise sources cannot be readily predicted and simulated.

4.3.3 Digital Signal Processing Methods

After the **start-up** company Masimo began publishing results of noise compensation using adaptive filtering in 1995 [10], other manufacturers began to utilize digital signal processing. Masimo was founded by in 1989 by two engineers: Massi Joe Kiani and Mohamed Diab. The Masimo algorithm is discussed in Section 4.4. In reaction to Massimo's algorithm, which can be described as an **adaptive comb filter,** Nellcor attempted to minimize noise artifact by applying an adaptive comb filter to digital pulse oximetry waveforms, based on the **harmonics** of the heart rate. An adaptive comb filter is a filter that selectively removes evenly spaced frequency bands. According to Nellcor's patent, the heart rate was determined through an empirical algorithm based on shape. Once filtered, the data were averaged using a **Kalman filter,** a recursive filter described by state equations. They were then processed to calculate the ratio, R, and a corresponding S_pO_2 value [12].

Even if it can be assumed that the heart rate and pulse oximetry spectra completely overlap, the noise spectra may also overlap in these frequency bands. It is possible that signal spectra may be removed while noise spectra may be retained during Nellcor's implementation of adaptive comb filtering. Further, averaging the remaining signal and noise by the Kalman filter may decrease the accuracy of calculated S_pO_2 values. The Nellcor algorithm, which is implemented in the Nellcor N-3000 monitor, is compared to the Masimo algorithm in Section 4.4.3.

4.3.4 Market Forces

By the mid 1980s, Ohmeda, the first manufacturer of calibration curve pulse oximeters, and Nellcor, the first manufacturer to expand pulse oximetry markets, dominated the pulse oximetry market. Nellcor was also the first manufacturer of **single-use finger probes** [5]. As the concern for sterility and cost containment increased, Nellcor convinced hospitals to use only their single-use probe within various hospital units. (Remember that oximeters from various manufacturers have different calibration curves and report different values. **Standardization** is necessary if data is to be shared within units and disposables are to be purchased in bulk at reduced prices.) With this strategy, Nellcor became the dominant pulse oximetry manufacturer. Nellcor manufactures its own monitors, and sells monitors to other manufacturers to market under other brand names. All their monitors use the Nellcor disposable probe.

In 1997 alone, hospital sales of pulse oximeters and their disposable probes generated $32,011,546 and $197,494,350, respectively, in the United States. These sales estimates exclude sales in federal hospitals and nursing homes. Nellcor accounted for 88.1% of total disposable sales [13]. Annual sales are expected to decline when the pulse oximetry probe patent expires in 2001 [14]. (Patents issued before 1995 are protected for 17 years from the date of issue.)

Nellcor's success from a small start-up company to a major corporation that purchased Puritan-Bennett, and was later purchased by Mallinkrodt, has reached mythical proportions in the **venture capital** community. Venture capital refers to investment money for new technologies in which the risk for loss and the potential for profit are considerable. 5½ years after inception, Nellcor became a public company with stock worth $250 million [15]. Every new medical instrumentation start-up wants to become "the next Nellcor," inventing the market for a new **vital sign** measurement.

4.4 ADAPTIVE NOISE CANCELLATION IN MASIMO SOFTWARE

As described in Chapter 3, an adaptive filter requires a reference noise source that is uncorrelated from the primary input signal. Unfortunately, a reference noise source is not readily present in pulse oximetry. In this section, the method by which Masimo derived a reference noise source is described. The use of the reference noise source in adaptive noise cancellation is highlighted.

4.4.1 Calculation of a Reference Noise Source

The derivation in Section 4.2.1 was simplified by omitting the discrete time dependence of all variables. Including this dependence, Eq. (4.11) can be rewritten to describe the ratio, R, of the input signal, $u_{\lambda_1}(k)$, for each wavelength, λ_1 and λ_2:

$$R(k) = \frac{\varepsilon_o(\lambda_1, k)c_o(k) + \varepsilon_d(\lambda_1, k)c_d(k)}{\varepsilon_o(\lambda_2, k)c_o(k) + \varepsilon_d(\lambda_2, k)c_d(k)} = \frac{u_{\lambda_1}(k)}{u_{\lambda_2}(k)} \quad (4.14)$$

It was also assumed in the derivation that noise was not present. Let us define the signals detected by the photodiodes, the primary signals $y_{\lambda_i}(k)$, as the combination of the input and noise, $n_{\lambda_i}(k)$, so that

$$y_{\lambda_1}(k) = u_{\lambda_1}(k) + n_{\lambda_1}(k) \quad (4.15)$$

$$y_{\lambda_2}(k) = u_{\lambda_2}(k) + n_{\lambda_2}(k) \quad (4.16)$$

Rearranging Eqs. (4.15) and (4.16) and substituting them into Eq. (4.14) yields

$$R(k) = \frac{y_{\lambda_1}(k) - n_{\lambda_1}(k)}{y_{\lambda_2}(k) - n_{\lambda_2}(k)} \quad (4.17)$$

Cross-multiplication and rearrangement results in

$$R(k)y_{\lambda_2}(k) - R(k)n_{\lambda_2}(k) = y_{\lambda_1}(k) - n_{\lambda_1}(k) \quad (4.18)$$

$$n_{\lambda_1}(k) - R(k)n_{\lambda_2}(k) = y_{\lambda_1}(k) - R(k)y_{\lambda_2}(k) \equiv n_1(k) \quad (4.19)$$

Note that we have defined the reference noise source, $n_1(k)$, as a linear combination of the detected noise sources and the ratio. The reference noise source will effectively vary with motion artifact and poor perfusion. However, without noise, $n_1(k)$ equals zero and

$$R(k) = \frac{y_{\lambda_1}(k)}{y_{\lambda_2}(k)} \quad (4.20)$$

Again using Eq. (4.19), the reference noise source is derived from a linear combination of the detected signals and the ratio.

Unfortunately, the reference noise source depends on the ratio, R, which is the parameter being monitored to calculated S_pO_2. As a consequence, 117 possible values of the ratio which correspond to uniformly spaced S_pO_2 values from 34.8 to 105% are used to calculate candidate reference noise sources. Each candidate reference noise source is then input to the adaptive noise canceller with the infrared primary input; a corresponding filter output is determined. It is assumed by Masimo that the peak of the output power at the highest saturation corresponds to the arterial saturation. In other words, the output power for each candidate ratio and corresponding S_pO_2 can be plotted as a function of S_pO_2. The peak on this plot at the highest saturation is assumed to correspond to the arterial saturation, as an uncorrelated primary input and reference noise source are assumed by Massimo to result in a higher output power and arterial blood has the highest saturation (Figure 4.9) [16]. A second peak at a lower saturation may correspond to venous saturation, but this has not been confirmed by experiment.

4.4.2 Data Acquisition and Processing

According to their patents, Masimo data originate with a red LED with wavelength 660 nm and an infrared LED with wavelength 940 nm [17]. The LEDs are driven by an **emit-**

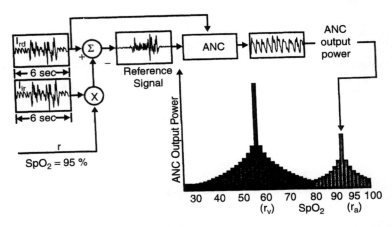

Figure 4.9 S_pO_2 selection, based on peak with highest saturation. ANC = adaptive noise cancellation. Courtesy of Masimo Corporation, Irvine, CA.

ter current driver to provide light transmission with **digital modulation** at 625 Hz. Within each cycle of 1/625 = 1.6 msec, the LEDs are sequenced such that the red LED is powered on for 0.4 msec, both LEDs are off for 0.4 msec, the infrared LED is powered on for 0.4 msec, and both LEDs are powered off for 0.4 msec. With both LEDs powered off, an estimate of the **ambient light** is obtained. The LEDs are driven at a power level that provides an acceptable intensity for detection.

In response to light from either LED, one photodetector signal is amplified and passed to an **antialiasing** single-pole lowpass filter with a corner frequency of 10 kHz. The output is then digitized to 16 bits/sample at a sample frequency of 20 kHz. After **demodulation** and further processing to obtain red and infrared data at 625 Hz, the data are decimated by a factor of 10 to 62.5 Hz. After bandpass filtering to obtain data within the heartbeat range of 34 to 250 beats per minute (bpm), 117 possible reference noise sources are calculated from 570 sample packets of red and infrared data. Each reference noise source is input with the corresponding infrared data to an FIR adaptive noise filter with 150 weights. The true arterial saturation is chosen as the output power peak with the highest saturation (Figures 4.9 and 4.10) [16]. The sequence of saturation calculations is repeated once per second. The mean of the most recent six calculated saturations is output to the monitor display [17].

4.4.3 Data Analysis

To validate this system, Masimo sponsored several clinical studies through support of equipment costs and volunteer fees. In one study, 10 healthy volunteers were monitored by three different pulse oximeters: the Nellcor N-200, with the C-Lock algorithm not used since it is known to degrade performance [10]; the Nellcor N-3000 that uses Kalman filtering; and a Masimo prototype that uses adaptive noise cancellation. A disposable sensor corresponding to each monitor was randomly placed on the index, middle, or ring finger of each subject's dominant hand. Additionally, a 20 g catheter was placed in the radial artery of the nondominant hand to obtain arterial blood samples for determination of **fractional** arterial oxygen saturation and the presence of MetHb and COHb using a co-oxime-

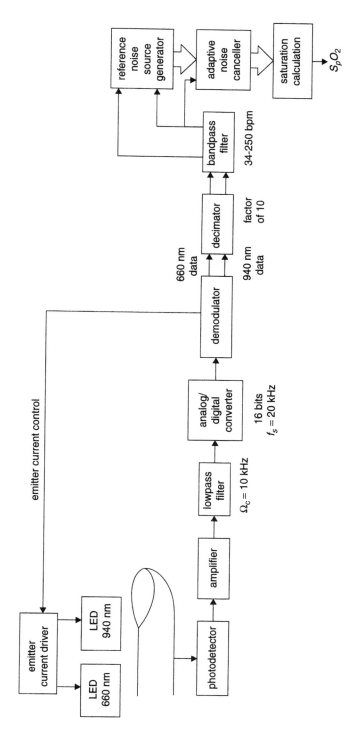

Figure 4.10 Masimo Data Acquisition System. Based on [16].

ter. (When MetHb and COHb are not present, the fractional and functional saturations are equal.) A sensor for a reference pulse oximeter was also placed on the nondominant hand.

Each subject inspired various oxygen fractions from an anesthesia machine through a tight-fitting face mask to simulate **normoxemia** (room air), steady-state hypoxemia ($S_pO_2 \approx 75\%$), and transient hypoxemia (S_pO_2 varied between 75 and 100%). During each oxygen state, the dominant hand with sensors was subjected to standardized "rubbing" and "tapping" motions generated by a motor-driven tilt table. Further, the sensors were disconnected and reconnected at various preselected times, forcing the instrument to reacquire data during the motion conditions.

Arterial blood co-oximetry analysis revealed insignificant levels of MetHb and COHb in any subject and that the reference oximeters performed within their specified uncertainty limits of 2% during steady-state conditions. The Masimo oximeter performed significantly better ($p < 0.05$) than the other monitors during motion in terms of **performance index** (PI), as shown in Table 1. In this study, performance index was defined as the time percentage during which the oximeter provided an S_pO_2 value within 7% of the control S_pO_2 value. The Masimo performance indices were significantly greater during motion, whether or not the oximeter had been functioning before motion began, and at the lowest true saturation point during rapid desaturation to 75%.

Sensitivity (correct detection percentage of an alarm condition) and **specificity** (correct detection percentage of a nonalarm condition) were also assessed using an alarm threshold of 90%. Including periods of **dropout** when the oximeter provided no display data, the Masimo oximeter possessed the highest sensitivity and specificity (Table 1). Typical S_pO_2 readings during rapid desaturation–resaturation and tapping motion are shown in Figure 4.11. In this figure, the N-200 and N-3000 indicate desaturation before it actually occurs. The N-200 also reads S_pO_2 values near 80% after the control S_pO_2 value has returned to normal. In general the N-200 had the greatest S_pO_2 errors and the N-3000 had the highest dropout rates [17]. In a similar study in which poor peripheral perfusion was induced in the test hands of six healthy volunteers, hypoxemia was induced to a S_pO_2 of 80%. In this study, the performance indices of all three monitors were similar (Table 4.1) [18].

Table 4.1 Pulse oximeter performance during motion and low perfusion. From [17] and [18].

		Nellcor N-200	Nellcor N-3000	Masimo prototype
Motion	Performance index (%) (oximeter connected before motion begins)	76	87	99
	Performance index (%) (oximeter connected after motion begins)	68	47	97
	Performance index (%) (oximetry readings at lowest point of rapid desaturation to 75%)	70	58	95
	Sensitivity	99	84	100
	Specificity	70	64	100
Low-perfusion performance index (%)		99.3	98.2	98.5

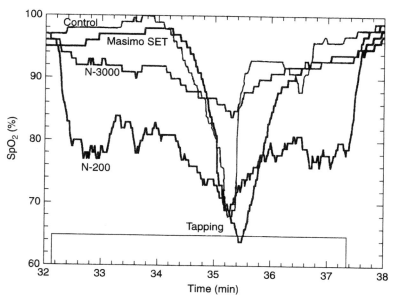

Figure 4.11 S_pO_2 versus time showing a rapid desaturation–resaturation occurring during tapping motion. Reprinted from [17] with permission from Lippincott Williams & Wilkins.

4.4.4 Implementation

As a method of gaining entry into the Nellcor-dominated pulse oximetry market, Masimo developed a marketing strategy of licensing its noise cancellation software to other pulse oximetry manufacturers (Figure 4.12). Monitors using Masimo software require Masimo disposable probes. According to the company's website, as of July, 2000, 27 manufacturers have licensed the Masimo software Signal Extraction Technology (SET) [19].

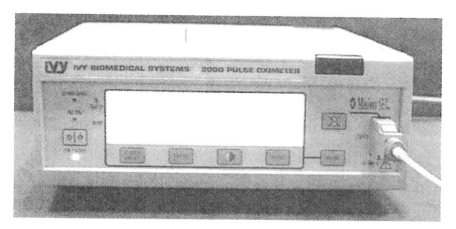

Figure 4.12 Ivy 2000 Pulse Oximeter with Masimo SET adaptive noise cancellation algorithm. Courtesy of Masimo Corporation, Irvine, CA.

4.5 SUMMARY

In this chapter, we have discussed how pulse oximetry can be used to monitor arterial oxygen saturation, which in turn can be used to infer tissue, anemia, or ischemic hypoxia. Pulse oximetry is loosely based on the Beer–Lambert law, which describes light transmission through a single absorbing medium of known length and concentration. Since light traveling through biologic tissue is much more complex, a ratio, R, of the two absorbance derivatives, which remains constant during arterial pulsing, is used to measure arterial oxygen saturation by pulse oximetry, or S_pO_2. An empirical relationship between R and S_pO_2 is determined for each unique pulse oximetry sensor design.

Pulse oximetry is widely used in monitoring patients during anesthesia, recovery, and critical care. S_pO_2 measurements are accuracy-limited by the signal-to-noise ratio, which is degraded by motion and poor peripheral perfusion. Empirical noise rejection algorithms have been developed that may actually decrease the accuracy of S_pO_2 measurements. In contrast, an adaptive noise cancellation algorithm developed by start-up company Masimo has been shown to effectively reject noise. With Masimo's method, a reference noise source is derived from data using two wavelengths. The Masimo oximeter performed significantly better than other monitors during motion testing. Masimo believes that its algorithm will enable it to significantly penetrate the Nellcor-dominated pulse oximetry market.

4.6 REFERENCES

[1]. Mackenzie, N. Comparison of a pulse oximeter with an ear oximeter and an in vitro oximeter. *J Clin Monit, 1,* 156–160, 1985.

[2]. Severinghaus, J. W. and Astrup, P. B. History of blood gas analysis: VI. Oximetry. *J Clin Monit, 2,* 270–288, 1986.

[3]. Kelleher, J. F. Pulse oximetry. *J Clin Monit, 5,* 37–62, 1989.

[4]. Severinghaus, J. W. and Honda, Y. History of blood gas analysis: VII. Pulse oximetry. *J Clin Monit, 3,* 135–138, 1987.

[5]. ———. Pulse oximeters. *Health Devices, 18,* 185–230, 1989.

[6]. ———. Standards for postanesthesia care. *ASA Directory of Members,* 56th ed., p. 672, 1991.

[7]. Pologe, J. A. Pulse oximetry: Technical aspects of machine design. *Int Anesthesiol Clin, 25,* 137–153, 1987.

[8]. Moller, J.T., Pedersen, T., Rasmussen, L. S., Jensen, P. F., Pedersen, B. D., Ravlo, O., Rasmussen, N. H., Espersen, K., Johannessen, N. W., Cooper, J. B., Gravenstein, J. S., Chraemmer-Jorgensen, B., Wiberg-Jorgensen, F., Djernes, M., Heslet, L., and Johansen, S. H. Randomized evaluation of pulse oximetry in 20,802 patients: I. *Anesth, 78,* 436–444, 1993.

[9]. Goodman, D. E. and Corenman, J. E. Method and apparatus for detecting optical signals. *U.S. patent 4,928,692.* June 19, 1990.

[10]. Elfadel, I. M., Weber, W., and Barker, S. J. Motion-resistance pulse oximetry. *J Clin Monit, 11,* 262, 1995.

[11]. Conlon, B., Devine, J. A., and Dittmar, J. A. ECG synchronized pulse oximeter. *U.S. patent 4,960,126.* October 2, 1990.

[12]. Baker, C. R., Jr., and Yorkey, T. J. Method and apparatus for estimating physiological parameters using model-based adaptive filtering. *U.S. Patent 5,853,364.* December 29, 1998.

[13]. IMS Health. *Hospital Supply Index.* IMS America: Plymouth Meeting, PA, 1997.

[14]. Pologe, J. A. Arterial blood monitoring probe. *U.S. patent 5,297,548*. March 29, 1994.
[15]. Nesheim, J. L. *High Tech Start Up*. John L. Neshiem, Saratoga, CA, 1997.
[16]. Diab, M. K., Kiani-Azarbayjany, E., Elfadel, I. M., McCarthy, R. J., Weber, W. M., and Smith, R. A. Signal processing apparatus. *U.S. patent 5,632,272*. May 27, 1997.
[17]. Barker, S. J., and Shah, N. K. The effects of motion on the performance of pulse oximeters in volunteers (revised publication). *Anesth*, 86, 101–108, 1997.
[18]. Barker, S. J., Novak, S., and Morgan, S. The performance of three pulse oximeters during low perfusion in volunteers. *Anesth*, 87(3A), A409, 1997.
[19]. Masimo website. July 22, 2000 update.
[20]. VitalWave Corporation, San Diego, CA. Research data, 1998.

Further Reading

Oxygen Transport Physiology: Guyton, A. C. and Hall, J. E., *Textbook of Human Physiology*, 10th ed. Saunders: Philadelphia, 2000.

Pulse Oximeters: Webster, J. G., ed. *Design of pulse oximeters*. Institute of Physics Publishing: Bristol, UK, 1997.

4.7 NONINVASIVE BLOOD PRESSURE EXERCISES

Engineer B at Startup Company has been asked to remove the motion artifact from a new continuous, noninvasive blood pressure monitor. The blood pressure waveforms are obtained from two sensors positioned over the **radial artery** at the wrist. Having recently read a review on adaptive filtering, Engineer B wishes to use data from one sensor as the primary input, and some combination of data from both sensors as the reference noise source. We recreate his analysis below.

4.7.1 Clinical Data

The two pressure files in Figure 4.13 were simultaneously recorded by Engineer B from an employee at Startup Company. The simulated noise source was sudden arm move-

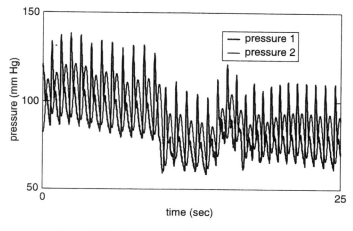

Figure 4.13 Pressure waveforms for exercise [20].

ment. For example, an instrumented arm may fall off its arm board during surgery and be repositioned a few seconds later. Twenty five seconds of data were recorded with a sampling frequency of 1.6 kHz and resolution of 12 bits/sample; the data were saved in units of mmHg. These data were generously provided by VitalWave Corporation [20].

Please move the *pressure1.dat* and *pressure2.dat* files from *ftp://ftp.ieee.org/uploads/press/baura* to your *Matlab\bin* directory. Input the data using "load file.dat-ascii".

4.7.2 Useful Matlab Functions

The following functions from the Signal Processing Toolbox are useful for completing this exercise: psd and randn.

4.7.3 Exercises

1. Check for waveform fidelity by comparing the power spectral densities of the waveforms. (a) Has one waveform been corrupted? (b) What is the likely source of error?
2. Even with a single waveform, adaptive filtering may be possible. Decimate the uncorrupted waveform by a factor of 16 to 100 Hz. Subtract the DC offset. (a) Implement a self-tuning filter using the LMS algorithm. Print the Matlab code. (b) Determine reasonable values for μ, the number of weights, and the delay. (c) Plot the primary input and self-tuning filter output. (d) What is the main assumption of a self-tuning filter? (e) Is this assumption valid?
3. Test the self-tuning filter by inputting the sum of a 500 Hz sine wave with a peak-to-peak amplitude of 10 plus normally distributed noise with a peak-to-peak amplitude of 10. (a) Determine reasonable values for μ, the number of weights, and the delay. (b) Plot the primary input and self-tuning filter output.
4. Empirical methods may also be used to remove motion artifact. (a) Describe one empirical method for removing the noise in these waveforms. (b) Would this method be reliable for more complicated noise sources such as eating with the instrumented arm?

5

TIME-FREQUENCY AND TIME-SCALE ANALYSIS

In Chapters 1 and 3, we reviewed methods for minimizing noise artifact. If the noise is restricted to a specific frequency band, we may reduce it through use of a frequency-selective filter. If the noise and signal bands overlap and we have a reference noise source and stationary system, the reference noise may be used to model and adaptively filter out the original noise. If none of these constraints are met, we still may be able to obtain a usable signal. In this case, the data may be transformed to recover important signal characteristics using a **time-frequency** or **time-scale representation.** A time-frequency representation is a two-dimensional mapping of the fraction of the energy of a one-dimensional signal at time, t, and frequency, Ω. A time-scale representation is a two-dimensional mapping of the fraction of the energy of a one-dimensional signal at time, t, and **scale**, a. Scale is a physical attribute representing compression.

Certainly, time-frequency and time-scale representations are not typically classified as filters. Their utility in minimizing noise artifact that is present in the time domain stems from the intrinsic nature of the representations. Rather than observing a signal as it changes with time, the time-frequency representation enables the user to observe a density in time and frequency simultaneously that indicates which frequencies are present in the signal and how they change in time. In contrast, a Fourier transform of the signal identifies the frequencies present in the signal, but not the times when these frequencies occurred.

For example, a Fourier transform may be calculated for the sum of three sine waves in Figure 5.1. The illustrated amplitude spectrum does not identify how the components change with time. If we window these three sine waves (the "gaps" in Figure 5.2), combine them sequentially rather than simultaneously, and take the Fourier transform of this new combination, we obtain the amplitude spectrum in Figure 5.2. The amplitude spectrum in Figure 5.2 does not differ substantially from the amplitude spectrum in Figure 5.1, even though the combination of sine waves is quite different. The time during which each sine wave is present cannot be recovered from the amplitude spectrum [1].

The ability to simultaneously track time and frequency is very useful in analyzing complex physiologic signals. For example, examining an **electrogastrogram** (EGG) in

88   TIME-FREQUENCY AND TIME-SCALE ANALYSIS

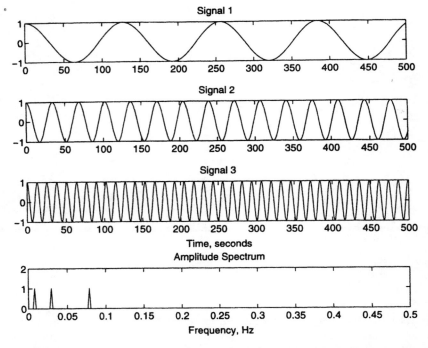

Figure 5.1   Three sine waves simultaneous in time, and the amplitude spectrum of their sum. From [1].

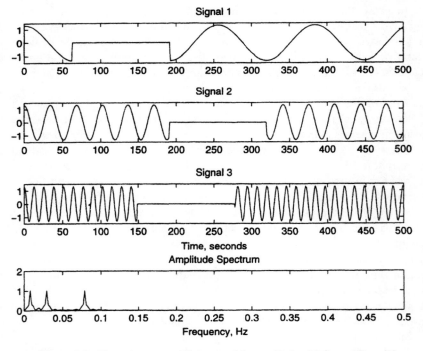

Figure 5.2   Three sine waves with gaps, and the amplitude of their sum. From [1].

the time domain is not very informative. An electrogastrogram is a recording of gastric **myoelectrical** (electrical impulses from muscles) activity obtained by placing electrodes on the abdomen. However, when the EGG in Figure 5.3 is separated simultaneously into time and frequency, it becomes apparent that this gastric signal has a frequency of 3 cycles per minute (cpm). Also, from this separation, it is shown that the accompanying interference has a frequency of 12 cpm, and does not affect the energy and frequency of the gastric signal. These frequencies do not vary substantially over the 30 minute interval [2].

Similarly, more insight into the **epicardial first heart sound** may be obtained from a time-frequency representation than from data plotted in the time or frequency domain. An epicardial first heart sound is the sound that occurs during initial cardiac contraction (specifically, during **isovolumetric contraction,** as described in Section 6.1.1), with the sound recorded from the inner layer of the sac surrounding the heart. A typical epicardial first heart sound is shown in Figure 5.4. Note that the time signal suggests two components, labeled I and II. From the corresponding power spectrum, it is observed that the first heart sound frequencies are concentrated below 150 Hz, and attenuate rapidly at approximately 20 dB per decade. In the power spectrum, the dynamics implied by the time signal are lost. In contrast, the time-frequency representation reveals that shortly after the ECG R-wave, a striking rise in frequency occurs. The frequency of component I rises from about 50 Hz to 120 Hz in 30 msec. Although the second component is less dynamic, it possesses slowly varying frequencies near 60 and 90 Hz [3]. This technique may be useful in diagnosing deficiencies in prosthetic heart valves [4].

In this chapter, we discuss some of the fundamental theories underlying time-frequency

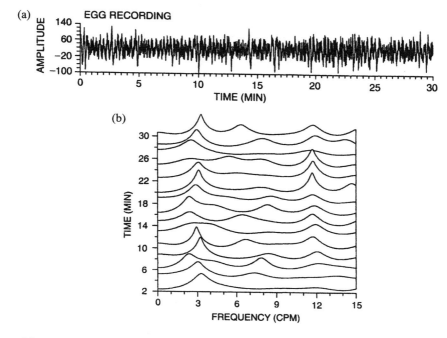

Figure 5.3 (a) An EGG recording with an interference of 12 cpm and (b) its frequency representation calculated by the adaptive spectral method. Each spectrum (from bottom to top) stands for 2 min of EGG data. From [2].

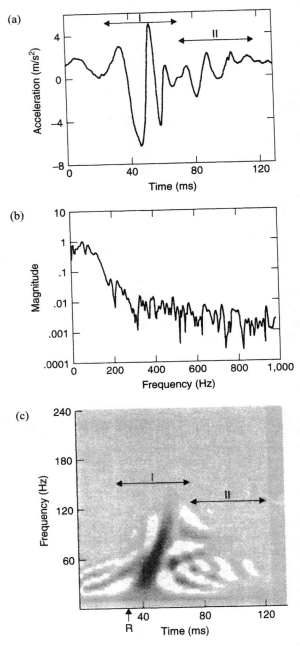

Figure 5.4 A typical (a) epicardial first heart sound, (b) its power spectrum, and (c) its time-frequency distribution. From [3].

and time-scale analysis. Although time-frequency analysis is currently a fashionable research topic, this methodology traces its roots to the invention of the sound **spectrogram, the short-time Fourier transform,** in the 1940s for analysis of human speech [5, 6]. To limit our discussion to a single chapter, only the Cohen and affine classes and some historically important time-frequency representations are presented. A more comprehensive list of time-frequency representations can be found in [7]. The time-frequency discussion in this chapter is based on Leon Cohen's textbook, *Time-Frequency Analysis* [8]. Further, examples of the affine class are limited to scalograms and wavelet transforms. References with more comprehensive wavelet discussions are given at the end of the chapter.

Unlike the other theoretical chapters in this textbook, continuous, rather than discrete, time is emphasized. Further, to simplify notation in this chapter, the Fourier transform of a continuous signal, $U(e^{j\Omega})$, is abbreviated as $U(\Omega)$, and the negative to positive infinity limits, $-\infty$ to $+\infty$, are omitted from integrals. With these modifications, the nomenclature in this chapter is consistent with other time-frequency nomenclature in the literature. However, to facilitate implementation, the discrete form of each time-frequency or time-scale representation is provided at the end of its section.

5.1 TIME-FREQUENCY REPRESENTATIONS

In time-frequency analysis, a function is constructed that describes the energy distribution of a signal simultaneously in time and frequency. Recall that the **instantaneous power** of a signal, $u(t)$, is the absolute value of the signal squared, $|u(t)|^2$, and that the **power density spectrum** of this signal is the absolute value of the Fourier transform squared, $|U(\Omega)|^2$. We assume the normalization that the total energy equals 1, such that

$$\int |u(t)|^2 \, dt = \int |U(\Omega)|^2 \, d\Omega = \text{total energy} = 1 \tag{5.1}$$

The function we seek is a **joint energy density,** $P(t, \Omega)$. Because these functions are also found in quantum mechanics, $P(t, \Omega)$ is commonly referred to as a **probability distribution, joint distribution,** or **distribution.** Methods for construction are described later in the chapter. In this section, we discuss desirable properties for time-frequency representations.

5.1.1 Marginals

Ideally, the sum of the distributions for all frequencies at a specified time should describe the instantaneous power. Further, the sum of the distributions for all times at a specified frequency should describe the power density spectrum. Therefore, a joint distribution in time and frequency should satisfy the time and frequency **marginal** conditions

$$\int P(t, \Omega) d\Omega = |u(t)|^2 \tag{5.2}$$

$$\int P(t, \Omega) dt = |U(\Omega)|^2 \tag{5.3}$$

The term "marginal" was coined by Leon Cohen, and is borrowed from probability theory [9].

5.1.2 Total Energy

The total energy of the distribution should be the total energy of the signal, such that

$$\text{Total energy} = \iint P(t, \Omega)d\Omega dt = \int |u(t)|^2 dt = \int |U(\Omega)|^2 d\Omega \quad (5.4)$$

Note that if the joint distribution satisfies the marginals, it automatically satisfies the total energy requirement; however, the converse is not true. Many time-frequency representations do not satisfy the marginals, but still provide a good representation of time-frequency structure.

5.1.3 Global Means

The **expected value**, or **mean**, of any function of time and frequency, $E[g(t, \Omega)]$, is calculated in the standard manner as

$$E[g(t, \Omega)] = \iint g(t, \Omega)P(t, \Omega)d\Omega dt \quad (5.5)$$

If the marginals are satisfied, the mean of the sum of a time and frequency function will be correctly calculated by substitution of the marginals in Eqs. (5.2) and (5.3), such that

$$E[g_1(t) + g_2(\Omega)] = \iint [g_1(t) + g_2(\Omega)]P(t, \Omega)d\Omega dt \quad (5.6)$$

$$= \int \left[\int g_1(t)P(t, \Omega)d\Omega\right]dt + \int \left[\int g_2(\Omega)P(t, \Omega)dt\right]d\Omega \quad (5.7)$$

$$E[g_1(t) + g_2(\Omega)] = \int g_1(t)|u(t)|^2 dt + \int g_2(\Omega)|U(\Omega)|^2 d\Omega \quad (5.8)$$

5.1.4 Local Means

By extension of Eq. (5.6), the **conditional mean** of a function at a given time or frequency is

$$E[g(\Omega)]_t = \frac{1}{P(t)} \int g(\Omega)P(t, \Omega)d\Omega \quad (5.9)$$

$$E[g(t)]_\Omega = \frac{1}{P(\Omega)} \int g(t)P(t, \Omega)dt \quad (5.10)$$

Note that Eq. (5.9), which is a calculation of the average frequency at a particular time, defines the **instantaneous frequency**.

Similarly, the **conditional standard deviations,** $\sigma_{i|j}^2$, are

$$\sigma_{\Omega|t}^2 = \frac{1}{P(t)} \int \{\Omega - E[g(\Omega)]_t\} P(t, \Omega) d\Omega \tag{5.11}$$

$$\sigma_{t|\Omega}^2 = \frac{1}{P(\Omega)} \int \{t - E[g(t)]_\Omega\} P(t, \Omega) dt \tag{5.12}$$

5.1.5 Time and Frequency Shift Invariance

For an ideal joint distribution, a **time shift** of t_0 in the signal, $u(t) \to u(t - t_0)$, translates into a time shift in the distribution. Similarly, a **frequency shift** of Ω_0 in the signal's Fourier transform translates into a frequency shift in the distribution. Recall from Fourier transform theory that a frequency shift in the frequency domain translates into multiplication in the time domain by the term $e^{j\Omega_0 t}$. Therefore, a desirable property of time-frequency representations is that

$$\text{if } u(t) \to e^{j\Omega_0 t} u(t - t_0), \text{ then } P(t, \Omega) \to P(t - t_0, \Omega - \Omega_0) \tag{5.13}$$

5.1.6 Scaling Invariance

It is desirable to keep the signal and its Fourier transform scaled so that the signal energy is normalized to a magnitude of 1 unit. This can be accomplished with the **scaling constant,** a. The scaled forms of the signal and its Fourier transform are then

$$u_{sc}(t) = \sqrt{a}\, u(at) \tag{5.14}$$

$$U_{sc}(\Omega) = \frac{1}{\sqrt{a}}\, U\!\left(\frac{\Omega}{a}\right) \tag{5.15}$$

Using these scaled forms, the spectrum is expanded when the signal is compressed, and vice versa. For these relations to hold for the joint distribution, the scaled joint distribution must be of the form

$$P_{sc}(t, \Omega) = P\!\left(at, \frac{\Omega}{a}\right) \tag{5.16}$$

The scaled joint distribution in Eq. (5.16) satisfies the marginals of the scaled signal

$$\int P_{sc}(t, \Omega) d\Omega = |u_{sc}(t)|^2 = a|u(at)|^2 \tag{5.17}$$

$$\int P_{sc}(t, \Omega) dt = |U_{sc}(\Omega)|^2 = \frac{1}{a}\left|U\!\left(\frac{\Omega}{a}\right)\right|^2 \tag{5.18}$$

5.1.7 Weak and Strong Finite Support

Ideally, a joint distribution equals zero for frequencies outside a specified frequency range if the Fourier transform of the signal is zero outside this frequency range. Similarly, an ideal joint distribution equals zero for times outside a specified time range if the signal is zero outside this time range. If each requirement is met, then the distribution has **weak finite frequency (or time) support**. If the joint distribution equals zero when the signal or spectrum is zero, then the joint distribution has **strong finite support**. Strong finite support implies weak finite support, but not vice versa.

5.1.8 Uncertainty Principle

An ideal joint distribution should observe the **uncertainty principle**. The uncertainty principle states that the product of the standard deviations, with respect to time and frequency, of a signal must be greater than or equal to ½:

$$\sigma_t \sigma_\Omega \geq \frac{1}{2} \tag{5.19}$$

where the standard deviations are defined in the standard manner with

$$\sigma_t^2 = \int \{t - E[t]\}^2 |u(t)|^2 \, dt \tag{5.20}$$

$$\sigma_\Omega^2 = \int \{\Omega - E[\Omega]\}^2 |U(\Omega)|^2 \, d\Omega \tag{5.21}$$

Adoption of this principle limits joint distribution construction, since a compromise must be made between resolution in the time and frequency domains. In other words, the standard deviation in both domains cannot be arbitrarily small.

5.1.9 Characteristic Functions

By definition, the joint **characteristic function**, $M(\theta, \tau)$, is

$$M(\theta, \tau) = \iint P(t, \Omega) e^{j\theta t + j\tau \Omega} \, dt \, d\Omega \tag{5.22}$$

The joint characteristic function is an effective method for analyzing energy distributions.

5.2 SPECTROGRAM

As stated above, the first time-frequency representation developed was the short-time Fourier transform (STFT), or spectrogram. Gabor introduced the STFT, which is basically a windowing process (see Chapter 1) over the input signal to extend the applicability of Fourier transforms [9]. To analyze a signal at a specified time, a small portion of the signal centered around that time is used to calculate its energy spectrum. It is assumed that the signal is stationary within the chosen time interval.

5.2.1 Derivation

Let us derive the STFT by using the continuous time version of the Fourier transform in Eq. (1.7) and normalizing to insure that the power of the signal equals 1, such that

$$U(\Omega) = \Im\{u(t)\} = \frac{1}{\sqrt{2\pi}} \int u(t) e^{-j\Omega t} dt \qquad (5.23)$$

Specifically, for a window function, $h(t)$, centered at t, we can calculate the spectrum of $u(\tau)h(\tau - t)$, where τ is a dummy integration variable, as

$$U_t(\Omega) = \frac{1}{\sqrt{2\pi}} \int u(\tau) h(\tau - t) e^{-j\Omega \tau} d\tau \qquad (5.24)$$

which is the short-time Fourier transform. The corresponding joint distribution is then

$$P_{STFT}(t, \Omega) = |U_t(\Omega)|^2 \qquad (5.25)$$

which is the power spectral density at samples t and Ω. Conversely, if we wish to study time properties at a specified frequency, the short-frequency time transform is

$$u_\Omega(t) = \frac{1}{\sqrt{2\pi}} \int U(\theta) H(\Omega - \theta) e^{-j\theta t} d\theta \qquad (5.26)$$

where $H(\Omega)$ is the Fourier transform of the window function, and θ is a dummy integration variable. The joint distribution can also be described as

$$P_{STFT}(t, \Omega) = |U_t(\Omega)|^2 = |u_\Omega(t)|^2 \qquad (5.27)$$

The window function controls the relative weight imposed upon different parts of the signal. The spectrogram can be used to estimate local quantities by choosing a window that weights the interval near the observation sample or frequency a greater amount than other samples or frequencies. Greater time resolution is achieved by choosing a more peaked window function; greater frequency resolution is achieved by choosing a more peaked window spectrum. Because the uncertainty principle is not followed, both $h(t)$ and $H(\Omega)$ cannot be narrowly peaked. Different windows can be used for estimating different properties.

The spectrogram is commonly displayed as a two-dimensional projection, where the energy is represented by different shades of gray.

5.2.2 General Properties

The basic properties and effectiveness of the STFT for a particular signal depend on the form of the window. If the energy of the window equals one, then the total energy of the spectrogram equals the total energy of the signal. The marginals of the spectrogram do not satisfy the marginal conditions in Eqs. (5.2) and (5.3) because the spectrogram scrambles the energy distributions of the window with those of the signal. Since the marginals are not satisfied, the global and local means do not satisfy Eqs. (5.5), (5.9), or (5.10).

While the spectrogram is time- and frequency-shift-invariant, it is not scale-invariant. It possesses neither strong nor weak finite support because the window may pick up some of the signal for a time, t, before the signal starts. It does not follow the uncertainty principle. The characteristic function is calculated directly from substitution of Eq. (5.25) into Eq. (5.22) as

$$M(\theta, \tau) = \iint |U_t(\Omega)|^2\, e^{j\theta t + j\tau\Omega}\, dt\, d\Omega \tag{5.28}$$

5.2.3 Example

For discrete signals, the spectrogram is calculated as

$$P_{STFT}(k,f) = \sum_{\tau=0}^{L-1} u(k+\tau)h(\tau)\exp(-j2\pi f\tau) \tag{5.29}$$

where $f = \omega/2\pi$, $h(k)$ = window, L = window length, and $0 \leq \tau \leq L - 1$. The resulting spectrogram for a test signal composed of two sinusoidal segments of differing time and frequency placement (Figure 5.5) is shown in Figure 5.6.

The spectrogram has been used to analyze **electrocorticograms** (ECoGs) for **ictal**, or seizure, activity in epileptic patients. An ECoG is a nonstationary signal taken from electrodes that are chronically implanted over lobes of the **cerebral cortex** after surgical incision of the skull. For a 500 ms, 250 sample ictal ECoG segment with two high-frequency bursts and a low-frequency component between them (Figure 5.7c), a spectrogram was calculated using a 160 ms Hanning window, shifted 2 ms between successive evaluations. The resultant spectrogram in Figures 5.7a and 5.7b places significant energy in time-frequency regions corresponding to three signal areas. However, the time and frequency definitions of the components are blurred as their energy is smeared across the time-frequency plane [11].

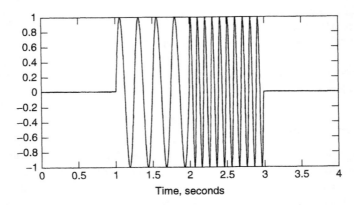

Figure 5.5 Test signal used to evaluate time-frequency distributions. From [1].

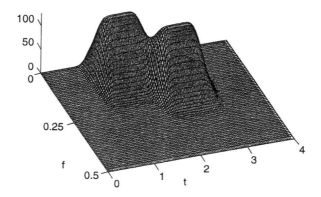

Figure 5.6 Spectrogram based on test signal in Figure 5.5. From [1].

5.3 WIGNER DISTRIBUTION

The **Wigner distribution,** $P_W(t, \Omega)$, is the prototype of distributions that are qualitatively different from the STFT. Wigner introduced this distribution in 1932 as a means of calculating deviations from the ideal gas law [12]. Around the same period during which the spectrogram was first investigated, Ville introduced the Wigner distribution into signal analysis [13]. For this reason, it is also referred to as the **Wigner–Ville distribution.**

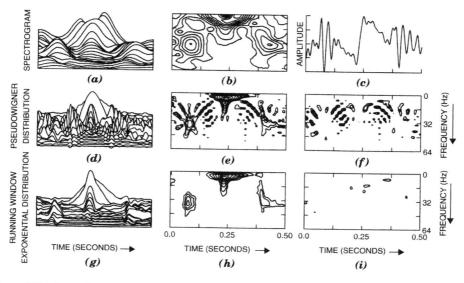

Figure 5.7 Comparison of the STFT, PWD, and CWD from the ictal phase ECoG shown in (c). The CWD in this figure is labeled as the running window exponential distribution. Surface plots are shown in (a), (d), and (g). Positive contour plots are shown in (b), (e), and (h). Negative contour plots of the representations shown in (d) and (g) are shown in (f) and (i), respectively. From [11].

5.3.1 Derivation

Wigner introduced his distribution "because it seems to be the simplest" [9]. The Wigner distribution is

$$P_W(t, \Omega) = \frac{1}{2\pi} \int u^*(t - 0.5\tau) u(t + 0.5\tau) e^{-j\Omega\tau} d\tau \tag{5.30}$$

$$P_W(t, \Omega) = \frac{1}{2\pi} \int U^*(\Omega - 0.5\theta) U(\Omega + 0.5\tau) e^{-jt\theta} d\theta \tag{5.31}$$

where * represents the complex conjugate. This distribution is **bilinear** in the signal because the signal enters twice in its calculation.

Note that to obtain the Wigner distribution at a particular time, we sum together pieces of the product of the signal at a past time multiplied by the signal at a future time. Therefore, any signal overlap at time, t, in the left and right pieces determines the properties of the Wigner distribution. This also holds true for signal overlap in the frequency domain. The Wigner distribution weighs local and distant times equally, and is therefore highly nonlocal. For an infinite duration signal, the Wigner distribution will be nonzero for all time, since all sums will be nonzero. Similarly, for a finite duration signal, the Wigner distribution will be zero before the start and after the end of a signal. For a bandlimited signal, the Wigner distribution will be zero for frequencies outside the band. The Wigner distribution is always real, even if the signal is complex.

One disadvantage of the Wigner distribution is the appearance of **interference** or **cross terms.** For a signal that disappears for a finite time interval and then reappears, the Wigner distribution may not be zero during this time interval. For a signal with noise present only during a small percentage of the total signal duration, the distribution will possess noise at times when the noise is not present. Again, this occurs because each part of the distribution is based on past and future time.

5.3.2 General Properties

The Wigner distribution satisfies the time and frequency marginals. Since these conditions are satisfied, the total energy condition is also satisfied and the global and local means can be calculated using Eqs. (5.5), (5.9), and (5.10). This distribution is time and frequency shift invariant since replacement of the original signal, $u(t)$, by $e^{j\Omega_0 t} u(t - t_0)$ leads to

$$\frac{1}{2\pi} \int e^{-j\Omega_0(t-0.5\tau)} u^*(t - t_0 - 0.5\tau) e^{j\Omega_0(t+0.5\tau)} u(t - t_0 + 0.5\tau) e^{-j\Omega\tau} d\tau$$

$$= P_W(t - t_0, \Omega - \Omega_0) \tag{5.32}$$

While the distribution is not scale-invariant, it does possess weak time and frequency finite support and follows the uncertainty principle. The characteristic function is calculated directly from substitution of Eq. (5.30) into Eq. (5.22) as

$$M(\theta, \tau) = \int u^*(\zeta - 0.5\tau) u(\zeta + 0.5\tau) e^{j\theta\Omega} d\Omega \tag{5.33}$$

This function and its variants have played a major role in signal analysis. This form is called the **symmetric ambiguity function**, $A(\theta, \tau)$.

5.3.3 Pseudo Wigner Distribution

To compensate for equal weighting of past and future times, a window, $h(t)$, may be added to the Wigner distribution. This addition is called the **pseudo Wigner distribution**, $P_{PW}(t, \Omega)$:

$$P_{PW}(t, \Omega) = \frac{1}{2\pi} \int h(\tau) u^*(t - 0.5\tau) u(t + 0.5\tau) e^{-j\Omega\tau} d\tau \tag{5.34}$$

The window enables emphasis of properties near the time of interest and study of **range limiting** (integration to finite limits) effects.

5.3.4 Example

For discrete signals, the pseudo Wigner distribution is calculated as

$$P_{PW}(k, f) = 2 \sum_{\tau=-L}^{+L} e^{-j4\pi f\tau/N} u^*(k - \tau) u(k + \tau) \tag{5.35}$$

where $f = \omega/2\pi$, $u(t)$ and its complex conjugate are sample-limited to $\{-K/2, +K/2\}$, K is even, and $N = K + 1$. A rectangular window is used, so that $L = K/2 - |k|$ [14]. For the test signal in Figure 5.5, the resulting Wigner distribution is given in Figure 5.8.

For the ictal ECoG segment described in Section 5.2.3, a pseudo Wigner distribution was calculated. The resultant distribution in Figures 5.7d and 5.7e resolves the three frequency components, rather than blurring them as the spectrogram did. However, significant cross terms are also generated between the frequency components [11]. In general, the spectrogram cannot resolve signal components effectively, while the Wigner distribution includes confusing artifacts between resolved components.

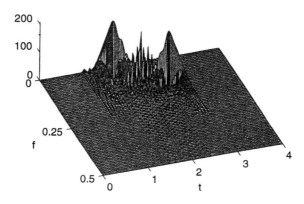

Figure 5.8 Wigner distribution based on test signal in Figure 5.5. From [1].

5.4 KERNEL METHOD

Time-frequency distributions in **Cohen's class,** including the spectrogram and Wigner distribution, can be generated using the **kernel method,** which was devised by Cohen in 1966 as a generalization of existing methods [5]. To use this method, a characteristic function, $M(\theta, \tau)$, is transformed to obtain Cohen's class, $C(\tau, \Omega)$, such that

$$C(\tau, \Omega) = \frac{1}{4\pi^2} \iint M(\theta, \tau) e^{-j\theta t - j\tau\Omega} \, d\theta \, d\tau \tag{5.36}$$

The characteristic function was defined in Eq. (5.22); an equivalent definition is

$$M(\theta, \tau) = \phi(\theta, \tau) \int u^*(\zeta - 0.5\tau) u(\zeta + 0.5\tau) e^{j\theta\zeta} \, d\zeta \tag{5.37}$$

$$M(\theta, \tau) = \phi(\theta, \tau) A(\theta, \tau) \tag{5.38}$$

where $\phi(\theta, \tau)$ is a two-dimensional function called the **kernel** and $A(\theta, \tau)$ is the symmetric ambiguity function.

The kernel determines a distribution and its properties. For the Wigner distribution, the kernel is equal to $\phi(\theta, \tau) = 1$. For the spectrogram, the kernel equals $\int h^*(\Psi - 0.5\tau) h(\Psi + 0.5\tau) e^{-j\theta\psi} \, d\psi$. If the kernel does not explicitly depend on the signal, the resulting distribution is said to be bilinear.

5.4.1 Basic Kernel Properties

A new distribution with desired properties can be generated by carefully choosing its kernel. In order to satisfy the time marginal, the kernel at $\tau = 0$, $\phi(\theta, 0)$, must equal 1. In order to satisfy the frequency marginal, the kernel at $\theta = 0$, $\phi(0, \tau)$, must equal 1. Once the marginals are satisfied, the total energy condition and uncertainty principle are also satisfied. Further, the global and local means can be calculated using Eqs. (5.5), (5.9), and (5.10). Note that total energy conservation by normalization to one can be met without satisfying the marginals. If the kernel at τ and θ equal 0, $\phi(0, 0)$ equals 1, and the energy is normalized to one.

For time invariance, the kernel must be independent of time. Similarly, for frequency invariance, the kernel must be independent of frequency. For scaling invariance, it is required that

$$\phi(a\theta, \tau/a) = \phi(\theta, \tau) \tag{5.39}$$

Assuming the kernel is independent of the signal, this can only occur if

$$\phi(\theta, \tau) = \phi(\theta\tau) \tag{5.40}$$

The kernel in Eq. (5.40) is called a **product kernel** because it is a function of the product of two variables.

Finite support can also be specified by the kernel. To obtain weak finite time support, the following integral must be satisfied for $|\tau| \leq 2|t|$:

$$\int \phi(\theta, \tau) e^{-j\theta t} d\theta = 0 \qquad (5.41)$$

Similarly, for weak finite frequency support, the following integral must be satisfied for $|\theta| \leq 2|\Omega|$:

$$\int \phi(\theta, \tau) e^{-j\tau\Omega} d\tau = 0 \qquad (5.42)$$

For bilinear distributions, strong finite support can also be specified. To obtain strong finite time support, Eq. (5.41) must be satisfied for $|\tau| \neq 2|t|$. Similarly, for strong finite frequency support, Eq. (5.42) must be satisfied for $|\theta| \neq 2|\Omega|$.

5.4.2 Choi–Williams Distribution

Choi and Williams were the first researchers to generate a new distribution, $P_{CW}(t, \Omega)$, using the kernel method that produced remarkably better results than the Wigner distribution [16]. The **Choi–Williams** or **exponential distribution**, $P_{CW}(t, \Omega)$, is based on the kernel

$$\phi(\theta, \tau) = e^{-\theta^2 \tau^2/\sigma} \qquad (5.43)$$

where σ is a parameter. Substituting the kernel into the general class and integrating over θ results in the energy distribution

$$P_{CW}(t, \Omega) = \iint \frac{1}{\sqrt{\frac{4\pi\tau^2}{\sigma}}} \exp\left[\frac{-(\zeta - t)^2}{\frac{4\tau^2}{\sigma}}\right] \qquad (5.44)$$

$$\times u^*(\zeta - 0.5\tau) u(\zeta + 0.5\tau) e^{-j\Omega\tau} d\zeta\, d\tau$$

With this kernel, the time and frequency marginals are satisfied. Thus, the total energy condition is also satisfied and the global and local means can be calculated using Eqs. (5.5), (5.9), and (5.10). Because the kernel is time- and frequency-independent, the distribution is time- and frequency-shift invariant. Because the kernel is a product kernel, the distribution is scale invariant. Finally, because Eqs. (5.41) and (5.42) are not satisfied, the distribution does not have weak or strong finite support.

When σ is large, the Choi–Williams distribution approaches the Wigner distribution, as the kernel approaches one. When σ is small, the kernel is peaked at the origin, is one along the θ and τ axes, and falls off rapidly away from the axes. This results in minimization of cross terms. In general, for cross term minimization, a kernel must be significantly less than one for $\theta\tau \gg 0$.

For discrete signals, the running windowed Choi–Williams distribution is calculated as

$$P_{CW}(k,f) = 2 \sum_{\tau=-\infty}^{+\infty} W_N(\tau) e^{-j2\pi f \tau/N} \qquad (5.45)$$

$$\times \left[\sum_{\mu=-\infty}^{+\infty} W_M(\zeta) \frac{1}{\sqrt{\frac{4\pi\tau^2}{\sigma}}} \exp\left(\frac{-\zeta^2}{\frac{4\tau^2}{\sigma}} \right) u^*(k + \zeta - \tau) u(k + \zeta + \tau) \right],$$

where $f = \omega/2\pi$, $W_N(\tau)$ is a symmetric window that has nonzero values in the range of $-N/2 \leq \tau \leq N/2$, and $W_M(\mu)$ is a rectangular window which has a value of 1 for the range of $-M/2 \leq \zeta \leq M/2$. For the test signal in Figure 5.5, the resulting distribution is shown in Figure 5.9.

For the ictal ECoG segment described in Section 5.2.3, a Choi–Williams distribution was calculated, using $\sigma = 1$. The resultant distribution is shown in Figures 5.7g and 5.7h. Unlike the spectrogram and Wigner distribution, the Choi–Williams distribution provides a clear display of the time-frequency structure of the signal features without any loss of definition due to smearing or cross terms. Note also that the negative components in the Choi–Williams distribution (Figure 5.7i) are significantly less than those in the Wigner distribution (Figure 5.7f) [11].

5.4.3 Zhao-Atlas-Marks Distribution

Another distribution that minimizes cross terms is the **Zhao–Atlas–Marks** or **cone kernel distribution**, $P_{ZAM}(t, \Omega)$ [17]. Instead of spreading the cross terms out in the time-frequency plane with low intensity, the cross terms are placed under the **self** terms (noncross terms). The cross terms will only fall exactly on the self terms if they lie exactly midway between the self terms [18]. The required kernel is of the form

$$\phi(\theta, \tau) = \begin{cases} g(\tau) & |\tau| \geq b|\theta| \\ 0 & \text{otherwise} \end{cases} \qquad (5.46)$$

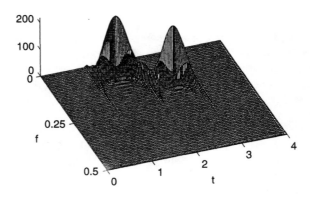

Figure 5.9 Choi–Williams distribution based on test signal in Figure 5.5. From [1].

where b is a constant that adjusts the shape of the cone. Substituting this kernel into Cohen's class results in the distribution

$$P_{ZAM}(t, \Omega) = \frac{1}{4\pi b} \int g(\tau) e^{-j\tau\Omega} \int_{t-b|\tau|}^{t+b|\tau|} u^*(\zeta - 0.5\tau) u(\zeta + 0.5\tau) d\zeta \, d\tau \qquad (5.47)$$

Depending on the choice of $g(\tau)$, the kernel may or may not satisfy other desirable properties such as the marginals.

For $b = 1$, the discrete form of the Zhao–Atlas–Marks distribution is:

$$P_{ZAM}(k, f) = 4 \sum_{\tau=0}^{L} \hat{g}(\tau) \cos(\tau f) \left[\sum_{\mu=-|\tau|}^{+|\tau|} u^*(k - \mu - \tau) u(k - \mu + \tau) \right] \qquad (5.48)$$

where

$$f = \frac{\omega}{2\pi}, \; L = \frac{\text{kernel window length} - 1}{2}$$

and

$$\hat{g}(n) = \begin{cases} 0.5 \, g(n) & n = 0 \\ g(n) & \text{otherwise} \end{cases} \qquad (5.49)$$

For the test result of Figure 5.5, the resulting distribution is shown in Figure 5.10.

This distribution has been used to model simulated speech. For two-tone signals of 3.0 and 3.08 kHz, the spectral profiles using the spectrogram, pseudo Wigner distribution, and Zhao–Atlas–Marks distribution were calculated, using window length of 128 samples and a sampling frequency of 20 kHz (Figure 5.11). Note that the spectrogram does not resolve the two spectral peaks, and that the pseudo Wigner has a spike of interference at 0 Hz. In contrast, the Zhao–Atlas–Marks distribution gives two distinct spectral peaks [17].

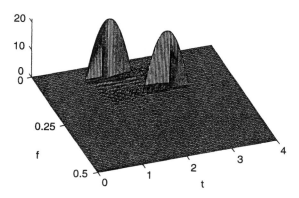

Figure 5.10 Zhao–Atlas–Marks distribution based on test signal in Figure 5.5. From [1].

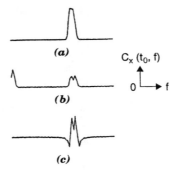

Figure 5.11 Spectral profiles of a signal with two tones at 3.0 kHz and 3.08 kHz. (a) Spectrogram, (b) pseudo Wigner distribution, (c) Zhao-Atlas-Marks distribution. From [17].

5.5 TIME-SCALE REPRESENTATIONS

In the same way that we used the kernel method to describe time-frequency distributions, we can define a time-scale distribution, $P(t, a)$. First, we define the scale transform and its inverse as

$$U(a) = \frac{1}{\sqrt{2\pi}} \int_0^\infty u(t) t^{-ja-0.5} \, dt \leftrightarrow u(t) = \int_0^\infty \frac{U(a) e^{-ja(\ln t)}}{\sqrt{t}} \, da; \; t \geq 0 \qquad (5.50)$$

As with the Fourier transform [see Eq. (5.1)], total energy is preserved within the scale function since integration of the squared time function is equal to integration of the squared scale function.

Now we use the kernel method to construct our time-scale distribution as

$$P(t, a) = \frac{1}{4\pi^2} \iint M(\theta, \sigma) e^{-j\theta t - j\sigma a} \, d\theta \, d\sigma \qquad (5.51)$$

where σ is a dummy integration variable and $M(\theta, \sigma)$ is the corresponding characteristic function defined by

$$M(\theta, \sigma) = \phi\left(a\theta, \frac{\sigma}{a}\right) \int u^*(\zeta - 0.5\tau) u(\zeta + 0.5\tau) e^{j\theta\zeta} \, d\zeta \qquad (5.52)$$

Here, $\phi(a\theta, (\sigma/a))$ is the time-scale kernel function, and the integrated function is the symmetric ambiguity function. Energy distributions satisfying Eq. (5.51) are members of the **affine class**.

5.5.1 General Properties

The time-scale distribution satisfies the time marginal condition,

$$\int P(t, a) \frac{da}{a^2} = |u(t)|^2 \qquad (5.53)$$

if

$$\int \phi\left(a\theta, \frac{\sigma}{a}\right) \frac{da}{a^2} = \delta(\sigma) \tag{5.54}$$

for any σ. Similarly, the time-scale distribution satisfies the frequency marginal condition,

$$\int P(t, a) = \left| U\left(\frac{\Omega}{a}\right) \right|^2 \tag{5.55}$$

if $\phi(0, \sigma) = e^{-j\Omega\sigma}$. Once the marginals are satisfied, the total energy condition is also satisfied.

For frequency invariance, the kernel must be independent of frequency. The time-scale distribution is always time- and scale-invariant.

5.6 SCALOGRAMS

A simple time-scale distribution is the **scalogram,** or squared modulus of the wavelet transform. A **wavelet,** $\psi(t)$, is a function with a mean of zero, such that

$$\int \psi(t)dt = 0 \tag{5.56}$$

Each wavelet function can be used to bandpass filter a signal. With this method, the scale plays the role of a local frequency. As a increases, wavelets are stretched and analyze low frequencies. As a decreases, wavelets are compressed and analyze high frequencies. The **continuous wavelet transform,** $WT(t, a)$, of a signal, $u(t)$, can be defined in the time and frequency domains as

$$WT(t, a) = \frac{1}{\sqrt{a}} \int u(\tau)\psi^*\left(\frac{\tau - t}{a}\right) d\tau \tag{5.57}$$

$$WT(t, a) = \sqrt{a} \int U(\Omega)\hat{\psi}^*(a\Omega)e^{j\Omega t} d\Omega \tag{5.58}$$

where $\hat{\psi}(\Omega)$ is the Fourier transform of $\psi(t)$. The center frequency and bandwidth of the transform vary inversely with scale, such that the ratio of the center frequency to the bandwidth, or **quality factor,** Q, is constant. For efficient processing, only dyadic (powers of two) scales may be used, leading to the **dyadic discrete wavelet transform,** $WT(k, 2^j)$:

$$WT(k, 2^j) = \frac{1}{\sqrt{2^j}} \sum_{i=0}^{N-1} u(i)\psi^*\left(\frac{i - k}{2^j}\right) \tag{5.59}$$

In discrete time, the scalogram is given by

$$P_S(k, a) = |WT(k, a)|^2 = \left| \frac{1}{\sqrt{a}} \sum_{i=0}^{N-1} u(i)\psi^*\left(\frac{i - k}{2a}\right) \right|^2 \tag{5.60}$$

where $a = 2^j$. This distribution is useful in detecting signal transients.

5.6.1 Example

The scalogram has been used to detect individual heartbeats. Looking back at Figure 4.8, the **QRS complex** of each heartbeat is a signal transient that can be detected in many scales. Using a cubic spline wavelet, Boudreaux-Bartels' group detected R waves in the QRS complex of beats from the American Heart Association (AHA) database. A beat was identified when local maxima at the scales 2^2 and 2^3 were present. 1950 beats from channel 1 of the 30 minute tape 3203 were analyzed, using this method and three traditional time-based methods. Error rate was defined as the sum of false positives and false negatives divided by the total number of beats. Using the scalogram, an error rate of 3% was obtained. In contrast, the time-based methods resulted in error rates of 6, 15, and 34%.

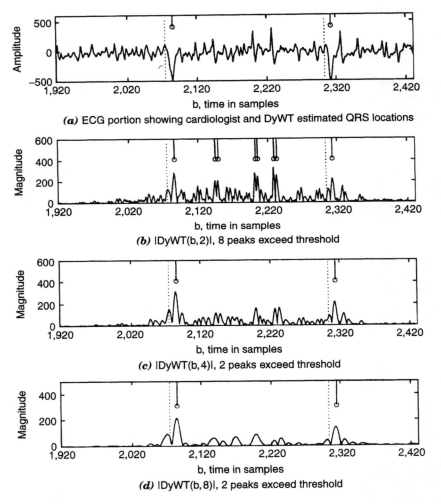

Figure 5.12 Electrocardiograms with external noise from Tape 3203 and corresponding square root scalograms. (a) ECG, with dashed lines indicating cardiologist estimate of QRS onset and tic marks indicating scalogram estimate of R wave locations. (b) Square root of scalogram at scale 2^1. (c) Square root of scalogram at scale 2^2. (d) Square root of scalogram at scale 2^3. From [19].

Scalograms for noisy ECGs in tape 3203 are shown in Figure 5.12 [19]. In a subsequent analysis of four AHA tapes (including tape 3203), respresenting 8598 beats with various arrhythmias, the scalogram method resulted in an error rate of 7%, whereas the time-based methods resulted in error rates of 12, 14, and 21% [20].

5.6.2 Wavelet Processing

The calculations in Eq. (5.59) may be classified functionally as filtering through convolution and **downsampling.** Through choice of $\psi(k)$, lowpass or highpass filtering may be accomplished. Common discrete filters and continuous functions are given in Table 5.1. When a scaling filter, $\psi_a(k)$, is used, convolution results in smoothing or lowpass filtering. The resulting wavelet transform is composed of **approximation coefficients.** As shown

Table 5.1 Common wavelet and scaling filters and functions

Haar wavelet filter $\psi_d(k) = \{-0.7071, +0.7071\}$	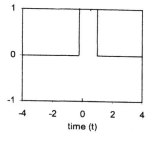
Haar scaling filter $\psi_a(k) = \{+0.7071, +0.7071\}$	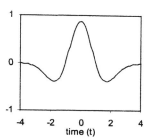
Mexican hat wavelet function $\psi(t) = \left(\dfrac{2}{\sqrt{3}}\ \pi^{-1/4}\right)(1 - t^2)e^{-t^2/2}$	

The Mexican hat wavelet and scaling filters do not exist.

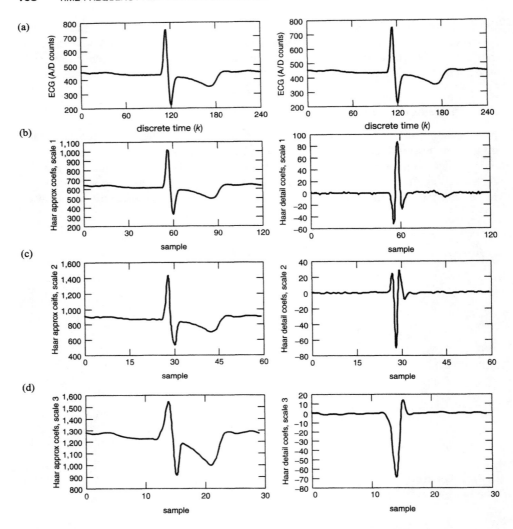

Figure 5.13 Electrocardiogram beat [21], with wavelet transforms. (a) Original beat. (b) Scale 1 Haar approximation and detail coefficients. (c) Scale 2 Haar approximation and detail coefficients. (d) Scale 3 Haar approximation and detail coefficients.

in Figure 5.13, an electrocardiogram beat degrades in subsequent scales of Haar approximation coefficients to emphasize the slower T wave, rather than the faster QRS complex. When a wavelet filter, $\psi_d(k)$, is used, convolution results in emphasis of discontinuities or highpass filtering. The resulting wavelet transform is composed of **detail coefficients.** As shown in Figure 5.13, the same electrocardiogram beat degrades in subsequent scales of Haar detail coefficients to emphasize the high-frequency QRS complex. The Haar wavelet and scaling filters are defined in Table 5.1.

As described in Eq. (5.59), each filtering operation is followed by downsampling, which refers to the retention of every other sample during transformation from one dyadic scale to the next. In Eq. (5.59), $u(i)$ refers to either the original signal at scale 0 or the approximation coefficients of the signal at scale j. These properties may be exploited to compress and reconstruct signals.

5.7 SUMMARY

In this chapter, we have reviewed the basic concepts underlying time-frequency analysis. A time-frequency representation is a two-dimensional mapping of the fraction of the energy of a one-dimensional signal at time, t, and frequency, Ω. Time-frequency distributions in Cohen's class are created using the kernel method. With this method, a kernel chosen to satisfy desirable properties is multiplied by the symmetric ambiguity function, and then transformed in time and frequency. The kernel of the simplest distribution, the Wigner distribution, equals one.

In order to satisfy the time marginal, the kernel at $\tau = 0$ must equal 1. In order to satisfy the frequency marginal, the kernel at $\theta = 0$ must equal 1. Once the marginals are satisfied, the total energy condition and uncertainty principle are also satisfied. If the kernel at τ and θ equals 0, $\phi(0, 0)$ equals 1, and the energy is normalized to one. For time invariance, the kernel must be independent of time. Similarly, for frequency invariance, the kernel must be independent of frequency. For scaling invariance, the kernel must be a product kernel.

In the same way that we use the kernel method to describe time-frequency distributions, we can describe time-scale distributions. First, we define the scale transform and its inverse, and realize that total energy is preserved within a scale function. A time-scale distribution is then created by mutiplying a time-scale kernel by the symmetric ambiguity function, and integrating with respect to time and scale. These distributions are members of the affine class. A simple time-scale distribution is the scalogram, or squared modulus of the wavelet transform. A wavelet is a function with a mean of zero. For efficient processing, only dyadic scales are used, leading to dyadic discrete wavelet transforms.

Both time-frequency and time-scale distributions may be used to recover important signal characteristics that are corrupted by noise in the time domain.

5.8 REFERENCES

[1]. Williams, W. J. Recent advances in time-frequency representations: Some theoretical foundations. In *Time Frequency and Wavelets in Biomedical Signal Processing*, edited by M. Akay, IEEE Press: Piscataway, NJ, 1998, pp. 3–43.

[2]. Lin, Z. and Chen, J. Z. Time-frequency analyses of the electrogastrogram. In *Time Frequency and Wavelets in Biomedical Signal Processing*, edited by M. Akay, IEEE Press: Piscataway, NJ, 1998. pp. 147–182.

[3]. Wood, J. C., Buda, A. J., and Barry, D. T. Time-frequency transforms: A new approach to first heart sound frequency dynamics. *IEEE Trans Biomed, 39,* 730–740, 1992.

[4]. Wood, J. C. and Barry, D. T. Time-frequency analysis of the first heart sound. *IEEE EMBS Mag, 14,* 144–151, 1995.

[5]. Koenig, R., Dunn, H. K., and Lacy, L. Y. The sound spectrograph. *J Acoust Soc Am, 18,* 19–49, 1946.

[6]. Potter, R. K., Kopp, G., and Green, H. C. *Visible Speech.* Van Nostrand: New York, 1947.
[7]. Boudreaux-Bartels, G. F., and Murray, R. Time-frequency signal representations for biomedical signals. In *The Biomedical Engineering Handbook,* edited by J. D. Bronzino. CRC Press: Boca Raton, FL, 1995, pp. 866–885.
[8]. Cohen, L. *Time-Frequency Analysis.* Prentice Hall: Englewood Cliffs, NJ, 1995.
[9]. Cohen, L. Introduction: A primer on time-frequency analysis. In *Time-Frequency Signal Analysis: Methods and Applications,* edited by B. Boashash, Longman Cheshire: Melbourne, Australia, 1992, pp. 3–42.
[10]. Gabor, D. Theory of communications. *J IEE (London), 93,* 429–457, 1946.
[11]. Zaveri, H. P., Williams, W. J., Iasemidis, L. D., and Sackellares, J. C. Time-frequency representation of electrocorticograms in temporal lobe epilepsy. *IEEE Trans Biomed, 39,* 502–509, 1992.
[12]. Wigner, E. P. On the quantum correction for thermodynamic equilibrium. *Phys Rev, 40,* 749–759, 1932.
[13]. Ville, J. Theorie et applications de la notion de signal analytique. *Cables et Transmissions, 2A,* 61–74, 1948.
[14]. Boashash, B. and Black, P. J. An efficient real-time implementation of the Wigner–Ville distribution. *IEEE Trans ASSP, 35,* 1611–1618, 1987.
[15]. Cohen, L. Generalized phase-space distribution functions. *Math Phys, 7,* 781–786, 1966.
[16]. Choi, H. I. and Williams, W. J. Improved time-frequency representation of multicomponent signals using exponential kernels. *IEEE Trans ASSP, 37,* 862–871, 1989.
[17]. Zhao, Y., Atlas, L. E., and Marks, R. J. The use of cone-shaped kernels for generalized time-frequency representations of nonstationary signals. *IEEE Trans ASSP, 38,* 1084–1091, 1990.
[18]. Loughlin, P. J., Pitton, J. W., and Atlas, L. E. Bilinear time-frequency representations: New insights and properties. *IEEE Trans Sig Proc, 41,* 750–767, 1993.
[19]. Murray, R., Kadambe, S., and Boudreaux-Bartels, G. F. Extensive analysis of a QRS detector based on the dyadic wavelet transform. In *Proceedings IEEE-SP International Symposium on Time-Frequency and Time-Scale Analysis,* Philadelphia, PA, pp. 540–543, 1994.
[20]. Kadambe, S., Murray, R., and Boudreaux-Bartels, G. F. Wavelet transform-based QRS complex detector. *IEEE Trans BME, 46,* 838–848, 1999.
[21]. Sugimachi, M. National Cardiovascular Center Research Institute, Osaka, Japan. Research data, 1992.

Further Reading

Hyperbolic and κth Power Classes:

Papandreou, A. Hlawatsch, F., and Boudreaux-Bartels, G. F. The hyperbolic class of quadratic time-frequency representations, Part I: Constant-Q warping, the hyperbolic paradigm, properties, and members. *IEEE Trans Sig Proc, 45,* 3425–3444, 1993.

Hlawatsch, F., Papandreou, A., and Boudreaux-Bartels, G. F. The power classes of quadratic time-frequency representations: a generalization of the affine and hyperbolic classes. In *Proceedings 27th Annual Asilomar Conference Signals Systems Computation,* Pacific Grove, CA, pp. 1265–1270, 1993.

Wavelet Transforms:

Rioul, O. and Flandrin, P. Time-scale energy distributions: A general class extending wavelet transforms. *IEEE Trans Sig Proc, 40,* 1746–1757, 1992.

Rioul, O. A discrete-time multiresolution theory. *IEEE Trans Sig Proc, 41,* 2591–2606, 1993.

5.9 RECOMMENDED EXERCISES

Spectrogram: See [1], Examples 7.1 to 7.10.
Wigner Distribution: See [1], Examples 8.1 to 8.12, 8.15, and 8.16.
Kernel Method: See [1], Examples 9.1 to 9.8.
Scalograms: See Akay, M. *Biomedical Signal Processing.* Academic Press: San Diego, 1994. Computer Experiments 8.1 and 8.2 for scalograms based on the Haar and Mexican hat wavelets.

6

IMPROVED IMPEDANCE CARDIOGRAPHY

In this chapter, we discuss an industrial application of time-scale and time-frequency analysis: compensation of noise artifact from **impedance cardiography** (ICG) measurements. During impedance cardiography, **cardiac output,** or total blood flow, is estimated by monitoring changes in thoracic impedance. Features within the impedance signal are used to estimate cardiac output. Unfortunately, respiratory, motion, and other artifacts in the impedance data may obscure these features. Transformation to the time-space or time-frequency domain enables robust pattern recognition of key parts of the impedance signal used in calculations.

6.1 PHYSIOLOGY OF CARDIAC OUTPUT

The mammalian heart can be considered as two hollow organs—the right and left halves—with muscular walls. Each half comprises an **atrium** and **ventricle.** The left atrium and right ventricle are part of the **pulmonary circulation,** the movement of blood from the right to left heart. The right atrium and left ventricle are part of the **systemic circulation,** the distribution of blood to, and return from, the rest of the body (Figure 6.1). Within the circulatory system, **veins** and smaller **venules** transport blood to the heart; **arteries** and smaller **arterioles** transport blood away from the heart. In the pulmonary circulation, oxygenated blood is transported by the veins. In the systemic circulation, oxygenated blood is transport by the arteries.

During each heartbeat, the ventricles are rhythmically relaxed and then contracted. During this relaxation or **diastole,** the ventricles fill with blood; during this contraction or systole, they expel the blood into the large arteries—the **aorta** and **pulmonary artery.** Before blood reenters the ventricles, it passes from the large veins—the **vena cava** and **pulmonary vein**—into the associated atria. The systole of each atrium precedes that of its ventricle, in order to boost flow to the ventricle. This continual circulation of fluid throughout the body serves as a route for the delivery and removal of substances.

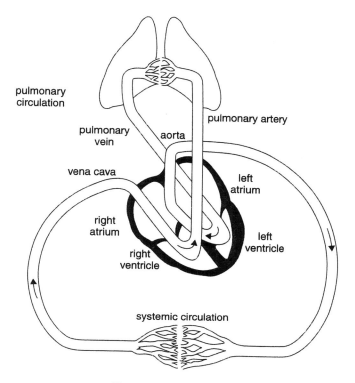

Figure 6.1 Circulation.

6.1.1 Heart Mechanics

Systole and diastole could not occur without the participation of **valves** covering the inlets and outlets of both ventricles. The **mitral valve** in the left ventricle and **tricuspid valve** in the right ventricle prevent regurgitation of blood into the atria during ventricular systole. These valves are called the **atrioventricular (AV) valves.** The **aortic** and **pulmonary valves** that are at the bases of the large arteries prevent regurgitation into the ventricles during diastole. Valve opening and closing occurs through pressure changes in the adjacent heart cavities or vessels. Valve motion therefore affects the mode of contraction of the heart muscle.

During systole, an **isovolumetric contraction period** and **ejection period** occur (Figure 6.2). At the onset of ventricular systole, the intraventricular pressure rises, causing immediate closure of the AV valves. At first, the arterial valves also remain closed, causing continual contraction around the incompressible contents.

When the intraventricular pressure exceeds the diastolic pressure of ~ 80 mmHg, the arterial valves open and blood begins to be expelled. Initially, the intraventricular pressure continues to rise, until it reaches a maximum of ~ 130 mmHg. Toward the end of systole, the pressure begins to fall. As the volume curve of Figure 6.2 illustrates, under resting conditions, the ventricle ejects only about half of the volume it contains. This ejected volume of ~ 60 ml is the **stroke volume** (SV). When the ejection phase is complete, an end-systolic volume of ~ 70 ml remains in the ventricle. Closure of the aortic valve marks

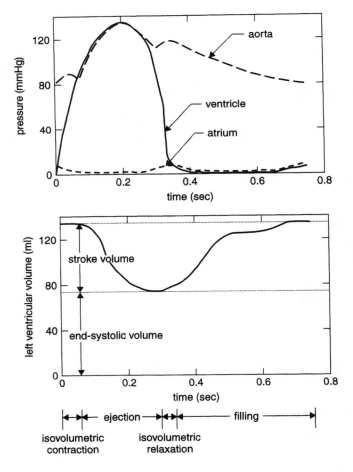

Figure 6.2 Pressure and volume changes during one heartbeat.

the end of systole. At normal resting level, the duration of this contraction period, the **left ventricular ejection time** (LVET), is ~ 150–300 ms in normal people.

During diastole, an **isovolumetric relaxation period** and **filling period** occur. Initially, for ~ 50 ms, all the valves remain closed. The resulting relaxation is thus isovolumetric, as the intraventricular pressure falls rapidly to almost zero. When the pressure is lower than the atrial pressure, the AV valves open, causing the ventricle to fill in preparation for the next systole. During the filling period, the intraventricular pressure rises only slightly.

In principle, the right heart encounters these same four periods. However, because the vascular resistance is lower in the pulmonary circulation, the pressure the right heart must develop in systole is considerably lower. The stroke volumes of the two ventricles are almost the same.

6.1.2 Systemic Circulation

Blood propelled during systole from the left ventricle, through the aorta, to arteries is divided among many regional vascular beds in parallel, each of which supplies a particular organ. Each artery branches numerous times, so that the total number of vessels increases while the individual vessels become progressively smaller in diameter. The smallest arterioles branch to form a **capillary bed,** a very dense network of narrow vessels with extremely thin walls. Across capillary walls, exchange between blood and cells in the surrounding tissue occurs. The capillaries merge to form venules, which in turn merge to form veins.

6.1.3 Pulmonary Circulation

Within the pulmonary circulation, blood passes from the right ventricle through the pulmonary trunk into the vascular system of the lungs (see Chapter 4). Four large pulmonary veins then transport the blood to the left atrium, where it moves to the left ventricle.

6.1.4 Cardiac Output

Cardiac output (CO) is the sum of all stroke volumes ejected within a given time interval. It is usually defined as the product of the mean stroke volume (in units of ml/beat) and heart rate (in units of beats/min):

$$CO = \text{stroke volume} \times \text{heart rate} \tag{6.1}$$

The **cardiac index** is cardiac output that has been normalized by body surface area. The stroke volume is primarily determined by the **ventricular preload, afterload,** and **contractility.** Preload is the passive load on the heart muscle that establishes the initial muscle length of the cardiac fibers before contraction. Afterload is the sum of all the loads against which the cardiac fibers must shorten during systole. Contractility is estimated by the maximum velocity of contraction of the cardiac muscle fibers. Cardiac output and the **systemic vascular resistance** of the systemic circulation are regulated by the central nervous system in order to maintain the blood pressure gradient required for flow through the vascular system.

Typically, blood pressure is used to assess overall cardiovascular performance. Readily measured noninvasively or intraarterially, blood pressure is clinically significant because blood flow to tissues may be inadequate when blood pressure is too low. However, a normal blood pressure does not always indicate optimal blood flow to the tissues. In severely compromised patients, the volume of blood flow, and therefore the quantity of oxygen and other vital substances delivered to the tissues, must also be considered. The expected result after intervention of increased cardiac output along with increased blood pressure does not always occur. For example, Shoemaker's group measured cardiac output in 52 emergency room patients after standard fluid therapy with crystalloids. In all cases, the therapy increased blood pressure. However, cardiac output and related hemodynamic changes deteriorated in 26%, were unchanged in 37%, and improved in only 28% of the patients [1].

Further, Forrester et al. demonstrated that the clinical status of patients with acute **myocardial infarction** (typically called a "heart attack") is related to the CO and that after myocardial infarction, prognosis worsens as CO declines. Almost half of the patients with acute myocardial infarction have subnormal CO when examined by a physician; 25% of infarctions are not recognizable by standard clinical criteria [2].

6.2 IN VIVO AND IN VITRO CARDIAC OUTPUT MEASUREMENTS

The gold standard for cardiac output monitoring is **thermodilution,** one of a group of **indicator-dilution methods** by which a detectable indicator is applied upstream in the circulation and detected downstream to determine the flow rate by which it was mixed. It is assumed that the indicator mixes with all the blood flowing through the central mixing pool. Other methods for estimating cardiac output include measuring stroke volume changes through impedance cardiography and aortic blood velocity changes through **Doppler ultrasound.**

6.2.1 Thermodilution

In preparation for thermodilution, a multiple lumen **pulmonary artery (PA) catheter** is passed through the skin into a central vein (Figure 6.3a). Once the catheter tip reaches a central venous location, a balloon at the tip is inflated, which causes the catheter tip to rapidly move from the right atrium, through the right ventricle, and into the pulmonary artery. During thermodilution, a bolus of room temperature or iced (0°C) 5% dextrose in water or 0.9% NaCl is injected through the catheter into the right atrium. The volume injected in adult patients without fluid restrictions is generally 10 ml. The resulting blood temperature transient is detected downstream by a thermistor in the pulmonary artery. A typical thermodilution curve is shown in Figure 6.4.

Cardiac output is then calculated using the **Stewart–Hamilton equation** as

$$CO = \frac{V_I(T_B - T_I)K_1 K_2}{A} \quad (6.2)$$

where V_I is the injectate volume, T_B is the blood temperature in the pulmonary artery, T_I is the injectate temperature, K_1 is the density factor (injectate/blood), K_2 is the catheter manufacturer's computation constant, and A is the area under the thermodilution curve. Due to heat loss through the catheter wall, several serial injections are needed to obtain a consistent value for cardiac output. If the cardiac output is low, the resulting curve will be very broad, decreasing the signal-to-noise ratio of the measurement. Respiratory-induced variations in pulmonary artery blood temperature may confound the dilution curve when it is of low amplitude. This in vivo measurement can only be obtained intermittently.

6.2.2 Continuous Thermodilution

Continuous thermodilution measurements can be made using a PA catheter that has been modified to include a filamentous heating element in the right ventricle (Figure 6.3b). Rather than using decreased temperature, increased temperature of approximately +0.02°C due to intermittent heating by the filament is used as the indicator. To filter out noise primarily due to respiratory-induced temperature fluctuations, the heating is applied as a pseudorandom binary sequence (see Chapter 1). The maximum heating coil temperature is limited to 44°C [3]. When the heating filament is positioned correctly in the right atrium and right ventricle, continuous cardiac output averaged over 3–7 minutes is accurately measured. In 404 pairs of intermittent and continuous thermodilution measurements from 35 intensive care unit (ICU) patients, a mean bias (systemic error) of 0.03 l/min and relative error (percentage error) of 3.6% were obtained [4]. This method was in-

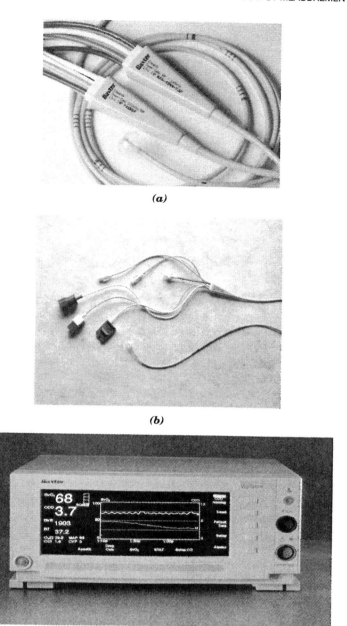

Figure 6.3 Cardiac output catheters and monitor. (a) Standard thermodilution Swan–Ganz catheter. (b) Continuous thermodilution CCOMBO catheter. (c) Continuous thermodilution Vigilance monitor. Courtesy of Baxter Edwards, Irvine, CA.

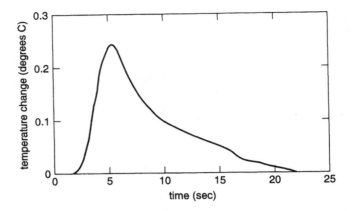

Figure 6.4 Typical thermodilution curve.

vented by Mark Yelderman while at Stanford University, and commercialized by Interflo Medical. Interflo was later purchased by Baxter (Figure 6.3c).

The feasibility of an alternate method for obtaining cardiac output once per minute has been investigated in six swine. In this system, cold saline is continuously cycled through a heat exchanger surrounding a special PA catheter. The heat exchanger cools blood in the right atrium; the resulting temperature change is measured by a thermistor in the pulmonary artery [5].

6.2.3 Fick Method

Oxygen, rather than temperature, may be employed as the indicator to measure cardiac output. Using the **Fick method,** oxygen entering the pulmonary circulation is measured in the pulmonary blood flow, since the oxygen concentration of venous blood increases as it passes through the lungs, along with the respiratory oxygen uptake. This method was first proposed by Adolph Fick in 1870. Based on this principle, cardiac output can be estimated as

$$CO = \frac{\dot{V}_{O_2} \times F}{C_{AO_2} - C_{VO_2}} \quad (6.3)$$

where $\dot{V}_{O_2}$ equals oxygen consumption, C_{AO_2} is the oxygen content in ml/dl in arterial blood, C_{VO_2} is the oxygen content in ml/dl in pulmonary artery blood, and F is a correction factor that compensates for oxygen measurement at room temperature versus oxygen consumption at body temperature. Oxygen consumption is measured at the airway, typically with an oxygen-filled **spirometer** containing a CO_2 absorber. A spirometer measures the volume of air during inspiration and expiration. Normal basal oxygen consumption divided by body surface area ranges from 110 to 150 ml/min-m². C_{AO_2} and C_{VO_2} are measured by blood gas analyzers from blood samples.

The Fick method does not require that fluid be added to the circulation. However, it may only be used during stable hemodynamic conditions, since measuring averaged oxygen consumption takes place over several minutes.

6.2.4 Impedance Cardiography

Noninvasive estimates of cardiac output can be obtained using impedance cardiography. Strictly speaking, impedance cardiography, also known as **thoracic bioimpedance** or **impedance plethysmography,** is used to measure the stroke volume of the heart. As shown in Eq. (6.1), when the stroke volume is multipled by heart rate, cardiac output is obtained. The basic method of correlating **thoracic,** or chest cavity, impedance, $Z_T(t)$, with stroke volume was developed by Kubicek et al. at the University of Minnesota for use by NASA [6, 7].

The total thoracic impedance, $Z_T(t)$, consists of a constant impedance, Z_0, and time-varying impedance, $\Delta Z(t)$. According to a parallel-column model of the thorax first described by Nyboer, this change in thoracic impedance, $\Delta Z(t)$, is related to the pulsatile blood volume change [8]. In this model, constant tissue impedances such as bone, muscle, and fat are modeled as a conducting volume, Z_0, in parallel with the pulsatile impedance of the blood, $\Delta Z(t)$. This second impedance is a time-varying fluid column with resistivity, ρ, cylindrical length, L, and a time-varying cross-sectional area that oscillates from zero to the stroke volume (Figure 6.5). When the pulsatile volume is at a minimum in the cardiac cycle, all the conducting tissues and fluids are represented by Z_0. During the cardiac cycle, the cylinder cross-sectional area increases from zero until its volume equals the blood volume change.

Because Z_0 is much greater than $\Delta Z(t)$, the following relationship holds:

$$SV = \rho \left(\frac{L^2}{Z_0^2} \right) LVET \frac{dZ}{dt}\bigg|_{max} \qquad (6.4)$$

where L is the distance between the inner band electrodes in cm, $LVET$ is the left ventricular ejection time in seconds, and dZ/dt_{max} is the magnitude of the largest negative derivative of the impedance change, $\Delta Z(t)$, occurring during systole in ohms/s. As shown in Figure 6.6, the impedance derivative has been purposely inverted so that the original negative minimum change will appear as a positive maximum, in a manner more familiar to clinicians. In this figure, **fiducial point** B is associated with aortic valve opening, and C represents a major upward deflection during systole. X is associated with aortic valve closure. The left ventricular ejection time occurs between B and X; dZ/dt_{max} is the maximum amplitude change at C. Typically, a value of 150 ohm-cm is used for the resistivity of the blood.

Because the impedance is assumed to be purely resistive, the pulsatile impedance may be estimated using Ohm's law. Current from a constant current source, $I_T(t)$, is applied, and the resulting voltage, $V_T(t)$, is measured in order to calculate the ratio

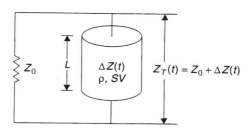

Figure 6.5 Thoracic impedance parallel-column model.

120 IMPROVED IMPEDANCE CARDIOGRAPHY

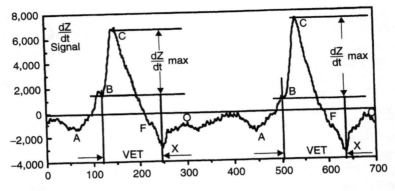

Figure 6.6 Basic features of $dZ(t)/dt$ signal used for application and diagnostic purposes. Because $-dZ/dt$ is plotted, $dZ(t)/dt_{min}$ is referred to as dZ/dt_{max}. From [9].

$$Z_T(t) = \frac{V_T(t)}{I_T(t)} \tag{6.5}$$

The applied current is in the range of 50 to 100 kHz, from 0.5 to 4 mA RMS, to minimize electric shock hazard.

At this frequency range, the typical skin impedance is 2 to 10 times the value of the underlying thoracic impedance. To eliminate the contribution from skin and tissue impedances, four band electrodes are used for measurement, as shown in Figure 6.7. Stimulation electrode 1 and sensing electrode 2 are applied above the thorax; stimulation electrode 4 and sensing electrode 3 are applied below the thorax. The current is supplied to the stimulation electrodes. Current flows from each electrode through each constant skin impedance, Z_{sk1} or Z_{sk4}, each constant body tissue impedance, Z_{b1} or Z_{b1}, and each constant skin impedance, Z_{sk2} or Z_{sk3}, to each sensing electrode. The voltages at electrodes

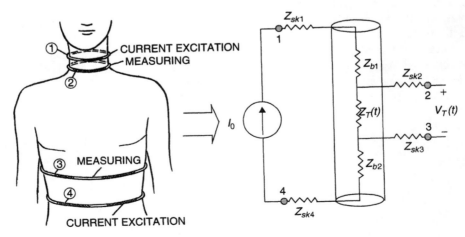

Figure 6.7 Thoracic impedance measurement using four electrodes. Based on [6].

2 and 3 are measured and input to a differential amplifier to obtain the differential voltage, $V_T(t)$. The desired thoracic impedance, $Z_T(t)$, is then obtained from Eq. (6.5).

In this original configuration, implemented as the **Minnesota impedance cardiograph,** cardiac output could only be estimated with impedance cardiography without motion and during **voluntary apnea,** during which time the patient held his breath. With apnea, ventilation changes that are typically an order of magnitude larger than cardiovascular impedance changes would not obscure cardiac output measurements. Unfortunately, cardiac output measurements obtained with this method in unhealthy individuals were inaccurate. Therefore, this original method is unsuitable for use with critically ill or unconscious patients.

6.2.5 Doppler Ultrasound

Noninvasive estimates of cardiac output can also be obtained using Doppler ultrasound. Again, stroke volume is estimated and multiplied by heart rate to obtain cardiac output. Lee Huntsman's group developed a method at the University of Washington, which was later commercialized by Lawrence Medical Systems [10], in which stroke volume was calculated as the product of the spatial average blood velocity in the aorta during systole, BV; the left ventricular ejection time; and the cross-sectional area of the aorta, A:

$$SV = BV \times LVET \times A \quad (6.6)$$

It is assumed that the aortic velocity is highest at the level of the narrowest aortic diameter.

For this measurement, an ultrasound transducer is positioned over the sternum to scan the **aortic root** (root of the aorta). The image selected for measurement is that which provides the narrowest aortic diameter, D. Assuming that the cross-sectional area is circular, it is calculated from the diameter as

$$A = \frac{\pi D^2}{4} \quad (6.7)$$

A second transducer is positioned over the **suprasternal notch** of the neck toward the ascending aorta, in order to measure the aortic velocity using the Doppler effect:

$$f_d = \frac{2 BV f_t \cos \theta}{c} \quad (6.8)$$

where f_d is the Doppler shift frequency, f_t is the transmitted frequency, c is the velocity of sound in blood, and θ is the angle between the ultrasound beam and the blood flow vector. The transmitted frequency is 2.5 MHz, and the velocity of sound in blood is assumed to be 1540 cm/sec. Because the angle is shallow, it is assumed to be 0°, so that $\cos \theta$ equals zero.

In a clinical validation of this technique, 110 measurements were compared to those obtained from thermodilution in 45 patients. All data were collected by a trained echocardiographic technician. The resulting correlation coefficient squared, r^2, was 0.88. Therefore, 88% of the variance in the thermodilution measurements could be accounted for by the ultrasound measurements [11]. However, because this technique was very sensitive to user positioning, it was not commercially successful. More recent systems have attempted to estimate aortic velocity and aortic cross-sectional area more accurately [12].

6.3 PROBLEM SIGNIFICANCE

In general, physiologic monitoring is moving from invasive to noninvasive methodology. Because the cardiac output gold standard of thermodilution increases cost and risk, it is desirable to find a noninvasive alternative. Recently, efforts have focused on improving the quality of measurements obtained from noninvasive impedance cardiography.

6.3.1 Clinical Significance

The pulmonary artery catheterization required for intermittent or continuous thermodilution exposes the patient to several risks. Possible complications include pulmonary artery rupture, pulmonary **infarction** (tissue death from obstruction of local blood supply), catheter-related **sepsis** (infection), balloon rupture, endocardial or valvular damage, and venous **thrombosis** (clots). Further, in a prospective study of five teaching hospitals in which 5735 adult cases were followed, it was determined that pulmonary artery catheterization was associated with increased mortality and increased utilization of resources. Patients with a higher baseline probability of surviving two months had the highest relative risk of death following pulmonary artery catheterization [13]. The results from this controversial study served as a catalyst for a national consensus meeting in 1997. The participants acknowledged that conclusive data showing benefit or harm did not yet exist, and recommended that a prospective, randomized trial be conducted [14].

Even if pulmonary artery catheterization does not increase mortality, it is still desirable to measure continuous cardiac output noninvasively. A study by Haller et al. suggested that the mean time to show a 75% response in actual changes in cardiac output using continuous thermodilution was 10.5 minutes. This response time was too slow for immediate detection of acute changes [15]. Moreover, continuous thermal measurement of cardiac output depends on relatively steady state measurement of pulmonary artery blood temperature. When large volumes of crystalloid and blood products are infused at cold or room temperatures during the resuscitation of a critically ill patient, significant variations in the continuously measured cardiac output value occur. Even when these products are warmed to near body temperature, administration of large volumes may still affect the reliability of the measurement [16].

6.3.2 Traditional Empirical Method

In 1984, Bohumir Sramek was issued a patent for a method that improved thoracic impedance measurements [17]. Rather than model the thorax as a cylinder, Sramek used a truncated cone, resulting in a modified stroke volume calculation of

$$SV = \frac{L^3}{4.25 Z_0} LVET \frac{dZ}{dt}_{max} \qquad (6.9)$$

Sramek also used a four-sample average of dZ/dt_{max} to reduce the effect of respiratory artifact, thus enabling measurement without voluntary apnea. He further simplified measurement by using pairs of spot electrodes rather than band electrodes. With these modifications, the accuracy of impedance cardiography measurements improved. A few years later, Donald Bernstein suggested a modification to Sramek's equation, which came to be known as the Sramek–Bernstein equation:

$$SV = \beta \left(\frac{\text{Weight}_{observed}}{\text{Weight}_{ideal}} \right) \left[\frac{(0.17\,H)^3}{4.2} \right] LVET \left(\frac{\frac{dZ}{dt}_{max}}{Z_0} \right) \quad (6.10)$$

where β is the relative blood volume index, H is the patient height, $\text{Weight}_{observed}$ is the observed weight, and Weight_{ideal} is the ideal weight [18].

Unfortunately, the correlation between these measurements from Sramek devices and those made using thermodilution was still low. In a study in which cardiac output was measured in 28 patients recovering from elective heart surgery, the squared correlation coefficient between the two methods initially after surgery was $r^2 = 0.30$ ($p = 0.002$), and decreased to $r^2 = 0.26$ ($p = 0.004$) two to four hours later [19].

6.3.3 Market Forces

Sramek commercialized his invention through his company BoMed Medical Manufacturing in the 1980s. The BoMed monitor was called the NCCOM3. BoMed experienced bankruptcy and reorganization in 1992–1993.

Also in the late 1980s, Xiang Wang improved impedance cardiography measurements for his dissertation under Hun Sun at Drexel University, using the spectrogram. Their patents [20, 21, 22], which are owned by Drexel, were licensed to Renaissance Technologies. The accuracy of the Renaissance IQ monitor, which is based on Drexel technology, is discussed in Section 6.4. In 2001, the technology of the IQ monitor was acquired by Wantagh Incorporated [23].

Wantagh's primary competition is CardioDynamics International Corporation, the reorganization of BoMed. CardioDynamics has begun to utilize digital signal processing throughout its instruments, including wavelet transforms for improved fiducial point processing [24]. CardioDynamics invests in technology and markets their monitor more aggressively than Renaissance/Wantagh by presenting papers and exhibiting their products at major medical conferences.

In 1997 alone, hospital sales of thermodilution catheters and their monitoring accessories generated $67,162,910 and $14,604,868, respectively, in the United States. These sales estimates exclude sales in federal hospitals and nursing homes. Baxter accounted for 54.9 and 58.8% of total catheter and accessories sales, respectively. [25] It has been predicted that cardiac output monitoring sales would increase significantly if a less expensive, accurate, and easy to use noninvasive alternative existed. Impedance cardiography may be this alternative. According to cost analysis conducted in 1991 for cardiac output materials and care, a single application of thermodilution cost $1528, whereas a single application of ICG cost only $923 [26]. The market for impedance cardiography has begun to increase, as this monitoring method was formally approved for full Medicare coverage on a nationwide basis in September, 1998. The decision was facilitated by CardioDynamics, who submitted "over 70 peer-reviewed studies on more than 5000 patients, conducted by 600 researchers at 275 institutions" [27].

6.4 SPECTROGRAM PROCESSING IN DREXEL PATENTS

As described in Chapter 5, the spectrogram, or short-time Fourier transform, is a time-frequency representation that can be used to identify signal features. In this section, the method by which Wang and Sun derived more accurate estimates of dZ/dt_{max} and $LVET$ using the spectrogram is described. The spectrogram approach was only utilized after experimentation with the pseudo Wigner distribution. As would be expected, the pseudo Wigner approach resulted in cross terms that blended different frequency components and widely distributed noise [9].

6.4.1 Data Acquisition and Processing

Using constant current stimulation similar to the Minnesota Impedance cardiograph [9], the analog signals Z_0, $\Delta Z(t)$, and dZ/dt, as well as an electrocardiogram (ECG), were acquired. The signals were digitized using a 12-bit A/D converter, with a 500 Hz sample rate. Additionally, dZ/dt was digitally filtered using a lowpass (55.5 Hz bandwidth, −27 dB sidelobes) and highpass (5 Hz passband, −14 dB sidelobes) filter to remove artifacts such as 60 Hz noise. The ECG was processed through a digital lowpass and highpass filter, a differentiator, and a nonlinear squaring filter to enhance the R wave into a very high R wave/P wave ratio (see Figure 4.8). Processing of each digital $\dot{Z}(k)$ waveform was triggered by the R wave (Figure 6.8).

As discussed in Chapter 5, the spectrogram can be calculated as

$$P_{STFT}(k,f) = \sum_{\tau=0}^{M-1} u(k+\tau)h(\tau)\exp(-j2\pi f\tau) \qquad (6.11)$$

where $f = \omega/2\pi$, $h(k)$ = window, M = window length, and $0 \leq \tau \leq M-1$. The window chosen for this application was a Hanning window (see Chapter 1):

$$h(k) = \begin{cases} 0.5 - 0.5\cos(2\pi k/M), & 0 \leq k \leq M \\ 0, & \text{otherwise} \end{cases} \qquad (6.12)$$

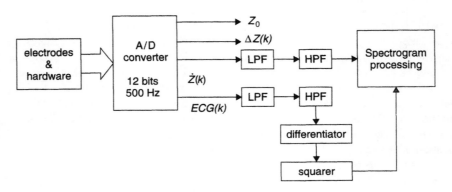

Figure 6.8 Drexel data acquisition system.

In actual implementation, only the magnitude of the **fast Fourier transform** (FFT) was used to generate the local frequency distribution by spectrogram function, $s(k)$. First, the location of $\dot{Z}(k)_{max}$ was located and stored as P. After windowing the Fourier transform of each $\dot{Z}(k)$ waveform and calculating its magnitude, the local frequency distribution by spectrogram function was calculated as

$$s(k) = \begin{cases} \dfrac{\sum_{n=f_1}^{f_2} FFT_{mag}(n)}{\sum_{n=f_2}^{f_3} FFT_{mag}(n)}, & 0 \leq k \leq P \\[2ex] \dfrac{\sum_{n=f_1}^{f_2} FFT_{mag}(n)}{\sum_{n=0}^{f_2} FFT_{mag}(n)}, & P \leq k \leq M + L \end{cases} \quad (6.13)$$

where L is any order of 2, $2L \leq M$, $f_1 \cong 30$ Hz, $f_2 \cong 45$ Hz, and $f_3 =$ sampling frequency/2.

Within this function, the aortic valve opening, B; upward deflection during systole, C; and aortic valve closure, X, were mapped to localized peaks (Figure 6.9). Detection of these peaks enabled noise-resistant estimation of the ventricular ejection time, the time interval between B and X, and estimation of dZ/dt_{max}, the maximum positive difference in derivative amplitude. Estimates of LVET and dZ/dt_{max} were then used to calculate stroke volume and cardiac output from the Sramek equation (Eq. 6.9).

This processing was implemented using integer math for fast signal processing with less required computational power [20, 21, 22]. The technology in the Drexel patents serve as the foundation for Renaissance Technologies' IQ System monitor (Figure 6.10).

6.4.2 Data Analysis

To validate this system, 842 pairs of simultaneously measured cardiac output by ICG and thermodilution in were analyzed in a multicenter trial [28]. The data were acquired from 68 severely ill patients who required thermodilution for management of their acute circulatory disorders. The mean cardiac output was 8.65 ± 3.82 l/min using ICG, and 8.74 ± 3.64 l/min using thermodilution. Using regression analysis, a squared correlation coefficient of $r^2 = 0.74$ ($p < 0.001$) was obtained. Cardiac output values estimated by both techniques are plotted against each other in Figure 6.11. In a follow-up study of 71 high-risk surgical patients measured before and after surgery, a squared correlation coefficient of $r^2 = 0.67$ ($p < 0.001$) was obtained [29]. Although the regression relationship decreased compared to results from the multicenter trial of more stable intensive care patients, the results are still a substantial improvement compared to the earlier BoMed system. Of the 71 patients, only 55 survived. During several hours of the post-op period, the decrease in cardiac index, measured by either method, was significantly greater ($p < 0.05$) in nonsurvivors than in survivors.

126   IMPROVED IMPEDANCE CARDIOGRAPHY

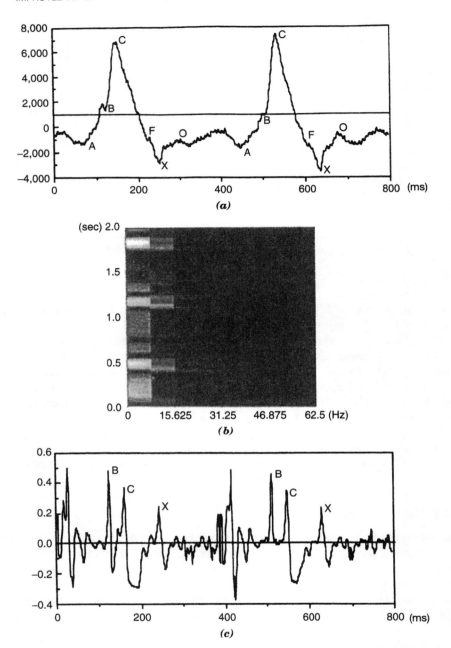

Figure 6.9 Spectrogram processing. (a) Raw $\dot{Z}(k)$, sampled at 500 Hz. (b) The spectrogram of $\dot{Z}(k)$. (c) The local frequency distribution by spectrogram, $s(k)$. From [9].

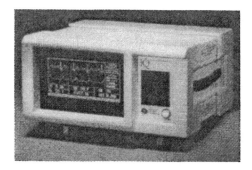

Figure 6.10 Renaissance Technologies' IQ System Model #2001. Courtesy of Renaissance Technologies, Newtown, PA.

6.5 WAVELET PROCESSING IN CARDIODYNAMICS SOFTWARE

It is the author's contention that improving the accuracy of ICG is a two-step process. First, the factors in the stroke volume equation should be calculated with minimal error. Second, the stroke volume equation should be reevaluated for accuracy, compared to thermodilution or another reference standard. The first journal article on Kubicek's equation (the first publications were U.S. Air Force technical reports) reports results from only "10 young, normal Caucasian male subjects" measured under 17 experimental conditions involving exercise or controlled rest [30]. The first journal article documenting Sramek's equation describes the Sramek system, but presents no data [31]. In his "theory and rationale" of modifications to the Sramek equation, which became known as the Sramek–

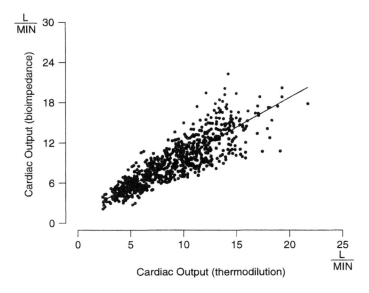

Figure 6.11 Bioimpedance cardiac output versus thermodilution cardiac output. 842 data pairs in 68 ICU patients. $r^2 = 0.74$ ($p < 0.001$). Reprinted from [28] with permission from Williams & Wilkins.

Bernstein equation, Bernstein presents theory but no data [18]. Only after both steps have been completed can accuracy comparisons be reasonably performed.

CardioDynamics accomplished this first step by calculating the fiducial points using wavelet transforms. A wavelet transform is distinctly separate from a time-frequency distribution because, as a type of time-scale distribution, it is based on scale, rather than on frequency [32]. In a similar manner to the time-frequency processing performed by Renaissance IQ software, $\Delta Z(t)$ and dZ/dt are processed in the time-scale domain, so that fiducial points may be more easily recognized in the face of noise. In contrast to the Renaissance software, the ECG is also processed in the time-scale domain for efficient R wave detection. The CardioDynamics monitor typically used in outpatient clinics, the BioZ.com, is illustrated in Figure 6.12.

6.5.1 Data Acquisition and Processing

Using a sine wave excitation of 2.5 mA RMS at 70 kHz, the analog signals $[Z_0, \Delta Z(t)]$ and ECG(t) were acquired. The signals were digitized using a 20 bit A/D converter, with a 1000 Hz sample rate, and decimated to 200 Hz. Additionally, $\Delta Z(t)$ was digitally filtered using a lowpass (20 Hz cutoff frequency) and highpass (0.9 Hz cutoff frequency) filter to remove artifacts such as 60 Hz noise and respiration. The ECG was also processed through a digital lowpass (45 Hz cutoff frequency) and highpass (0.9 Hz cutoff frequency) filter. Processing of each digital $\dot{Z}(k)$ or $\Delta Z(k)$ waveform was triggered by detection of an R wave using wavelet transforms. Detected B, C, and X points were used to calculate $LVET$ and dZ/dt_{max}, the derived parameters that are used to calculate stroke volume.

The C point calculation was based on the scale 1 approximation coefficients of $\dot{Z}(k)$, using the Mallet scaling filter (Figure 6.13). The global maximum was determined in the search range of ⅓ and ½ the length between the first and second Q points. The first occurring of these global maxima was designated the C point (Figure 6.14). The B point calculation was based on the scale 2 detail coefficients of $\Delta Z(k)$, using the Symlet 2 wavelet filter (Figure 6.13). Search for the most recent local maximum in the detail coefficients was

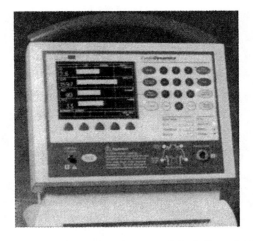

Figure 6.12 CardioDynamics BioZ.com Monitor. Courtesy of CardioDynamics International Corporation, San Diego, CA.

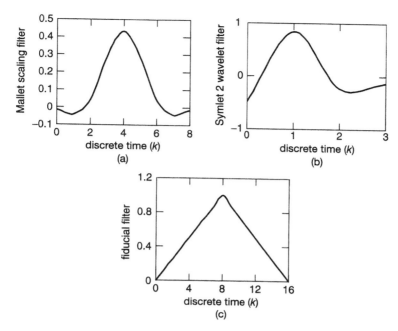

Figure 6.13 Scaling and wavelet filters used in fiducial point detection. (a) Mallet scaling filter; (b) Symlet 2 wavelet filter; (c) Fiducial filter. From [24, 33].

limited between $[(R + 1)/4] + 1$ and $[(C - 2)/4 + 1]$. The B point was detected as the global maximum of $\Delta Z(k)$ in the range bracketed by this local maximum and (C-2) (Figure 6.15).

The X point calculation was based on the scale 1 approximation coefficients of $\dot{Z}(k)$, using a newly created fiducial filter (Figure 6.13). Search for the X point was limited between the C point and the next Q wave. The X point was detected as the first-occurring local minimum of the scale 1 approximation coefficients (Figure 6.16).

This detection algorithm was developed from 48 training waveform sets. The data in Figures 6.14–6.16 were acquired from a 62 year old, 82 kg male pacemaker patient. The local minimum in $\dot{Z}(k)$ immediately after the Q point was caused by a pacemaker stimulation. Note from these figures that $\Delta Z(k)$ frequently cannot be represented by the "typical" $\Delta Z(k)$ illustrated in Figure 6.6. With the addition of noise artifact, identification of the fiducial points becomes even more difficult.

6.5.2 Data Analysis

Validation of this detection algorithm was performed in a manner similar to the validation of QRS complexes during arrhythmias. An in-house CardioDynamics reviewer not associated with the detection project chose 13 test waveform sets randomly from a database of 266 waveforms, not used in training. Many of these waveforms possessed significant noise artifact. He also chose two waveforms that exhibited **bundle branch block,** which occurs when the right and left ventricles do not simultaneously depolarize. When this occurs, two R waves are present. None of the ECGs in the training data possessed bundle branch block. The reviewer then annotated 15 beats within each waveform set for each

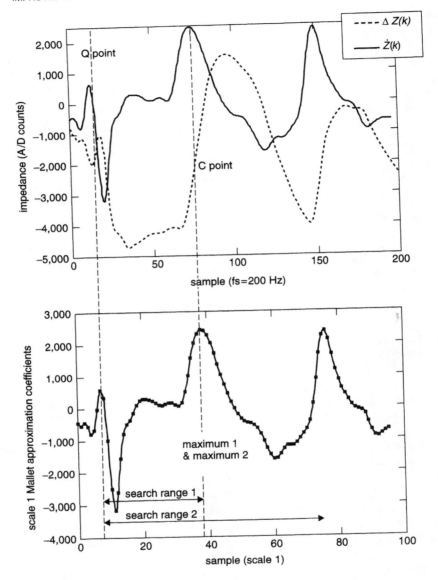

Figure 6.14 Detection of C point. Q point 2 = 200. From [24, 33].

fiducial point. The annotations were used as a reference for computing detection errors in the wavelet detection algorithm and in a previously implemented empirical detection algorithm. The empirical detection algorithm was based on curve fitting.

A typical group of beats selected for annotation is shown in Figure 6.17. Note that these beats are much noisier than the typical examples that appear in the literature, such as Figure 6.6. Accurate identification of the QRS complex is critical to limiting the search range for B, C, and X.

The empirical and wavelet detected beats were compared to the annotated beats. The

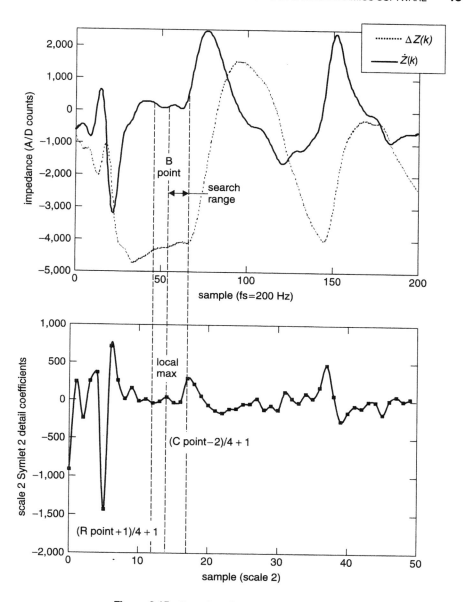

Figure 6.15 Detection of B point. From [24, 33].

mean absolute value error percentages were then computed for the stroke volume parameters. Based on 225 beats, *LVET* mean absolute error percentage decreased 48%, from 19.3 to 10%, and dZ/dt_{max} mean absolute error percentage decreased 37%, from 3.0 to 1.9%, using the wavelet, rather than empirical, detection [24, 33].

Unfortunately, Renaissance never validated their detection with annotated beats, so the time-frequency and wavelet methods of detection cannot be directly compared. Both methods rely on accurate R wave detection to trigger detection of the fiducial points.

132 IMPROVED IMPEDANCE CARDIOGRAPHY

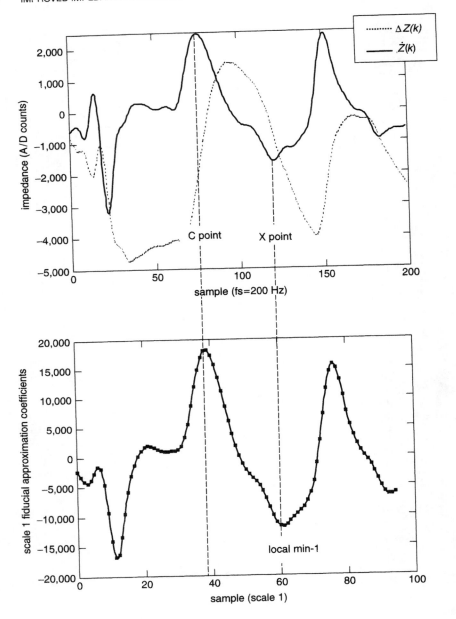

Figure 6.16 Detection of X point. From [24, 33].

However, in contrast to the Drexel technology that Renaissance licensed, which uses empirical R wave detection in the ECG, the CardioDynamics technology also processes the ECG in the time-scale domain for robust R and Q wave detection. Further, wavelet processing requires only simple additions and multiplications of real numbers to be performed, whereas time-frequency processing requires complex mathematics. Use of the spectrogram, rather than a time-frequency distribution such as the Choi–Williams distrib-

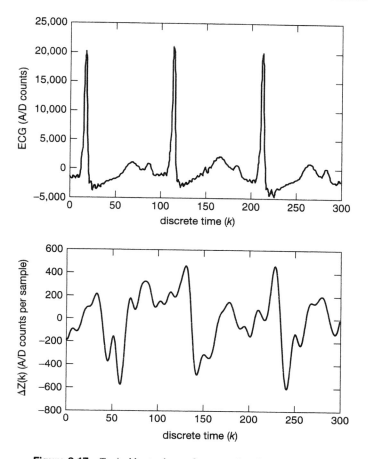

Figure 6.17 Typical beats chosen for annotation. From [24, 33].

ution, in the Renaissance approach results in cross terms (see Chapter 5) that may reduce detection accuracy. Finally, application of the spectrogram requires an assumption of short-term stationarity within the specified time window, which is not valid for the non-stationary impedance waveform.

6.6 SUMMARY

In this chapter, we have discussed many methods of monitoring cardiac output, the product of the mean stroke volume and heart rate. The gold standard for cardiac output monitoring is thermodilution. Using a pulmonary artery catheter, a bolus of room temperature or iced 5% dextrose in water or 0.9% NaCl is injected into the right atrium. The resulting blood temperature transient is detected downstream by a thermistor in the pulmonary artery, and is proportional to cardiac output. Thermodilution measurements may now be made continuously by heating blood in the right atrium, and measuring the decrease in temperature in the pulmonary artery. Alternatively, oxygen entering the pulmonary circulation may be measured to obtain cardiac output using the Fick method. However, the Fick method is only used during stable hemodynamic conditions, since measuring averaged oxygen consumption takes place over several minutes.

Noninvasive methods have been developed for monitoring stroke volume, which is then multiplied by heart rate to obtain cardiac output. These noninvasive methods eliminate the risks associated with pulmonary arterial catheters. With Doppler ultrasound, the stroke volume is measured as the product of the average blood velocity in the aorta, the ventricular ejection time, and the cross-sectional area of the aorta. The area and velocity are derived from ultrasound measurements. With impedance cardiography, the stroke volume is derived from changes in thoracic impedance, since this impedance is related to pulsatile blood volume change.

In its original configuration in the 1960s, impedance cardiography could only be used to monitor cardiac output without motion and with voluntary apnea. Sramek improved the technique in the 1980s by modifying the volume model of the thorax, using a simple four sample average of impedance changes to reduce the effect of respiratory artifact, and using spot, rather than band, electrodes. A typical squared correlation coefficient between cardiac output measurements based on thermodilution and ICG was $r^2 = 0.30$. Sramek's company BoMed Medical Manufacturing experienced bankruptcy and reorganization in 1992–1993.

Independently in the late 1980s, Wang and Sun at Drexel University used the spectrogram to improve fiducial point detection in the impedance cardiography signal. Their patents were licensed to Renaissance Technologies, which incorporated the spectrogram technology into its IQ System monitor. Renaissances' assets were later acquired by Wantagh Incorporated. Renaissance's primary competition is CardioDynamics International Corporation, the reorganization of BoMed. CardioDynamics utilizes wavelet transforms to calculate the fiducial points.

Unfortunately, Renaissance never validated their detection with annotated beats, so the time-frequency and wavelet methods of detection cannot be directly compared. Both methods rely on accurate R wave detection to trigger detection of the fiducial points. In contrast to the Drexel technology that Renaissance licensed, which uses empirical R wave detection in the ECG, the CardioDynamics software also processes the ECG in the time-scale domain for robust R and Q wave detection. Further, the wavelet processing requires only simple additions and multiplications of real numbers to be performed, whereas time-frequency processing requires complex mathematics. Use of the spectrogram, rather than a time-frequency distribution such as the Choi–Williams distribution, in the Renaissance approach results in cross terms that may reduce detection accuracy. Finally, application of the spectrogram requires an assumption of short-term stationarity within the specified time window, which is not valid for the nonstationary impedance waveform.

Impedance cardiography was formally approved for full Medicare coverage on a nationwide basis in September, 1998. With this approval, the market for impedance cardiography has begun to increase.

6.7 REFERENCES

[1]. Lopez-Saucedo, A., Hirt, M., Appel, P. L., Curtis, D. L., Harrier, H. D., and Shoemaker, W. C. Feasibility of noninvasive physiologic monitoring in resuscitation of trauma patients in the emergency department. *Crit Care Med, 17,* 567–568, 1989.

[2]. Forrester, J. S., Diamond, G., Chatterjee, K., and Swan, H. J. C. Medical therapy of acute my-

ocardial infarction by application of hemodynamic subsets. *N Engl J Med, 295,* 1404–1413, 1976.

[3]. McKown, R., Yelderman, M., and Quinn, M. Method and apparatus for continuously measuring volumetric flow. *U.S. Patent 5,146,414.* Sept. 8, 1992.

[4]. Boldt, J., Menges, T., Wollbruck, M., Hammermann, H., and Hempelmann, G. Is continuous cardiac output measurement using thermodilution reliable in the critically ill patient? *Crit Care Med, 22,* 1913–1918, 1994.

[5]. Jansen, J. R. C., Johnson, R. W., Yan, J. Y., and Verdouw, P. D. Near continuous cardiac output by thermodilution. *J Clin Mon, 13,* 233–239, 1997.

[6]. Kubicek, W.G., Kinnen, E., Patterson, R.P., and Witsoe, D.A. Impedance plethysmograph. *U.S. Patent Re. 30,101.* Sept. 25, 1979.

[7]. Patterson, R., Kubicek, W. G., and Kinnen, E. Development of an electrical impedance plethysmography system to monitor cardiac output. In *Proceedings 1st Annual Rocky Mountain Bioengineering Symposium.* USAF Academy, Colorado Springs, CO, pp. 56–71, 1964.

[8]. Nyboer, J. *Electrical Impedance Plethysmography,* 2nd ed. Charles C. Thomas: Springfield, IL, 1970.

[9]. Wang, X., Sun, H. H., and Van De Water, J. M. An advanced signal processing technique for impedance cardiography. *IEEE Trans Biomed, 42,* 224–230, 1995.

[10]. Huntsman, L. L., Leard, R. S., Tarbox, G. L., Barnes, S. R., and McLaren, B. D. Methods and apparatus for monitoring cardiac output. *U.S. Patent 4,796,634.* Jan. 10, 1989.

[11]. Huntsman, L. L., Stewart, D. K., Barnes, S. R., Franklin, S. B., Colocousis, J. S., and Hessel, E. A. Noninvasive doppler determination of cardiac output in man: Clinical validation. *Circ, 67,* 593–602, 1983.

[12]. Cariou, A., Monchi, M., Joly, L., Bellenfant, F., Claessens, Y., Thebert, D., Brunet, F., and Dhainaut, J. Noninvasive cardiac output monitoring by aortic blood flow determination: evaluation of the Sometec Dynemo–3000 system. *Crit Care Med, 26,* 2066–2072, 1998.

[13]. Connors, A. F., Jr., Speroff, T., Dawson, N. V., Thomas, C., Harrel, F. E. Jr., Wagner, D., Desbiens, N., Goldman, L., Wu, A. W., Califf, R. M., Fulkerson, W. J., Jr., Vidaillet, H., Broste, S., Bellamy, P., Lynn, J., and Knaus, W. A. The effectiveness of right heart catheterization in the initial care of critically ill patients. *JAMA, 276,* 889–897, 1996.

[14]. Pulmonary artery catheter consensus conference participants. Pulmonary artery catheter consensus conference: consensus statement. *Crit Care Med, 25,* 910–925, 1997.

[15]. Haller, M., Zollner, C., Briegel, J., and Forst, H. Evaluation of a new continuous thermodilution cardiac output monitor in critically ill patients: A prospective criterion standard study. *Crit Care Med, 23,* 860–866, 1995.

[16]. Nelson, L. D. The new pulmonary arterial catheters: right ventricular ejection fraction and continuous cardiac output. *Crit Care Clin, 12,* 795–818, 1996.

[17]. Sramek, B. Noninvasive continuous cardiac output monitor. *U.S. Patent 4,450,527.* May 22, 1984.

[18]. Bernstein, D. P. A new stroke volume equation for thoracic electrical bioimpedance: theory and rationale. *Crit Care Med, 14,* 904–909, 1986.

[19]. Yakimets, J., and Jensen, L. Evaluation of impedance cardiography: Comparison of NCCOM3-R7 with Fick and thermodilution methods. *Heart & Lung, 24,* 194–206, 1995.

[20]. Wang, X. and Sun, H. H. Apparatus and method for measuring cardiac output. *U.S. Patent 5,423,326.* June 13, 1995.

[21]. Wang, X. and Sun, H. H. System and method of impedance cardiography monitoring. *U.S. Patent 5,443,073.* Aug. 22, 1995.

[22]. Wang, X. and Sun, H. H. System and method of impedance cardiography and heartbeat determination. *U.S. Patent 5,309,917.* May 10, 1994.

[23]. Wantagh Incorporated website. April 1, 2002. www.wantagh-inc.com.
[24]. Baura, G. D., and Ng, S. K. Method and apparatus for hemodynamic assessment including fiducial point detection. *U.S. Patent pending.*
[25]. IMS Health, *Hospital Supply Index.* IMS America: Plymouth Meeting, PA, 1997.
[26]. Clancy, T. V., Norman, K., Reynolds, R., Covington, D., and Maxwell, J. G. Cardiac output measurement in critical care patients: Thoracic electrical bioimpedance versus thermodilution. *J Trauma, 31,* 1116–1120, 1991.
[27]. ———. Cardiac monitoring by electrical bioimpedance, issue #CAG-00001: decision memorandum dated 9/22/98. *Health Care Financing Administration:* website, Feb. 11,1999.
[28]. Shoemaker, W. C., Wo, C. C. J., Bishop, M. H., Appel, P. L., Van de Water, J. M., Harrington, G. R., Wang, X., and Patil, R. S. Multicenter trial of a new thoracic electrical bioimpedance device for cardiac output estimation. *Crit Care Med, 22,* 1907–1912, 1994.
[29]. Shoemaker, W. C., Wo, C. C. J., Bishop, M. H., Asensio, J., Demetriades, D., Appel, P. L., Thangathurai, D., and Patil, R. S. Noninvasive physiologic monitoring of high-risk surgical patients. *Arch Surg, 131,* 732–737, 1996.
[30]. Kubicek, W. G., Karnegis, J. N., Patterson, R. P., Witsoe, D. A., and Mattson, R. H. Development and evaluation of an impedance cardiac output system. *Aerosp Med, 37,* 1208–1212, 1966.
[31]. Sramek, B. B. Cardiac output by electrical impedance. *Med Electron, 13,* 93–97, 1982.
[32]. Cohen, Leon. Time-Frequency Analysis. Prentice Hall: Upper Saddle River, NJ. 1995.
[33]. CardioDynamics International Corporation. 2002.
[34]. Kadambe, S., Murray, R., and Boudreaux-Bartels, G. F. Wavelet transform-based QRS complex detector. *IEEE Trans BME, 46,* 838–848, 1999.
[35]. Sugimachi, M. National Cardiovascular Center Research Institute, Osaka, Japan. Research data, 1992.

Further Reading

Cardiac Output Physiology: Guyton, A. C. and Hall, J. E. Textbook of Human Physiology, 10th edition. Saunders: Philadelphia, 2000.

Cardiac Output Monitoring: Gravenstein, N., Good, M. L., and Banner, T. E. Assessment of cardiopulmonary function. In *Critical Care,* edited by J. M. Civetta, R. W. Taylor, and R. R. Kirby, 3rd ed. Lippincott-Raven: Philadelphia, 1997, pp. 867–898.

6.8 ELECTROCARDIOGRAM QRS DETECTION EXERCISES

Engineer C at Startup Company has been asked to detect heartbeat onsets for a new heart rate monitor. She is familiar with the work from Boudeaux-Bartels' laboratory [34], and wishes to implement their QRS detector. We recreate her analysis below.

6.8.1 Clinical Data

The ECG data file in Figure 6.18 was recorded from a patient with **atrial fibrillation** (erratic heart rhythm due to continuous, rapid discharges from areas of the atria) due to **rheumatic heart disease.** Ten seconds of data were recorded with a sampling frequency of 200 Hz, cutoff of 1 MHz, and resolution of 12 bits/sample; they were saved in units of A/D counts. These data were simultaneously recorded with the pressure data used in the exercises in Chapter 2. They were generously provided by Dr. Masaru Sugimachi [35].

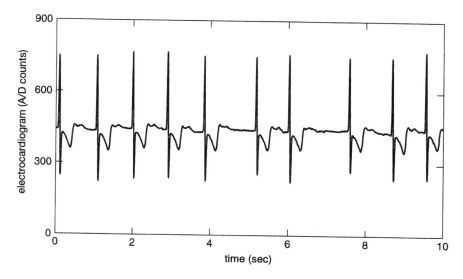

Figure 6.18 Electrocardiogram for exercise [35].

Please move the *ecg.dat* file from *ftp://ftp.ieee.org/uploads/press/baura* to your *Matlab\bin* directory. Input the data using "load ecg.dat -ascii."

6.8.2 Useful Matlab Functions

The following functions from the Wavelet Toolbox are useful for completing this exercise: wfilters, dwt, and rand.

6.8.3 Exercises

1. Boudeaux-Bartels' laboratory used the cubic spline wavelet in their work. The Haar wavelet is also very useful for detecting discontinuities. Calculate the wavelet and scaling filters for the Haar wavelet. Hint: these are the decomposition filters.
2. Compute the Haar detail coefficients for scales $a = 2^1$, 2^2, and 2^3. Remember that the input function is the original ECG or approximation coefficients of the previous scale.
3. (a) What thresholding scheme could be implemented to detect simultaneous peaks at scales 2^1 and 2^2? (b) List the detected beat onset samples. (c) What portion of the electrocardiogram is actually detected? (d) The "gold standard" for beat detection is visual detection by a cardiologist. Visually detect each beat. What are the mean and standard deviation of the beat detection errors?
4. Add uniformly distributed noise in the range {0,2000} to the ECG. Repeat Question 3 with this noisy waveform as the input.
5. How do you explain the results in scale 3?

II

MODELS FOR REAL TIME PROCESSING

Once a signal of interest has been sufficiently filtered, it may be modeled. Modeling enables signal classification, prediction, control, and investigation of underlying physiologic mechanisms. The process of the determination of dynamic models from experimental data is called system identification.

A linear model is defined by a linear differential equation. Within the framework of a linear time-invariant single input, single output system, the autoregressive moving average exogenous input (ARMAX) model and its variations are discussed in Chapter 7. The parameters of the versatile ARMAX model may be determined in batch mode or through recursive iterations. The application of this model to prediction of the effects of various waveform shapes during external defibrillation is discussed in Chapter 8.

If we relax our system constraints to that of a multiple input, multiple output system that is nonlinear but still time-invariant, we may describe the system operator as an artificial neural network. As its name implies, an artificial neural network refers to a mathematical model of human brain processing. Indeed, in the 1940s, physiologists and electrical engineers worked together toward this goal. Over time, it was discovered that these models did not simulate human neuron processing. However, because these models are very useful, engineers continue to investigate them. Various neural network architectures and methods of parameter identification are discussed in Chapter 9. The application of a feedforward neural network to classification of abnormal cervical cells in Pap smears is discussed in Chapter 10.

If we further relax our constraints on the multiple input, multiple output system so that our system operator is nonlinear, time-invariant, and sufficiently complex, it may not be easily described by mathematical equations. Given such a system, the system operator may be described using fuzzy logic. As its name implies, fuzzy logic is the logic underlying modes of reasoning that are approximate rather than exact. Conventional approaches to knowledge representation are based on bivalent (two states: true/false) logic. However, these approaches are unable to deal with the issues of uncertainty and precision. In contrast, fuzzy logic is derived from the fact that most modes of human reasoning, especially common sense, are approximate in nature. The original fuzzy model and its application to fuzzy control are discussed in Chapter 11. The application of a fuzzy model to blood pressure monitoring is described in Chapter 12.

In the discussion of each model, the process of determining the optimum model is highlighted. This process of model validation involves analysis of the coefficients of variation associated with each identified parameter, goodness of fit, and residual statistics. Model validation also encompasses model plausibility, which refers to the flexibility and simplicity of the model.

7

LINEAR SYSTEM IDENTIFICATION

In Chapter 1, we reviewed the concept of the **system operator,** that is, the mathematical representation of the relationship between the inputs and outputs. If we restrict this discussion to that of a **single input, single output (SISO) system** that is **linear** and **time-invariant,** then the input is $u(k)$, the output is $y(k)$, and the system operator is an **impulse response,** $h(k)$ (Figure 7.1). The relationship between the three variables is

$$y(k) = u(k) * h(k) \qquad (7.1)$$

where * represents convolution.

In this chapter, we wish to mathematically describe the impulse response. This process of the determination of dynamic models from experimental data is called **system identification.** We purposely restrict our discussion to the **autoregressive moving average exogenous input (ARMAX) model** and its variations. The ARMAX model is linear and quite versatile. It was first introduced in this textbook in Section 1.2.

7.1 THE ARMAX MODEL AND VARIATIONS

The ARMAX model consists of the linear constant coefficient difference equation

$$y(k) = -\sum_{n=1}^{N} a_n y(k-n) + \sum_{n=0}^{M} b_n u(k-n) + \sum_{n=0}^{P} c_n e(k-n) \qquad (7.2)$$

where a_n are **feedback** coefficients, b_n are **feedforward** coefficients, N is the model order, and $(M+1)$ is the number of feedforward coefficients. Because a measurement of the output may include noise, the noise is represented by $e(k)$, which is assumed to be **white noise** (sequence of independent random variables with zero mean). The $(P+1)$ noise coefficients are represented by c_n. It is also assumed that $a_0 = c_0 = 1$.

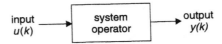

Figure 7.1 Linear time invariant single input, multiple output system.

This model is autoregressive because the output, $y(k)$, looks back on past values of itself:

$$y(k) = -\sum_{n=1}^{N} a_n y(k-n) + \sum_{n=0}^{M} b_n u(k-n) + \sum_{n=0}^{P} c_n e(k-n) \quad (7.3)$$

The model also possesses a moving average, $\sum_{n=0}^{P} c_n e(k-n)$, and an exogenous, or external, input, $u(k)$.

Alternatively, Eq. (7.2) may be represented using the argument q^{-1}, which denotes the **backward shift operator**. Using this argument, a delay of one sample in the input, $u(k)$, may be represented as

$$u(k-1) = q^{-1} u(k) \quad (7.4)$$

Using the backward shift operator, we may rewrite Eq. (7.2) as

$$A(q^{-1})y(k) = B(q^{-1})u(k) + C(q^{-1})e(k) \quad (7.5)$$

where

$$A(q^{-1}) = 1 + a_1 q^{-1} + \ldots + a_N q^{-N} \quad (7.6)$$

$$B(q^{-1}) = b_0 + b_1 q^{-1} + \ldots + b_M q^{-M} \quad (7.7)$$

$$C(q^{-1}) = 1 + c_1 q^{-1} + \ldots + c_P q^{-P} \quad (7.8)$$

The parameter vector, θ, for this model consists of

$$\theta = [a_1 \ldots a_N \; b_0 \ldots b_M \; c_0 \ldots c_P]^T \quad (7.9)$$

During system identification, we seek to estimate the parameter vector accurately. A block diagram of the ARMAX model is given in Figure 7.2.

7.1.1 Variations

Several important special cases of the ARMAX model exist. First, when $M = P = 0$ and $b_0 = 0$, it is assumed that an input signal is not present. For this **autoregressive model**,

$$A(q^{-1})y(k) = e(k) \quad (7.10)$$

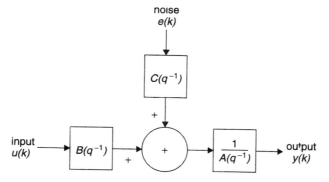

Figure 7.2 ARMAX system.

The corresponding parameter vector for this pure time series model is

$$\theta = [a_1 \ldots a_N]^T \qquad (7.11)$$

Second, when $N = M = 0$ and $b_0 = 0$, it is assumed that an input signal is not present and that the current output may be modeled by past values of noise. For this **moving average model**,

$$y(k) = C(q^{-1})e(k) \qquad (7.12)$$

The corresponding parameter vector for this model is

$$\theta = [c_1 \ldots c_P]^T \qquad (7.13)$$

Third, when $N = 0$ and noise is not present, it is assumed that the current output may be modeled by past values of the input. For this **finite impulse response (FIR) model** that was discussed extensively in Chapter 1,

$$y(k) = B(q^{-1})u(k) \qquad (7.14)$$

The corresponding parameter vector for this common model is

$$\theta = [b_0 \ldots b_M]^T \qquad (7.15)$$

Finally, when $P = 0$, it is assumed that only a single sample of noise is reflected in the output. For this **autoregressive with exogeneous input model** (ARX), which is also known as the **controlled autoregressive model**,

$$A(q^{-1})y(k) = B(q^{-1})u(k) + e(k) \qquad (7.16)$$

The corresponding parameter vector for this model is

$$\theta = [a_1 \ldots a_N \; b_0 \ldots b_M]^T \qquad (7.17)$$

Please note that without the white noise term, the ARX model becomes the **infinite impulse response (IIR) model.** Eq. (7.16) may be rewritten as

$$y(k) = \boldsymbol{\phi}^T(k)\boldsymbol{\theta} + e(k) \tag{7.18}$$

where the **regression vector,** $\boldsymbol{\phi}^T(k)$, is

$$\boldsymbol{\phi}^T(k) = [-y(k-1) \ldots -y(k-N) \: u(k) \ldots u(k-M)] \tag{7.19}$$

7.1.2 Input Signals

The input signal to the ARMAX model and its variations may be deterministic or random. To simplify the context of this chapter, we restrict the input signal to only deterministic signals.

7.2 UNIQUENESS PROPERTIES

The choice of an appropriate model structure for a system is determined by factors such as **flexibility, parsimony, algorithm complexity,** and **properties of the performance function.** In terms of flexibility, it should be possible to use the chosen model structure to describe most of the different system dynamics that can be expected in the application. Both the number of parameters and the way they enter into the model are important. In terms of parsimony, the chosen model should contain the smallest number of parameters required to represent the true system adequately. According to the **parsimony principle,** the simpler of two possible model structures will on average result in better accuracy if three assumptions can be made:

1. The two identifiable model structures fit the data (identifiability is discussed in Section 7.3).
2. The two model structures under consideration are **hierarchical** (one may be obtained by constraining the other structure in some way).
3. The method of system identification used is a **prediction error method** (prediction error methods are discussed in Section 7.4) [1].

Because algorithm complexity directly influences the amount of computation required, the chosen model should be as simple as possible. As illustrated in Chapter 3, the desired parameters [in Chapter 3, the weight vector, $\mathbf{w}(k)$] are often calculated through extensive iterations. Further, the performance function is affected by the chosen model structure. During the **optimization** process, parameters in the chosen model are estimated based on a performance function. In Chapter 3, a least mean squares equation was used as the performance function. For the adaptive linear combiner developed by Widrow for adaptive filtering, which is in essence an FIR model, this performance function is quadratic, with only a single global minimum. The existence of local minima as well as nonunique global minima is very much dependent on the model structure used.

Recalling the aorta and radial artery data from Figure 2.10 that were used in the Matlab exercises in Chapter 2, let us determine the impulse response between the aortic input and radial artery output over the course of this chapter. A model structure that is flexible, parsimonious, and relatively simple for this task is the ARX model. We discuss the choice of performance function in Section 7.4.

7.3 MODEL IDENTIFIABILITY

Model identifiability refers to the possibility of theoretically and practically obtaining unique estimates of all the unknown model parameters. The ARMAX, FIR, IIR, and ARX models are all theoretically identifiable. However, the practical process of parameter estimation may not yield acceptable results. Problems may arise from different types of experimental error, the number of data points, and the true system, leading to a loss of practical identifiability.

A model may be uniquely identifiable, **nonuniquely identifiable,** or **nonidentifiable.** If it is uniquely identifiable, the parameters may be uniquely determined. If it is nonuniquely identifiable, one or more of the parameters possesses more than one, but a finite, number of possible values. If it is nonidentifiable, one or more parameters possesses an infinite number of solutions.

7.4 PREDICTION ERROR METHODS

Given an identifiable model, the model parameters may be calculated through a performance function, $\xi(\theta)$. As an example, the error between a system and its prediction may be calculated, and a performance function may be constructed that minimizes a form of this error. When a performance function based on prediction error is utilized, the resulting system identification is classified as a prediction error method.

7.4.1 Performance Function

The error, $\varepsilon(k)$, between an output, $y(k)$, and its estimate, $\hat{y}(k)$, is

$$\varepsilon(k) = y(k) - \hat{y}(k) \qquad (7.20)$$

Since the estimate may be modeled as

$$\hat{y}(k) = \phi^T(k)\theta, \qquad (7.21)$$

the error may be calculated as

$$\varepsilon(k) = y(k) - \phi^T(k)\theta. \qquad (7.22)$$

Over time, the error vector containing these **residuals** is defined as

$$\mathbf{\varepsilon}(k) = [\varepsilon(1) \ldots \varepsilon(K)] \qquad (7.23)$$

where K is $> (N + M + 1)$, the dimension of the parameter vector. Since there are $(N + M + 1)$ parameters in θ, it should be theoretically possible to solve for θ from $(N + M + 1)$

146 LINEAR SYSTEM IDENTIFICATION

measurements. However, $K > (N + M + 1)$ measurements are used to account for noise, disturbances, and model misfit.

Let us define the **covariance**, $V(\theta)$, as

$$V(\theta) = \frac{1}{K} \sum_{k=1}^{K} \varepsilon^2(k) \tag{7.24}$$

The **performance functions** utilized in prediction error methods possess the form

$$\xi(\theta) = h[V(\theta)] \tag{7.25}$$

where $h(k)$ is a scalar-valued function that must satisfy certain conditions. In particular, the **least squares estimate** of θ is defined as the vector, $\hat{\theta}$, that minimizes the **mean squared error performance function**

$$\xi(\theta) = \frac{1}{2} \cdot \frac{1}{K} \sum_{k=1}^{K} \varepsilon^2(k) \tag{7.26}$$

The factor of ½ has been added to simplify calculation of the performance function derivative.

Similarly, the maximum likelihood estimate of θ is defined as the vector, $\hat{\theta}$, that minimizes the maximum likelihood performance function. The maximum likelihood refers to the probability distribution function of the observations conditioned on the parameter vector θ. Although the proof of the maximum likelihood performance function is beyond the range of our discussion, it may be shown that the maximum likelihood estimate of θ is the vector, $\hat{\theta}$, that minimizes the performance function

$$\xi(\theta) = V(\theta) \tag{7.27}$$

which is essentially the mean squared error performance function.

7.4.2 Derivation

The minimum of the mean squared error performance function is calculated by setting its derivative equal to zero:

$$\frac{d\xi(\theta)}{d\theta} = \frac{d}{d\theta}\left\{ \frac{1}{2} \cdot \frac{1}{K} \sum_{k=1}^{K} \varepsilon^2(k) \right\} = 0 \tag{7.28}$$

Substituting Eq. (7.22) into Eq. (7.28) yields

$$\frac{d\xi(\theta)}{d\theta} = \frac{d}{d\theta}\left\{ \frac{1}{2} \cdot \frac{1}{K} \sum_{k=1}^{K} [y(k) - \phi^T(k)\theta]^2 \right\} = 0 \tag{7.29}$$

$$\frac{1}{K} \sum_{k=1}^{K} [\phi(k)y(k) - \phi(k)\phi^T(k)\theta] = 0 \tag{7.30}$$

Moving the negative terms to the other side of the equation results in

$$\frac{1}{K}\sum_{k=1}^{K}\phi(k)y(k) = \frac{1}{K}\sum_{k=1}^{K}\phi(k)\phi^T(k)\theta \qquad (7.31)$$

We can then solve for $\hat{\theta}$, the least squares estimate, as

$$\hat{\theta} = \left[\frac{1}{K}\sum_{k=1}^{K}\phi(k)\phi^T(k)\right]^{-1}\frac{1}{K}\sum_{k=1}^{K}\phi(k)y(k) \qquad (7.32)$$

The general steps of the prediction error method are illustrated in Figure 7.3.

7.4.3 Example

For example, let us determine the impulse response between the second aortic and radial waveforms of Figure 2.10 [2], $u(k)$ and $y(k)$, respectively (Figure 7.4). The waveforms in Figure 7.4 have been lowpass filtered using a fourth order Butterworth filter with a cutoff of 20 Hz. The dc offset values have also been subtracted. Note that a delay of 14 or 15 samples ($f_s = 200$ Hz) exists between the input and output, representing the distance between the aorta and radial artery. We utilize the ARX model with $N = 6$ feedback parameters and 5 (b_0 to b_4) feedforward parameters. A delay, D, of 15 samples is input to the model as leading zeros in the $B(q^{-1})$ polynomial. Therefore, the feedforward parameters ($M = 4 + 15 = 19$) may be represented as

$$B(q^{-1}) = 0+0+0+0+0+0+0+0+0+0+0+0+0+0+0+b_0+b_1+b_2+b_3+b_4 \qquad (7.33)$$

It follows that $K = (N + M + 1) = 26$. The regression vector, $\phi(k)$, is constructed from Eq. (7.19) as

$$\phi^T(k) = [-y(k-1)\ldots-y(k-6)\ u(k)\ldots u(k-19)] \qquad (7.34)$$

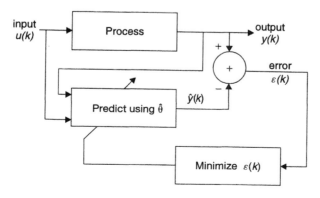

Figure 7.3 Prediction error method.

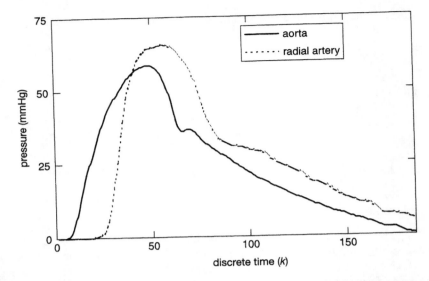

Figure 7.4 The second aortic and radial waveforms of Figure 2.10. The waveforms were obtained from a subject with atrial fibrillation due to rheumatic heart disease. These waveforms were lowpass filtered with a fourth order Butterworth filter with a cutoff of 20 Hz. The dc offsets were also subtracted. $f_s = 200$ Hz [2].

Solving for Equation (7.32) yields

$$\hat{\boldsymbol{\theta}}^T = [-3.81\ 5.86\ -4.38\ 1.34\ 0.136\ -0.13\ 0\ \ldots\ 0\ 0.37\ -1.27\ 1.75\ -1.14\ 0.30]^T \quad (7.35)$$

This equation may be easily solved using the Matlab function arx in the System Identification Toolbox. The resulting waveform prediction, using the input aortic waveform and the parameter vector in Eq. (7.35), is shown in Figure 7.5a.

7.4.4 Assumption of White Noise

Recalling Eq. (7.32), the least squares estimate of $\boldsymbol{\theta}$ is

$$\hat{\boldsymbol{\theta}} = \left[\frac{1}{K}\sum_{k=1}^{K}\boldsymbol{\phi}(k)\boldsymbol{\phi}^T(k)\right]^{-1}\frac{1}{K}\sum_{k=1}^{K}\boldsymbol{\phi}(k)y(k) \quad (7.36)$$

If we assume that the true system may be described as

$$y(k) = \boldsymbol{\phi}^T(k)\boldsymbol{\theta} + v(k) \quad (7.37)$$

where $v(k)$ is a **stochastic** (random) disturbance, the true parameter vector, $\boldsymbol{\theta}$, may be calculated by setting the derivative of the mean squared error performance function equal to zero:

$$\frac{d\xi(\boldsymbol{\theta})}{d\boldsymbol{\theta}} = \frac{d}{d\boldsymbol{\theta}}\left\{\frac{1}{2}\cdot\frac{1}{K}\sum_{k=1}^{K}\varepsilon^2\right\} = \frac{d}{d\boldsymbol{\theta}}\left\{\frac{1}{2}\cdot\frac{1}{K}\sum_{k=1}^{K}[y(k)-\boldsymbol{\phi}^T(k)\boldsymbol{\theta}-v(k)]^2\right\} = 0 \quad (7.38)$$

7.4 PREDICTION ERROR METHODS 149

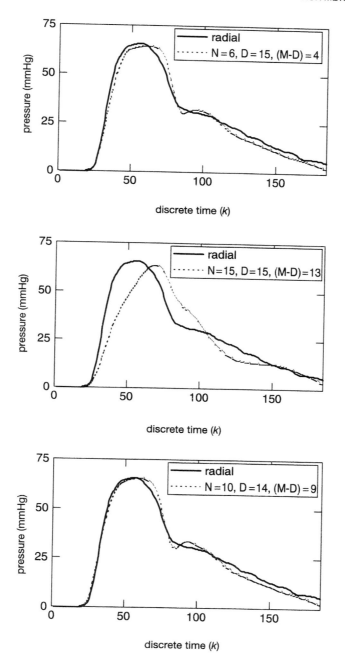

Figure 7.5 Least squares predictions of a radial waveform, with various combinations of feedback parameter number, N, sample delay, D, and feedforward parameter number, M. The input is an aortic waveform.

$$\frac{1}{K}\sum_{k=1}^{K}[\phi(k)y(k) - \phi(k)\phi^T(k)\theta - \phi(k)v(k)] = 0 \qquad (7.39)$$

$$\frac{1}{K}\sum_{k=1}^{K}\phi(k)\phi^T(k)\theta = \frac{1}{K}\sum_{k=1}^{K}\phi(k)y(k) - \frac{1}{K}\sum_{k=1}^{K}\phi(k)v(k) \qquad (7.40)$$

$$\theta = \left[\frac{1}{K}\sum_{k=1}^{K}\phi(k)\phi^T(k)\right]^{-1}\left[\frac{1}{K}\sum_{k=1}^{K}\phi(k)y(k) - \frac{1}{K}\sum_{k=1}^{K}\phi(k)v(k)\right] \qquad (7.41)$$

Combining Eqs. (7.36) and (7.41), the difference between the estimated and true parameter vectors, $(\hat{\theta} - \theta)$, can be determined by

$$\hat{\theta} - \theta = \left[\frac{1}{K}\sum_{k=1}^{K}\phi(k)\phi^T(k)\right]^{-1}\left[\frac{1}{K}\sum_{k=1}^{K}\phi(k)v(k)\right] \qquad (7.42)$$

When K tends to infinity, then the mean calculations may be replaced by estimated values, $E\{\ \}$, as

$$\hat{\theta} - \theta = [E\{\phi(k)\phi^T(k)\}]^{-1}[E\{\phi(k)v(k)\}] \qquad (7.43)$$

Thus, the least squares estimate, $\hat{\theta}$, will possess an asymptotic bias (or not be consistent) unless

$$E\{\phi(k)v(k)\} = 0 \qquad (7.44)$$

which can only occur if $v(k)$ is white noise. This very restrictive assumption is one reason other methods, such as **instrumental variable methods**, are utilized.

7.5 INSTRUMENTAL VARIABLE METHODS

Let us derive the instrumental variable (IV) methods as a generalization of the prediction error methods we have already discussed. First, we define a new performance function with

$$\xi(\theta) = \frac{1}{2} \cdot \frac{1}{K}\sum_{k=1}^{K}z(k)\varepsilon(k) = 0 \qquad (7.45)$$

where $z(k)$ is a vector of dimension $(M + N + 1) \times 1$ called an **instrumental variable**. It is assumed that $z(k)$ is uncorrelated with $v(k)$. Substitution of Equations (7.20) and (7.21) yields

$$\xi(\theta) = \frac{1}{K}\sum_{k=1}^{K}z(k)[y(k) - \phi^T(k)\theta] = 0 \qquad (7.46)$$

The basic IV estimate, $\hat{\theta}$, for a single input–single output system can then be calculated as

$$\frac{1}{K}\sum_{k=1}^{K} \mathbf{z}(k)y(k) = \frac{1}{K}\sum_{k=1}^{K} \mathbf{z}(k)\boldsymbol{\phi}^T(k)\boldsymbol{\theta} \qquad (7.47)$$

$$\hat{\boldsymbol{\theta}} = \left[\frac{1}{K}\sum_{k=1}^{K} \mathbf{z}(k)\boldsymbol{\phi}^T(k)\right]^{-1}\left[\frac{1}{K}\sum_{k=1}^{K} \mathbf{z}(k)y(k)\right] \qquad (7.48)$$

We can choose $\mathbf{z}(k)$ in different ways, subject to certain conditions, to guarantee the consistency of the estimate in Eq. (7.48). For $\mathbf{z}(k) = \boldsymbol{\phi}(k)$, the basic IV estimate reduces to the least squares estimate. The instrumental variable method is illustrated in Figure 7.6.

7.5.1 Assumptions

For IV methods, the following assumptions are made:

1. The system is strictly causal and asymptotically stable.
2. The input, $u(k)$, is persistently exciting of a sufficiently high order.
3. The disturbance, $v(k)$, is a stationary stochastic process with rational spectral density. It is thus uniquely described by

$$v(k) = \beta e(k) \qquad (7.49)$$

4. The input, $u(k)$, and disturbance, $v(k)$, are independent.
5. The model [Eq. (7.21)] and true system [Eq. (7.37)] possess the same transfer function *if and only if* $\hat{\boldsymbol{\theta}} = \boldsymbol{\theta}$.

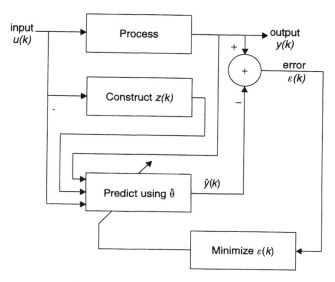

Figure 7.6 Instrumental variable method.

152 LINEAR SYSTEM IDENTIFICATION

These assumptions are much milder than the least squares estimation assumption that $v(k)$ must equal white noise $[v(k) = \beta e(k)]$.

7.5.2 Instrument Choices

A natural method for choosing instruments for a single input–single output system is to generate instruments similar to the ARX model, but not let them be influenced by $v(k)$. This leads to

$$\mathbf{z}(k) = K(q^{-1})[-x(k-1) \ldots -x(k-N) \; u(k-1) \ldots u(k-N-M)]^T \quad (7.50)$$

where $K(q^{-1})$ is a linear filter and $x(k)$ is generated from the input through a linear system

$$x(k) = M(q^{-1})u(k) \quad (7.51)$$

and

$$M(q^{-1}) = m_0 + m_1 q^{-1} + \ldots m_R q^{-R} \quad (7.52)$$

Most instruments used in practice are generated this way.

7.6 RECURSIVE LEAST SQUARES ALGORITHM

Alternatively, we may estimate model parameters recursively, rather than in batch fashion. Recursive identification possesses the benefit of decreased memory allocation, and may be derived from the least squares estimate in Eq. (7.32), provided that the inverse exists.

7.6.1 Derivation

We start by defining $\hat{\theta}(k)$, based on Eq. (7.32), as

$$\hat{\theta}(k) = \left[\sum_{i=1}^{k} \phi(i)\phi^T(i)\right]^{-1} \sum_{i=1}^{k} \phi(i)y(i) \quad (7.53)$$

Let us define the matrix, $\mathfrak{R}(k)$, as

$$\mathfrak{R}(k) = \sum_{i=1}^{k} \phi(i)\phi^T(i) \quad (7.54)$$

From the definition of $\mathfrak{R}(k)$, it follows that

$$\mathfrak{R}(k-1) = \sum_{i=1}^{k} \phi(i)\phi^T(i) - \phi(k)\phi^T(k) \quad (7.55)$$

$$\mathfrak{R}(k-1) = \mathfrak{R}(k) - \phi(k)\phi^T(k) \quad (7.56)$$

Substituting Eq. (7.54) into (7.53) yields

$$\hat{\theta}(k) = \underline{\mathfrak{R}}^{-1}(k) \sum_{i=1}^{k} \phi(i) y(i) \tag{7.57}$$

which can be rewritten as

$$\hat{\theta}(k) = \underline{\mathfrak{R}}^{-1}(k) \left[\sum_{i=1}^{k-1} \phi(i) y(i) + \phi(k) y(k) \right] \tag{7.58}$$

Now, let us rewrite Eq. (7.53) in terms of $(k-1)$ as

$$\hat{\theta}(k-1) = \left[\sum_{i=1}^{k-1} \phi(i) \phi^T(i) \right]^{-1} \sum_{i=1}^{k-1} \phi(i) y(i) \tag{7.59}$$

Substituting Eq. (7.54) and multiplying both sides of the equation by $\underline{\mathfrak{R}}(k-1)$ yields

$$\underline{\mathfrak{R}}(k-1) \hat{\theta}(k-1) = \sum_{i=1}^{k-1} \phi(i) y(i) \tag{7.60}$$

Eq. (7.60) can be substituted into Eq. (7.58) to obtain

$$\hat{\theta}(k) = \underline{\mathfrak{R}}^{-1}(k)[\underline{\mathfrak{R}}(k-1) \hat{\theta}(k-1) + \phi(k) y(k)] \tag{7.61}$$

Substituting Eq. (7.56) yields

$$\hat{\theta}(k) = \underline{\mathfrak{R}}^{-1}(k)\{[\underline{\mathfrak{R}}(k) - \phi(k)\phi^T(k)] \hat{\theta}(k-1) + \phi(k) y(k)\} \tag{7.62}$$

$$\hat{\theta}(k) = \underline{\mathfrak{R}}^{-1}(k)[\underline{\mathfrak{R}}(k) \hat{\theta}(k-1) - \phi(k)\phi^T(k) \hat{\theta}(k-1) + \phi(k) y(k)] \tag{7.63}$$

$$\hat{\theta}(k) = \hat{\theta}(k-1) + \underline{\mathfrak{R}}^{-1}(k)\phi(k)[y(k) - \phi^T(k)\hat{\theta}(k-1)] \tag{7.64}$$

We may prefer to work with an alternate matrix, $\underline{\mathbf{R}}(k)$, which is defined as

$$\underline{\mathbf{R}}(k) \equiv \frac{1}{k} \underline{\mathfrak{R}}(k) \tag{7.65}$$

$$\underline{\mathbf{R}}(k-1) \equiv \frac{1}{k-1} \underline{\mathfrak{R}}(k-1) \tag{7.66}$$

Let us rearrange Eq. (7.56) and multiply both sides by $1/k$ to obtain

$$\frac{1}{k} \underline{\mathfrak{R}}(k) = \frac{1}{k} \underline{\mathfrak{R}}(k-1) + \frac{1}{k} \phi(k) \phi^T(k) \tag{7.67}$$

Substituting Eqs. (7.65) and (7.66) yields

$$\underline{\mathbf{R}}(k) = \frac{1}{k} \underline{\mathfrak{R}}(k-1) + \frac{1}{k} \phi(k) \phi^T(k) \tag{7.68}$$

$$\underline{R}(k) = \frac{1}{k}[(k-1)\,\underline{R}(k-1)] + \frac{1}{k}\,\phi(k)\,\phi^T(k) \qquad (7.69)$$

$$\underline{R}(k) = \frac{k-1}{k}\,\underline{R}(k-1) + \frac{1}{k}\,\phi(k)\,\phi^T(k) \qquad (7.70)$$

$$\underline{R}(k) = \underline{R}(k-1) + \frac{1}{k}[\phi(k)\,\phi^T(k) - \underline{R}(k-1)] \qquad (7.71)$$

Similarly, the parameter vector can be recalculated using Eqs. (7.64) and (7.65) as

$$\hat{\theta}(k) = \hat{\theta}(k-1) + \frac{1}{k}\,\underline{R}^{-1}(k)\,\phi(k)\,[y(k) - \hat{\theta}(k-1)\phi(k)] \qquad (7.72)$$

Eqs. (7.71) and (7.72) describe a recursive algorithm for calculation of the parameter vector. At iteration, k, only $\hat{\theta}(k)$, $\underline{R}(k)$, $y(k)$, and $\phi(k)$ need to be kept in memory. However, this algorithm is not well suited to computation because it requires matrix inversion for each iteration. Instead, let us introduce the matrix, $\underline{P}(k)$, where

$$\underline{P}(k) \equiv \underline{R}^{-1}(k) \equiv \frac{1}{k}\,\mathfrak{R}^{-1}(k) \qquad (7.73)$$

We wish to update $\underline{P}(k)$, rather than $\underline{R}(k)$, directly. According to the **matrix inversion lemma**,

$$[\underline{A} + \underline{B}\,\underline{C}\,\underline{D}]^{-1} = \underline{A}^{-1} - \underline{A}^{-1}\,\underline{B}\,[\underline{D}\,\underline{A}^{-1}\,\underline{B} + \underline{C}^{-1}]^{-1}\,\underline{D}\,\underline{A}^{-1} \qquad (7.74)$$

where $\underline{A}$, $\underline{B}$, $\underline{C}$, and $\underline{D}$ are matrices of compatible dimensions. Let us define

$$\underline{A} = \underline{P}^{-1}(k-1) \qquad (7.75)$$

$$\underline{B} = \phi(k) \qquad (7.76)$$

$$\underline{C} = 1 \qquad (7.77)$$

$$\underline{D} = \phi^T(k) \qquad (7.78)$$

Substitution of Eq. (7.73) into (7.67) gives

$$\underline{P}(k) = [\underline{P}^{-1}(k-1) + \phi(k)\phi^T(k)]^{-1} \qquad (7.79)$$

Applying the matrix inversion lemma yields

$$\underline{P}(k) = \underline{P}(k-1) - \underline{P}(k-1)\phi(k)[\phi^T(k)\underline{P}(k-1)\phi(k) + 1]^{-1}\phi^T(k)\underline{P}(k-1) \qquad (7.80)$$

$$\underline{P}(k) = \underline{P}(k-1) - \frac{\underline{P}(k-1)\phi(k)\phi^T(k)\underline{P}(k-1)}{1 + \phi^T(k)\underline{P}(k-1)\phi(k)} \qquad (7.81)$$

Let us define the vector, $\mathbf{L}(k)$, as

$$\mathbf{L}(k) = \underline{\mathbf{P}}(k)\, \boldsymbol{\phi}(k) \tag{7.82}$$

$\mathbf{L}(k)$ may be calculated by multiplying both sides of Equation (7.81) by $\boldsymbol{\phi}(k)$ as

$$\mathbf{L}(k) = \underline{\mathbf{P}}(k)\,\boldsymbol{\phi}(k) = \underline{\mathbf{P}}(k-1)\,\boldsymbol{\phi}(k) - \frac{\underline{\mathbf{P}}(k-1)\boldsymbol{\phi}(k)\,\boldsymbol{\phi}^T(k)\underline{\mathbf{P}}(k-1)\boldsymbol{\phi}(k)}{1 + \boldsymbol{\phi}^T(k)\underline{\mathbf{P}}(k-1)\boldsymbol{\phi}(k)} \tag{7.83}$$

$$\mathbf{L}(k) = \frac{\underline{\mathbf{P}}(k-1)\boldsymbol{\phi}(k) + \underline{\mathbf{P}}(k-1)\boldsymbol{\phi}(k)\boldsymbol{\phi}^T(k)\underline{\mathbf{P}}(k-1)\boldsymbol{\phi}(k) - \underline{\mathbf{P}}(k-1)\boldsymbol{\phi}(k)\boldsymbol{\phi}^T(k)\underline{\mathbf{P}}(k-1)\,\boldsymbol{\phi}(k)}{1 + \boldsymbol{\phi}^T(k)\underline{\mathbf{P}}(k-1)\boldsymbol{\phi}(k)}$$

$$\tag{7.84}$$

$$\mathbf{L}(k) = \frac{\underline{\mathbf{P}}(k-1)\boldsymbol{\phi}(k)}{1 + \boldsymbol{\phi}^T(k)\underline{\mathbf{P}}(k-1)\boldsymbol{\phi}(k)} \tag{7.85}$$

The parameter vector is updated from Eqs. (7.72) and (7.73) as

$$\hat{\boldsymbol{\theta}}(k) = \hat{\boldsymbol{\theta}}(k-1) + \underline{\mathbf{P}}(k)\boldsymbol{\phi}(k)[y(k) - \hat{\boldsymbol{\theta}}(k-1)\boldsymbol{\phi}(k)] \tag{7.86}$$

Substitution of Equation (7.82) yields

$$\hat{\boldsymbol{\theta}}(k) = \hat{\boldsymbol{\theta}}(k-1) + \mathbf{L}(k)[y(k) - \hat{\boldsymbol{\theta}}(k-1)\boldsymbol{\phi}(k)] \tag{7.87}$$

Eqs. (7.85), (7.87), and (7.81) define the **recursive least squares algorithm.** This algorithm is robust and easily implemented. Note that, as with the batch least squares estimate, it is assumed that the disturbance, $v(k)$, is white noise. The recursive least squares algorithm is illustrated in Figure 7.7.

7.6.2 Initial Conditions

To execute the recursive least squares algorithm, we require initial values of $\underline{\mathbf{P}}(0)$ and $\hat{\boldsymbol{\theta}}(0)$. Luckily, the relative importance of the initial values decays with time, as the magnitudes of the sums increase. Also, as $\underline{\mathbf{P}}^{-1}(0) \to 0$, the recursive estimate approaches the batch estimate. Therefore, a common choice of initial values is to take $\underline{\mathbf{P}}(0) = C \cdot \underline{\mathbf{I}}$ ($\underline{\mathbf{I}}$ = the **identity matrix**) and $\hat{\boldsymbol{\theta}}(0) = 0$, where C is some large constant.

7.6.3 Forgetting Factor

Thus far in our discussion of recursive algorithms, we have assumed that the parameter vector is constant, i.e., that the system is time-invariant. However, an important reason for using recursive identification is that the dynamics may be changing with time, requiring tracking. One way to attack this problem in a heuristic way is to discount old measurements.

Let us rewrite $\hat{\boldsymbol{\theta}}(k)$ in Eq. (7.53) to discount old measurements by weighting measurements by the **forgetting factor,** $\lambda(k)$:

156 LINEAR SYSTEM IDENTIFICATION

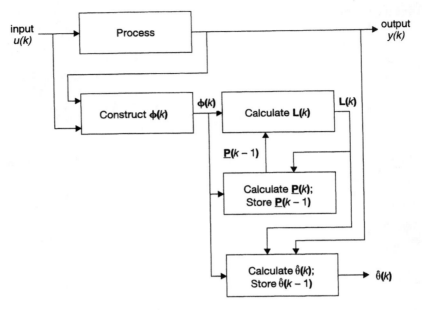

Figure 7.7 Recursive least squares algorithm.

$$\hat{\theta}(k) = \left[\sum_{i=1}^{k}\lambda(i)\phi(i)\phi^{T}(i)\right]^{-1}\sum_{i=1}^{k}\lambda(i)\phi(i)y(i) \tag{7.88}$$

This equation results from a performance function that is weighted by the forgetting factor. Proceeding through a derivation similar to that of Section 7.6.1, we obtain the **forgetting factor recursive least squares algorithm:**

$$\underline{P}(k) = \frac{1}{\lambda(k)}\left[\frac{\underline{P}(k-1) - \underline{P}(k-1)\phi(k)\phi^{T}(k)\underline{P}(k-1)}{\lambda(k) + \phi^{T}(k)\underline{P}(k-1)\phi(k)}\right] \tag{7.89}$$

$$\mathbf{L}(k) = \frac{\underline{P}(k-1)\phi(k)}{\lambda(k) + \phi^{T}(k)\underline{P}(k-1)\phi(k)} \tag{7.90}$$

$$\hat{\theta}(k) = \hat{\theta}(k-1) + \mathbf{L}(k)[y(k) - \hat{\theta}(k-1)\phi(k)] \tag{7.91}$$

The effect of the forgetting factor in Equation (7.89) is that $\underline{P}(k)$, and hence $\mathbf{L}(k)$, are kept large, even in the face of a sudden change in dynamics, δ. If $\lambda(k) \leq (1 - \delta)$, and $\delta > 0$, then $\underline{P}(k)$ will not tend to zero and the algorithm will always be alert to track changing dynamics. Obviously, with $\lambda(k) = 1$, we obtain the recursive least squares algorithm.

7.6.4 Example

For example, the arterial impulse response between the aortic and radial waveforms in Figure 7.4 may be approximated using recursive least squares. We choose $\underline{P}(0)$ as

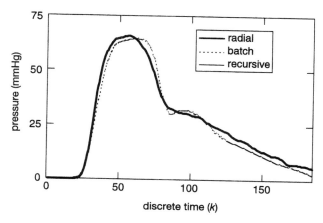

Figure 7.8 Radial waveform predictions using batch and recursive least squares algorithms. In both cases, $N = 6$, $D = 15$, and $(M - D) = 4$. The recursive prediction overlaps the observed radial waveform.

$10,000 \cdot \underline{\mathbf{I}}$ and $\hat{\boldsymbol{\theta}}(0)$ as 0. As before, let $N = 6$ and $M = 19$, which accounts for a delay of 15 samples. Using Eqs. (7.85), (7.87), and (7.81), we obtain the following parameter vector at $k = 185$ samples:

$$\hat{\boldsymbol{\theta}}^T = [-3.19 \; 3.71 \; -1.50 \; -0.48 \; 0.62 \; -0.16 \; 0 \ldots 0 \; 0.55 \; -1.82 \; 2.47 \; -1.60 \; 0.41]^T \quad (7.92)$$

This equation may be easily solved using the Matlab function rarx in the System Identification Toolbox. A comparison of the waveform predictions using batch least squares and recursive least squares is shown in Figure 7.8.

7.7 MODEL VALIDATION

Given various models for the same data set, how do we choose the best model? Model validation involves analyzing the results of parameter identification and model plausibility in order to select an "optimum" model.

7.7.1 Results of Parameter Identification

Once parameters have been identified for a chosen model from experimental data, the results may be analyzed to determine the model's utility using three validity measures. The first measurement is the **covariance matrix**, $\underline{\mathbf{V}}(K)$, which is the expected value of the regression vector multiplied by its transpose:

$$\underline{\mathbf{V}}(K) = \frac{1}{K} \sum_{i=1}^{K} \boldsymbol{\phi}(i) \boldsymbol{\phi}^T(i) \quad (7.93)$$

The diagonal elements of the covariance matrix contain estimates of the variance associated with each identified parameter. The square roots of these variances are used to calculate the standard deviations and therefore associated **coefficient of variation** (CV) for each parameter estimate. The coefficient of variation is also known as the **fractional standard deviations** (FSD), and is merely the standard deviation divided by its mean value. When the CVs of estimated parameter values are unreasonably large (i.e., much greater than 100%), the model may considered suboptimal. Large CVs may arise from limitations in the experimental data such as a small number of measurements or large measurement errors. Large CVs may also arise from utilization of a model that is too complex for the available experimental data. As the CVs become larger, the covariance matrix tends toward nonsingularity (nonunique identifiability).

A second validity measure is the **goodness of fit.** The best of several candidate models may be determined using Akaike's **final prediction error criterion** (FPE) [3], which is closely related to his **information criterion,** AIC. Because a more complicated model with more parameters may better fit experimental data, the number of model parameters is weighed against the number of data points and **mean squared error,** which refers to the average of the squared difference between the observed and estimated data:

$$FPE = \left[\frac{NN + p}{NN - p}\right]\left\{\frac{1}{NN}\sum_{k=1}^{NN}[y(k) - \hat{y}(k)]^2\right\} \qquad (7.94)$$

where NN is the number of data points and p is the number of parameters. In terms of this criterion, the best model is that which yields the lowest value of FPE. FPE is only used for linear, unbiased models. In Chapter 13, when we discuss nonlinear least squares estimation, we will use the more general AIC to measure goodness of fit.

The third validity measure is **residual statistics.** The residuals are an estimate of the noise in the system. If this noise is assumed to be white, Gaussian, and of zero mean, then the residuals should display these properties. If the residuals do not meet the assumptions, a systematic error in model identification may be present.

7.7.2 Model Plausibility

After the model fit has been analyzed, the model plausibility should be examined. First, the model should flexible, that is, capable of describing most of the different system dynamics that can be expected in the application in question. Both the number of parameters and the way they enter the model are important. Second, the model should be as simple as possible, avoiding unneeded complexity.

7.7.3 Example

For example, let us choose the best of three least squares estimates of the impulse response between the aorta and radial artery. The first model was given in Section 7.4.3 with $N = 6$, $D = 15$, and $(M - D) = 4$. Alternatively, we also fit the data in Figure 7.4 to the models $N = 15$, $D = 15$, $(M - D) = 13$ and $N = 10$, $D = 14$, $(M - D) = 9$. The three sets of estimated parameters are given in Table 7.1, along with the corresponding mean CVs and FPEs. The three predicted waveforms are given in Figure 7.5; the three residuals from these predictions are given in Figure 7.9.

All three models seem plausible. With regard to parameter identification results, the highest order model obviously suffers from a lack of fit, as evidenced by the large residuals. Although the lowest order model possesses lower mean CVs, it also possesses a higher FPE. However, the residuals of the lowest order model seem less random (alternation

Table 7.1 Models of the aorta-radial artery impulse response

Parameter	$N=6, D=15,$ $(M-D)=4$	$N=15, D=15,$ $(M-D)=13$	$N=10, D=14,$ $(M-D)=9$
$a_1 \ldots a_N$	−3.8, 5.8, −4.4, 1.3, 0.1, −0.1	−6.1, 18.2, −36.4, 55.4, −69.3, 74.1, −69.5, 57.1, −40.8, 24.6, −11.9, 4.0, −0.6, −0.1, 0.01	−5.4, 13.8, −21.8, 23.9, −18.8, 10.4, −3.8, 0.7, −0.01, −0.01
$b_{D+1} \ldots b_M$	0.4, −1.3, 1.7, −1.1, 0.3	0.3, −1.8, 4.8, −8.6, 11.3, −11.4, 9.2, −6.1, 3.5, −2.0, 1.0, −0.3, 0.04, 0.01	0.4, −1.7, 3.6, −4.8, 4.4, −2.9, 1.3, 0.2, −0.1, 0.06
mean CV_a	23%	11%	48%
mean CV_b	21%	268%	222%
FPE	4.3×10^{-4}	3.1×10^{-5}	6.5×10^{-5}

between positive and negative values) than the residuals of the medium order model. In other words, the differences between the observed and lowest order modeled data can not be attributed to noise. Therefore, the medium order model [$N = 10, D = 14, (M - D) = 9$] is the best model choice.

7.8 SUMMARY

In this chapter, we have reviewed the basic concepts underlying system identification, the determination of dynamic models from experimental data, of linear time-invariant single

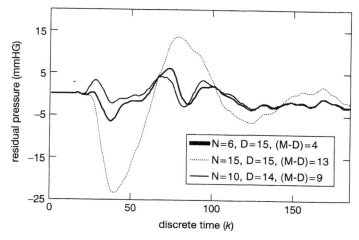

Figure 7.9 Residuals from least square predictions of radial waveforms using various combinations of feedback parameter number, N, sample delay, D, and feedforward parameter number, M. The input is an aortic waveform.

input, single output systems. We purposely restricted our discussion to the autoregressive moving average exogeneous input model because it is linear and quite versatile. We further restricted our discussion to deterministic input signals.

The ARMAX model consists of a linear constant coefficient difference equation containing N feedback coefficients multiplied by delayed samples of the output, $(M + 1)$ feedforward coefficients multiplied by delayed samples of the input, and $(P + 1)$ noise coefficients multiplied by delayed samples of white noise. It is autoregressive because the output looks back on past values of itself. It also possesses a moving average (the noise terms) and an exogenous, or external, input.

Several important special cases of the ARMAX model exist. First, when $M = P = 0$, it is assumed that an input signal is not present. This model is the autoregressive model. Second, when $N = M = 0$, it is assumed that an input signal is not present and that the current output may be modeled by past values of noise. This model is the moving average model. Third, when $N = 0$ and noise is not present, it is assumed that the current output may be modeled by past values of the input. This model is the finite impulse response model that was discussed extensively in Chapter 1. Finally, when $P = 0$, it is assumed that only a single sample of noise is reflected in the output. This model is the autoregressive with exogeneous input model, which is also known as the controlled autoregressive model. Without the white noise term, the ARX model becomes the infinite impulse response model.

The choice of an appropriate model structure for a system is determined by factors such as flexibility, parsimony, algorithm complexity, and properties of the performance function. In terms of flexibility, it should be possible to use the chosen model structure to describe most of the different system dynamics that can be expected in the application. In terms of parsimony, the chosen model should contain the smallest number of parameters required to represent the true system adequately. Because algorithm complexity directly influences the amount of computation required, the chosen model should be as simple as possible. During the optimization process, parameters in the chosen model are estimated based on a performance function.

Model identifiability refers to the possibility of theoretically and practically obtaining unique estimates of all the unknown model parameters. The ARMAX, FIR, IIR, and ARX models are all theoretically identifiable. However, the practical process of parameter estimation may not yield acceptable results. Problems may arise from different types of experimental error, the number of data points, and the true system, leading to a loss of practical identifiability. Depending on the capability of uniquely determining the parameters, a model may be uniquely identifiable, nonuniquely identifiable, or nonidentifiable.

Given an identifiable model, the model parameters may be calculated through a performance function, $\xi(\theta)$. As an example, the error between a system and its prediction may be calculated, and a performance function may be constructed that minimizes a form of this error. When a performance function based on prediction error is utilized, the resulting system identification is classified as a prediction error method. In particular, the least squares estimate of θ is defined as the vector, $\hat{\theta}$, that minimizes the mean squared error performance function. Similarly, the maximum likelihood estimate of θ is defined as the vector, $\hat{\theta}$, that minimizes the maximum likelihood performance function. The maximum likelihood refers to the probability distribution function of the observations conditioned on the parameter vector θ.

The least squares estimate, **θ**, will possess an asymptotic bias (or not be consistent) unless the noise input is white noise. This very restrictive assumption is one reason other methods, such as instrumental variable methods, are utilized. To derive the instrumental variable methods, we define a performance function with the error terms weighted by a vector of dimension $(M + N + 1) \times 1$, **z**(k), called an instrument. We can choose **z**(k) in different ways, subject to certain conditions, to guarantee the consistency of the basic IV estimate. For $\mathbf{z}(k) = \boldsymbol{\phi}(k)$, the basic IV estimate reduces to the least squares estimate.

Alternatively, we may estimate model parameters recursively, rather than in batch fashion. Recursive identification possesses the benefit of decreased memory allocation, and may be derived from the least squares estimate, provided that the inverse exists. The resulting recursive least squares algorithm is robust and easily implemented. Note that, as with the batch least squares estimate, it is assumed that the disturbance, $v(k)$, is white noise. A common choice of initial values for this algorithm is to take $\underline{\mathbf{P}}(0) = C \cdot \mathbf{I}$ and $\boldsymbol{\theta}(0) = 0$, where C is some large constant.

In the recursive least squares algorithm, we assume that the parameter vector is constant, i.e., that the system is time-invariant. However, an important reason for using recursive identification is that the dynamics may be changing with time, requiring tracking. One way to attack this problem in a heuristic way is to discount old measurements. This is accomplished by weighting the performance function by the forgetting factor, $\lambda(k)$. Proceeding through a derivation similar to that of the recursive least squares algorithm, we obtain the forgetting factor recursive least squares algorithm. The effect of the forgetting factor is that $\underline{\mathbf{P}}(k)$, and hence $\mathbf{L}(k)$, are kept large, even in the face of a sudden change in dynamics, δ. If $\lambda(k) \leq (1 - \delta)$, and $\delta > 0$, then $\underline{\mathbf{P}}(k)$ will not tend to zero and the algorithm will always be alert to track changing dynamics. With $\lambda(k) = 1$, we obtain the recursive least squares algorithm.

Given various models for the same data set, we use model validity, which involves analyzing the results of parameter identification and model plausibility, in order to select an "optimum" model. Model utility may be determined from coefficients of variation, goodness of fit, and residual statistics. Model plausibility refers to the flexibility and simplicity of the model.

7.9 REFERENCES

[1]. Box, G. E. P. and Jenkins, G. W. *Time Series Analysis: Forecasting and Control,* 2nd ed. Holden-Day: San Francisco, 1976.

[2]. Sugimachi, M. National Cardiovascular Center Research Institute, Osaka, Japan. Research data, 1992.

[3]. Akaike, H. Statistical predictor identification. *Ann Inst Stat Math, 22,* 203–217, 1970.

Further Reading

Linear System Identification:

Soderstrom, T. and Stoica, P. *System Identification.* Prentice Hall: Englewood Cliffs, NJ, 1989.

Ljung, L. and Soderstrom, T. *Theory and Practice of Recursive Identification.* MIT Press: Cambridge, MA, 1983.

Ljung, L. *System Identification: Theory for the User,* 2nd ed. Prentice Hall: Upper Saddle River, NJ, 1999.

7.10 RECOMMENDED EXERCISES

Prediction error method: see Soderstrom, T. and Stoica, P. *System Identification*. Problems 7.17 and 7.18.

Instrumental variable method: see Soderstrom, T. and Stoica, P. *System Identification*. Problem 8.12.

Recursive least squares: see Ljung, L. *System Identification: Theory for the User*. 11E.2 and 11E.3.

8

EXTERNAL DEFIBRILLATION WAVEFORM OPTIMIZATION

In this chapter, we discuss the application of linear system identification to **external defibrillation.** During external defibrillation, a strong current of short duration is administered across the thorax via paddles or electrode adhesive pads to convert rapid twitching of the ventricles, or **ventricular fibrillation,** to a slower rhythm that allows the heart to pump blood. The current may be administered as one of many possible waveform shapes. It is shown that the effects of various waveform shapes may be predicted by modeling the thoracic impedance as an autoregressive with exogenous input (ARX) model.

8.1 PHYSIOLOGY

In order to understand defibrillation, we must first understand electrical conduction within the heart. In Section 6.1, we discussed the mechanical function of the heart. The mechanical work of pumping is performed by the fibers of the **working myocardium** of atria and ventricles, which make up the main mass of the heart. However a second type of fiber exists—the fibers of the pacemaker and conducting system. These fibers are specialized to generate excitatory impulses and send them to working cells. In this section, we discuss the electrical function of the heart, which is based on the **action potential** of cardiac muscle cells. We then describe how the conduction of an action potential across the heart stimulates contraction, and how this excitation conduction may be observed in an **electrocardiogram** (ECG). ECGs are used to diagnose abnormal cardiac rhythms. The most lethal rhythm, ventricular fibrillation, is a type of cardiac arrest that may only be stopped through defibrillation.

8.1.1 Action Potential

The working myocardium action potential begins with a rapid reversal of the myocardial cell membrane potential, from a **resting potential** of ~ −90 mV to the initial peak of ~

+30 mV. This rapid phase of **depolarization** lasts 100–200 ms and is followed by a prolonged plateau and then **repolarization** to the resting potential (Figure 8.1). The cardiac action potential lasts ~200–400 ms. It is generated through a combination of membrane-potential changes, changes in ionic conductivity, and ion currents.

During the action potential's **absolute refractory period,** until the time when the membrane has repolarized to ~ –40 mV, the cell is inexcitable. During the **relative refractory period,** the cell may be reexcited. If an action potential is generated early during the relative refractory period, it will not rise as sharply, obtain the same maximum voltage, nor last as long as a normal action potential. This prolonged refractory period protects the myocardium from too-rapid reexcitation, which could impair mechanical function, and from recycling of excitation in the muscle, which would interfere with the rhythmic alternation of contraction and relaxation. The refractory period is normally longer than the period during which the excitation spreads over the atria or ventricles. Thus, a wave of positive excitation spreads across the heart only once and dies out as it encounters refractory tissue everywhere.

In the working myocardium of atria and ventricles, action potentials are generated by

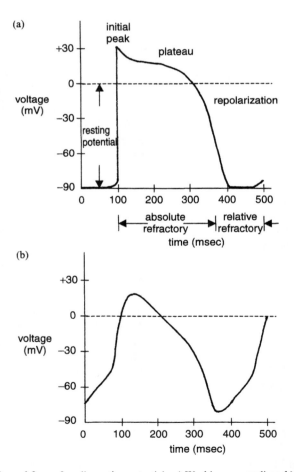

Figure 8.1 General form of cardiac action potentials. a) Working myocardium, b) dominant pacemaker.

conduction of excitation. In contrast, the cardiac cells capable of automaticity (those not responsible for pumping) possess the ability to depolarize spontaneously until a threshold potential of ~ −40 mV is reached, which generates a new action potential (Figure 8.1). These **dominant pacemaker** cells are present in the **sinoatrial (SA) node** and control the timing of heart contractions.

8.1.2 Spread of Excitation

A heartbeat is normally initiated in the SA node, in the wall of the right atrium at the opening of the superior vena cava (Figure 8.2). At rest, the SA node paces the heart at ~ 70 beats per min (bpm). From the SA node, the excitation is conducted over the working myocardium of both atria, initiating atrial contraction. The excitation is then briefly delayed at the **atrioventricular (AV) node.** Should the SA node not initiate a heartbeat or an excitation not be conducted to the atria, the AV node may substitute as a **secondary pacemaker** with a frequency of 40–60 bpm. This brief pause allows the atrial blood to enter the ventricles.

The excitation then quickly moves at a velocity of ~ 2 m/s through the remainder of the system, from the **bundle of His,** through the left and right bundle branches to the subendocardial endings of the **Purkinje fibers.** From the subendocardial endings of the Purkinje fibers, the excitation is conducted at a velocity of ~ 1m/s over the ventricular musculature, initialing ventricular contraction. In the case of **complete heart block,** when the conduction from the atria to the ventricles is completely interrupted, a **tertiary** center in the ventricular conducting system (that is normally suppressed) may take over as a pacemaker for ventricular contraction, with a frequency of ~ 30–40 bpm.

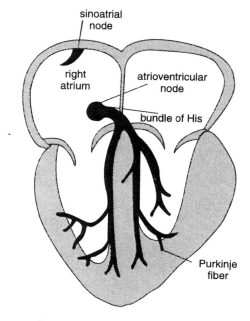

Figure 8.2 Cardiac conduction system.

8.1.3 Electrocardiogram

This excitation conduction may be observed in an electrocardiogram. An ECG is a body surface recording of the electrical activity generated by the heart, based on the differential voltage between two points on the body. In the early 1900s, the Dutch scientist Willem Einthoven defined three differential voltage (V) recordings, or **leads** [1], that are still used today:

$$\text{lead I} = V_{\text{left arm}} - V_{\text{right arm}} \tag{8.1}$$

$$\text{lead II} = V_{\text{left leg}} - V_{\text{right arm}} \tag{8.2}$$

$$\text{lead III} = V_{\text{left leg}} - V_{\text{left arm}} \tag{8.3}$$

Because the body is assumed to be purely resistive at ECG frequencies, the four limbs may be thought of as wires attached to the torso. It then follows that lead I may be recorded from the respective shoulders without loss of cardiac information. Please note that lead II is the sum of lead I and lead III.

A typical normal lead II ECG is shown in Figure 8.3. ECG signals are typically in the range of ± 2 mV, and require a bandwidth of 0.05 to 150 Hz. The features of the ECG were named by Einthoven as waves P to U. The **P wave** represents the depolarization and simultaneous contraction of both atria. The short, relatively isoelectric segment following the P wave represents the slowing of depolarization within the AV node and conduction to the subendocardial endings of the Purkinje fibers. The **QRS complex,** which is composed of the **Q, R,** and **S waves,** represents ventricular depolarization and the beginning of ventricular contraction. The Q wave is an initial downward deflection, the R wave is an initial upward deflection, and the S wave is the terminal downward deflection. Following the QRS complex, the **ST segment** is another short relatively isoelectric segment; it represents the plateau before ventricular repolarization. The T wave represents ventricular repolarization. A U wave may also be present; its origin has never been fully established.

8.1.4 Ventricular Fibrillation

The appearance of abnormal ECG waveforms is used to diagnose cardiac disease, including cardiac **arrhythmias,** or abnormal rhythms. One pathological type of arrhythmia is ventricular fibrillation, which is shown in Figure 8.4. During ventricular fibrilla-

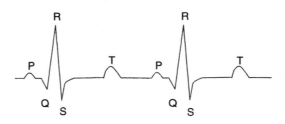

Figure 8.3 Representation of a typical normal lead II electrocardiogram.

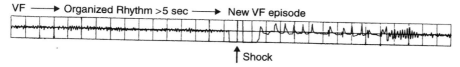

Figure 8.4 Ventricular fibrillation (VF) ECG before and after defibrillation shock. Amplitude = 1 mV/division. Time = 1 second/division. Reprinted from [2] with kind permission from the Association for the Advancement of Medical Instrumentation, Arlington, VA.

tion, the conduction of cardiac excitation is transiently or permanently blocked. Conduction block may be caused by **myocardial ischemia** (insufficient blood flow), congestive heart failure, hypothermia, or electric shock. In response, potential pacemaker cells in the ventricle that are normally silent become irritable and simultaneously rapidly discharge. The resulting rhythm is so chaotic that distinct, complete waves are indistinguishable. The ventricles no longer contract, but only twitch rapidly. Ventricular fibrillation is a type of **sudden cardiac arrest** (SCA), and requires immediate defibrillation to prevent sudden death.

8.1.5 Defibrillation

Defibrillation refers to the application of a strong electric shock in an effort to convert fibrillation to a slower rhythm that allows the heart to pump blood. The shock may be applied externally through paddles or **electrode adhesive pads** positioned on the thorax using an **external defibrillator,** or internally through electrodes in contact with heart tissue using an **internal defibrillator.** Successful defibrillation stimulates cells by passing an adequate current intensity through the cells for an adequate period of time.

The mechanism of defibrillation remains a subject of ongoing research. According to the **upper limit of vulnerability hypothesis** of defibrillation, a defibrillation shock must accomplish two goals. First, the shock must halt conduction fronts propagating through tissue by directly exciting or prolonging the refractoriness of the myocardium just in front of these conduction fronts. Second, it must prevent new conduction fronts at the border of the directly excited region that reinitiate fibrillation.

8.2 EXTERNAL DEFIBRILLATION WAVEFORMS

External defibrillation was first reported in 1899, when Prevost and Batelli discovered that a capacitor discharge abolished ventricular fibrillation [3]. This observation was later confirmed by Gurvich and Yuniev, who also showed that damping and prolonging the discharge by adding a small inductance to the circuit increased its effectiveness (Figure 8.5) [4]. However, the first widely used waveform for ventricular defibrillation was the standard household 60 Hz ac current. The delivered shock was a sine wave that alternated between positive and negative polarity every 17 ms, and lasted up to 1 second. Later, in the early 1960s, to decrease the potential for significant cardiac damage, Lown et al. [5] and Edmark [6] introduced the damped capacitor discharge into clinical use to replace AC defibrillators. This discharge delivered a **damped sinusoid waveform** to the patient (Figure 8.6A).

168 EXTERNAL DEFIBRILLATION WAVEFORM OPTIMIZATION

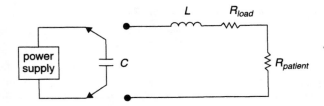

Figure 8.5 External defibrillation circuit that shapes damped sinusoid waveform. R_{load} refers to internal defibrillator resistance.

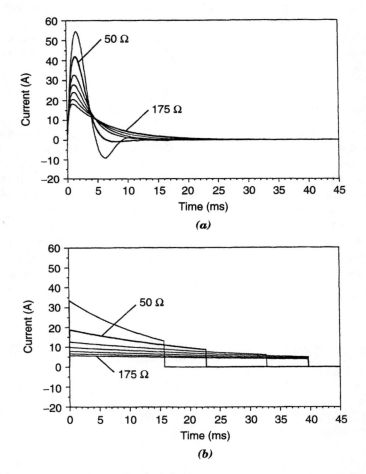

Figure 8.6 Representative current plots of (a) damped sinusoid and (b) monophasic truncated exponential waveforms as a function of load impedance from 25 to 175 Ω in 25 Ω increments. Reprinted from [2] with kind permission from the Association for the Advancement of Medical Instrumentation, Arlington, VA.

8.2 EXTERNAL DEFIBRILLATION WAVEFORMS

8.2.1 Damped Sinusoid Waveform

The damped sinusoid waveform was delivered by most external defibrillators until the mid 1990s (Figure 8.7). To obtain this waveform, the voltage stored by a capacitor is discharged across a **resistor–inductor–capacitor** (RLC) circuit. The patient resistance forms part of the resistive contribution to the circuit (Figure 8.5). Typically, such a defibrillator stored a maximum of 435 J of energy and 5200 V, and delivered about 80% of the stored energy as 360 J into a 50 Ω resistive load, with a pulse duration of 3 to 5 msec. Typical capacitance and inductance values were 16 μF and 50 mH, respectively. The charge was delivered through steel paddles/applied gel or electrode adhesive pads to the thorax. Because the patient resistance varied from an assumed 50 Ω, the duration and damping characteristics of the waveform varied. Although the industry standard for calibration is 50 Ω, the average patient resistance is about 80 Ω, with a clinical range of about 35 to 170 Ω [2]. In these defibrillators, the waveform was overdamped in high-resistance patients and underdamped in low-resistance patients (Figure 8.6). The resulting waveform variability impacted defibrillation effectiveness, since **defibrillation efficacy** with this waveform decreases at higher impedances [7]. Defibrillation efficacy refers to the percentage of successful defibrillation attempts.

8.2.2 Monophasic Truncated Exponential Waveforms

Other defibrillators manufactured during the same timeframe as those that delivered the damped sinusoid utilized the **monophasic truncated exponential waveform**. The current from these defibrillators was shaped only by a **resistance–capacitor** (RC) circuit, eliminating the inductor. Such defibrillators were designed to deliver the selected energy

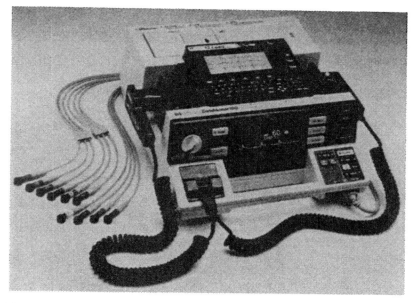

Figure 8.7 Agilent Technologies Codemaster 100 external defibrillator. This defibrillator uses a damped sinusoid waveform. Courtesy of Agilent Technologies, Healthcare Solutions Group, Andover, MA.

170 EXTERNAL DEFIBRILLATION WAVEFORM OPTIMIZATION

to all load resistances by extending the waveform duration until the desired energy was delivered. For high resistances, the delivered waveform was low in current and excessively long in duration (Figure 8.6). Animal studies have shown that the efficacy of this waveform decreases for waveform durations greater than 20 msec [8].

8.2.3 Biphasic Truncated Exponential Waveforms

In the early 1990s, many researchers began to investigate the utility of the **biphasic truncated exponential** (BTE) waveform for external defibrillation (Figure 8.8). Originally, this waveform had been demonstrated to increase defibrillation efficacy, as compared to the monophasic truncated exponential, during **internal defibrillation** (direct application of current) [9]. For successful internal defibrillation, less energy and voltage were required with a biphasic, rather than monophasic, truncated exponential.

Utilization of this biphasic waveform for external defibrillation requires the selection of several parameters. Researchers investigated the defibrillation efficacy of voltages discharged from a single capacitor with varying **total duration, phase duration,** and **waveform tilt.** Total duration ignores the switching time between shock phases. In Figure 8.8, the total duration is 10 msec, with 5 msec during each phase and the switching time of 0.5 msec neglected. Phase duration refers to the percentage of the total waveform duration occupied by the first shock phase. In Figure 8.8, the phase duration is 50%. For a BTE waveform with initial positive voltage, V_0, and terminal negative voltage, V_3, tilt is calculated as

$$\text{tilt} = \left[\frac{V_0 - (-V_3)}{V_0} \right] \times 100\% \quad (8.4)$$

For example, the former startup company Heartstream investigated BTE waveforms with 12 msec total duration, based on the parameter E_{50}. E_{50} refers to the total energy required for 50% successful defibrillation (subject resuscitated 50% of trials). For 12 msec dura-

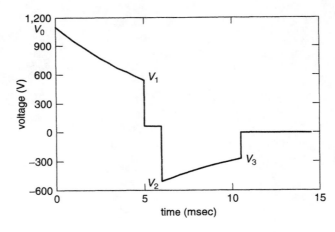

Figure 8.8 Biphasic truncated exponential waveform example. The phase duration is 50%, and excludes the 0.5 msec switching time.

tion, E_{50} was 169 ± 101 J for the BTE waveform compared with 414 ± 114 J for the monophase trucated exponential waveform, which is significantly different (p = 0.003). Optimization of phase duration to 60% and tilt to 76% reduced the defibrillation requirements of the 12 msec BTE waveform to 129 ± 36 J [8].

8.3 PROBLEM SIGNIFICANCE

Research conducted on the biphasic truncated exponential waveform for external defibrillation was directed toward the goal of a better waveform for **automatic external defibrillators** (AEDs). An AED is a device that recognizes and treats rapid ventricular arrhythmias in patients with cardiac arrest without requiring interpretation of the rhythm by medical personnel. The AED analyzes a measured ECG and prompts the user to deliver (or self-deliver) a shock if it detects ventricular fibrillation. The first use of an AED was reported in 1979 [10]. By 1988, the International Association of Fire Chiefs had adopted a proposal to equip every fire truck in the United States with an AED [11].

During this timeframe, AEDs were as bulky as their hospital counterparts. Damped sinusoid waveforms discharged from these defibrillators required large wave-shaping inductors that accounted for approximately 150 cc of volume and 300 g of mass, thus hindering device miniaturization. In contrast, BTE waveforms provided the opportunity of equal or better defibrillation efficacy without the use of a bulky inductor, and with lower voltages/smaller batteries. Using the BTE waveform, the physical size and weight of an AED could be reduced from a typical 8–22 pounds to less than 5 pounds, enabling it to be better suited for transport.

8.3.1 Clinical Significance

The AED is an important factor in increasing the odds of survival from sudden cardiac arrest because patients can be defibrillated much faster. Over one million deaths each year in the United States are attributed to cardiovascular disease [12]. Approximately one-fourth of these deaths occur suddenly outside the hospital, the majority of these caused by ventricular fibrillation [13]. The likelihood of survival of SCA decreases by approximately 10% every minute from the time of collapse to defibrillation [14]. In large metropolitan areas such as Chicago [15] and New York City [16], survival rates are near 2%, largely due to extensive delays in the availability of defibrillators. In contrast, in cities with mature early defibrillation programs like Seattle [14] and Rochester, MN [17], survival rates are in the range of 25 to 45%. In particular in Seattle, the survival rate was recently increased to 30% by providing 90 sec of cardiopulmonary resuscitation (CPR) before AED shock delivery [18]. Access to AEDs in these early defibrillation programs increased the chance of survival from SCA.

8.3.2 Market Forces

The survival of SCA victims through external defibrillation is cost-efficient for society. As estimated by one study, the cost-per-life by emergency medical systems staffed by emergency medical technicians certified to provide external defibrillation or paramedics in 1988 dollars was $2100–2300 [19]. This figure compares favorably to an expenditure of $35,000–45,000 for renal dialysis per year of useful life, $50,000 per year of life saved for primary prevention of coronary heart disease by cholesterol reduction treatment with lovastatin (except in very high risk patients), and $15,000–30,000 per year of life saved for screen and generic drug treatment of high blood pressure [20].

In November, 1996, the FDA approved a compact AED utilizing a BTE waveform for treatment of SCA on airlines. The $3000 Forerunner AED (Figure 8.9) was manufactured by the startup company Heartstream, which was later acquired by external defibrillation hospital market leader Hewlett Packard. (In late 1999, Hewlett Packard spun off its medical division to the newly formed Agilent Technologies.) Shortly after FDA approval, American Airlines ordered this AED; United and Delta Airlines soon announced they would also equip their airplanes with AEDs [21].

Heartstream's AED spawned comparable models from other companies. By 1999, half a dozen firms debuted their own AED. All weighed about 4 pounds, utilized biphasic truncated exponential waveforms, and cost from $2500–$4000. Approximately 50,000 were sold by September, 1999 to customers that included airlines, police departments, hospitals, and home consumers. About 40% of these sales were made by Hewlett Packard/Agilent [22].

If we assume an average cost of $3250 per AED and 25,000 sales per year, the annual market around 1999 was $81,250,000, excluding disposable electrode adhesive pads. In contrast, in 1997, hospital sales of external defibrillators and their disposable external adhesive pads generated only $10,863,681 and $9,440,189, respectively, in the United States. These sales estimates exclude sales to federal hospitals and nursing homes. Hewlett Packard accounted for 43.1% of total device sales; Zoll Medical, another defibrillator manufacturer, accounted for 84.7% of disposable sales [23]. Models of hospital defibrillators from the top three hospital manufacturers (Agilent, Medtronic Physio Control, and Zoll) are now available using BTE waveforms with various waveform parameters.

During the early 1990s and continuing to this day, various waveform shapes, many of them variations of the BTE, have been postulated for use in automatic and hospital external defibrillators. The approach to this research has often been quite scattershot, as an overall mechanism for determination of an optimal waveform is not known. A clinical

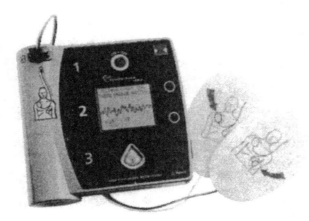

Figure 8.9 Agilent Technologies Heartstream ForeRunner automatic external defibrillator. Courtesy of Agilent Technologies, Healthcare Solutions Group, Andover, MA.

need exists to determine an underlying mechanism for optimization, in order to efficiently allocate resources for this research.

8.4 PREVIOUS STUDIES

An optimal waveform may be determined by understanding the **transthoracic impedance**. Because the defibrillation energy level is preset before shock administration, the transthoracic impedance uniquely determines the quantity of current that flows to the myocardium. Transthoracic impedance consists of three parts:

1. Impedance from extratissue sources, including the defibrillator, leads, and electrodes
2. Impedance from tissue sources, which include intracardiac and extracardiac tissue
3. Impedance from the interface between the electrode and tissue

Impedance itself is comprised of two parts:

1. The resistance, which is the real component of impedance
2. The reactance, which is the frequency dependent complex component

In this section, we review the pure resistance model of transthoracic impedance and early evidence for a more complicated model.

8.4.1 Geddes' Pure Resistance Model

Traditionally, transthoracic impedance has been measured as a pure resistance. The original measurement method was proposed by Leslie Geddes' group at Purdue University [24, 25], and was later validated in humans by Richard Kerber's group at the University of Iowa [26]. Two variations of the method exist. In the first variation, the ratio of peak voltage to peak current is calculated in response to a transthoracic defibrillation pulse. In the second variation, the ratio of peak voltage to peak current is calculated in response to transthoracic application of a sinusoidal current with an approximate frequency of 30 kHz. This frequency is used because the resulting ratio approximates the ratio obtained with a defibrillation pulse. Based on Kerber's mean impedance in 19 patients of 78.1 ± 19.4 Ω, as well as earlier impedance measurements with a mean of 58 Ω by Machin in 175 patients [27], human transthoracic impedance has been estimated for over two decades as a "typical" quantity of 50 Ω. This 50 Ω quantity has been used as a standard for external defibrillation and electrode calibration [28]. The measurement has also been incorporated into Hewlett-Packard defibrillators [29, 30].

8.4.2 Evidence for a Model Based on Resistance and Reactance in Dogs

With this peak ratio method, impedance is calculated as a purely resistive quantity, whereas the reactive component of impedance is ignored. However, as early as 1976, studies demonstrated that impedance is not purely resistive, as plots of simultaneous voltage versus current during defibrillation discharge resulted in ellipses, rather than the expected straight line with approximate slope of 50 Ω (Figure 8.10) [27, 31]. More recently,

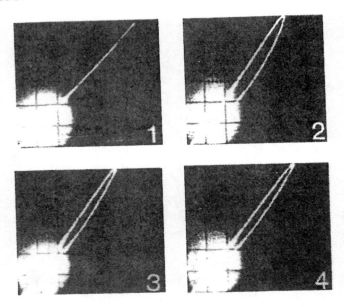

Figure 8.10 Simultaneous X–Y plot of current (10 A per division) and voltage (500 V per division) from dc defibrillator discharge. Dog G: (1) Discharge into artificial resistive load of 50 ohms; (2) first transthoracic discharge; (3) second transthoracic discharge; (4) tenth transthoracic discharge. Reprinted from [31] with kind permission from the Association for the Advancement of Medical Instrumentation, Arlington, VA.

Tang et al. have shown that the ratio of voltage to current increases through each phase of a biphasic truncated exponential discharge [32]. In his original papers, Geddes himself showed that the peak ratio of voltage to current is inversely proportional to frequency [23, 24] rather than independent of frequency, as would be required by a pure resistance. This evidence for a reactive component has been disregarded by citing the "virtual" lack of phase difference between the defibrillation current and voltage [33, 34].

8.5 APPLICATION OF THE ARX MODEL TO PREDICTION OF TRANSTHORACIC IMPEDANCE

The ARX model discussed in Chapter 7 may be used to estimate the resistive and reactive components of transthoracic impedance. This model was first validated by estimating known resistance and capacitance values in a series R–C circuit. It was then used to estimate transthoracic resistance and capacitance during external defibrillation in swine and during voltage discharges across electrode adhesive pads. The following discussion is based on work conducted by the author in 1995–1996 at Cardiotronics Systems Inc [35, 36, 37, 38].

8.5.1 Resistance Estimation Method

The transthoracic impedance was modeled as a resistor, R_t, and capacitor, C_t, in series. This impedance was connected to either a 1 Hz, 0.2 second pulse duration, 2 $V_{p\text{-}p}$ square

8.5 APPLICATION OF THE ARX MODEL TO PREDICTION OF TRANSTHORACIC IMPEDENCE

wave voltage source (National Instruments LabPC+ board) or defibrillator damped sinusoid waveform (HP43110 defibrillator), $V_{in}(k)$; and load resistor, R_l (Figure 8.11). A 1983 Ω resistor was used with the square wave; a 50 Ω resistor was used with the damped sinusoid. The transfer function between $V_{in}(z)$ and $V_{out}(z)$, the voltage across R_t and C_t, was derived from circuit analysis as:

$$\frac{V_{out}(z)}{V_{in}(z)} = \frac{\dfrac{R_t + \dfrac{1}{C_t}}{R_t + R_l + \dfrac{1}{C_t}} + \dfrac{\dfrac{1}{C_t} - R_t}{R_t + R_l + \dfrac{1}{C_t}} z^{-1}}{1 + \dfrac{\dfrac{1}{C_t} - R_t - R_l}{R_t + R_l + \dfrac{1}{C_t}} z^{-1}} \tag{8.5}$$

To derive an equivalent transfer function, an ARX model was used to model the impulse response between the measured voltage source and impedance as

$$\tilde{V}_{out}(k) = -a_1 \tilde{V}_{out}(k-1) + b_0 \tilde{V}_{in}(k) + b_1 \tilde{V}_{in}(k-1) \tag{8.6}$$

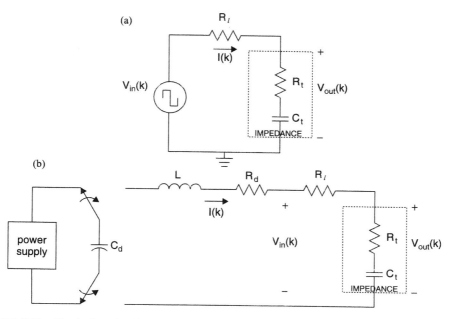

Figure 8.11 Circuits for estimation of transthoracic impedance. The transthoracic impedance is modeled as a resistor, R_t, and capacitor, C_t, in series. $V_{out}(k)$ is the voltage across the transthoracic impedance. $I(k)$ is the current. a) For low-current estimation, $V_{in}(k)$ is a 1 Hz, 0.2 second duration, 2 Vp-p square wave. b) for high-current estimation, $V_{in}(k)$ is the voltage across C_d, L, R_l, and R_d. From [38].

The model coefficients were obtained from measured $\tilde{V}_{in}(k)$ and $\tilde{V}_{out}(k)$ using least squares estimation. Substitution of Equation (8.6) into Equation (7.32) yields

$$\hat{\theta} = [a_1 \ b_0 \ b_1]^T = \left[\frac{1}{K}\sum_{k=0}^{K-1}\phi(k)\phi^T(k)\right]^{-1} \frac{1}{K}\sum_{k=0}^{K-1}\phi(k)\tilde{V}_{out}(k) \quad (8.7)$$

where K = number of samples per waveform period and

$$\phi^T(k) = [-\tilde{V}_{out}(k-1) \ \tilde{V}_{in}(k) \ \tilde{V}_{in}(k-1)]^T \quad (8.8)$$

1000 samples of each voltage were digitized to 12 bits with sampling frequency 1 kHz for the square wave or 20 kHz for the damped sinusoid (National Instruments LabPC+ board), lowpass filtered with a corner frequency of 40 Hz, and decimated by 8 samples before applying least squared estimation. Transformation of (8.6) to the z domain and rearrangement yielded the transfer function:

$$\frac{\tilde{V}_{out}(z)}{\tilde{V}_{in}(z)} = \frac{V_{out}(z)}{V_{in}(z)} = \frac{b_0 + b_1 z^{-1}}{1 + a_1 z^{-1}} \quad (8.9)$$

Equating both transfer functions (8.5) and (8.9) yielded an estimate of the transthoracic resistance:

$$R_t = \frac{(b_0 - b_1)R_l}{2(1 - b_0)} \quad (8.10)$$

For each low-current measurement, the mean of estimated resistance values from three square wave periods was used. For each high-current measurement, the resistance value from a single defibrillation pulse was used.

8.5.2 Proposed Capacitance Estimation Method

Although capacitance could also be obtained from these transfer functions, it was not determined, as this method does not have sufficient resolution for accurate capacitance estimates on the order of 10^{-5} Farads. Instead, using R_t as an input, capacitance was estimated by fitting the measured decay of $\tilde{V}_{out}(k)$ or $\tilde{I}(k)$ from its peak value (Figure 8.12) to the appropriate exponential(s). Using circuit analysis, a single exponential of form $e^{-kt/(R_t + R_l)C_t}$ represents the voltage decay in response to a square wave. Similarly, two exponentials of the form $e^{m_1 t} + e^{m_2 t}$ represent the current decay in response to a damped sinusoidal defibrillator waveform, where

$$m = \frac{-(R_d + R_l + R_t) \pm \sqrt{(R_d + R_l + R_t)^2 - 4L\left(\frac{1}{C_t} + \frac{1}{C_d}\right)}}{2L} \quad (8.11)$$

where R_d = defibrillator resistance, L = defibrillator inductance, and C_d = defibrillator capacitance (Figure 8.11). As before, Eq. (8.11) was determined from circuit analysis. For the HP 43110 defibrillator, $R_d = 11\ \Omega$, $L = 20$ mH, and $C_d = 53.4\ \mu$F [29]. To accurately

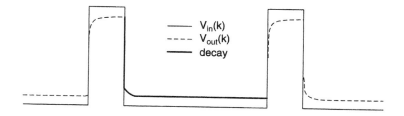

Figure 8.12 Decay portion of output voltage used for capacitance estimation. From [38].

estimate capacitance, the squared error between the observed and calculated decay in the data used for resistance estimation was minimized using the optimization technique called Powell's successive quadratic estimation [39].

For Powell's method, let us assume that the squared error is a unimodal function that is continuous in an interval. Therefore, the squared error can be approximated by a quadratic polynomial based on three capacitor samples c_0, c_1, and c_2:

$$\text{decay\_error}(C_t) = d_0 + d_1(C_t - c_0) + d_2(C_t - c_0)(C_t - c_1) \quad (8.12)$$

Substituting c_0, c_1, and c_2 yielded values of d_0, d_1, and d_2:

$$d_0 = \text{decay\_error}(c_0) \quad (8.13)$$

$$d_1 = \frac{\text{decay\_error}(c_1) - \text{decay\_error}(c_0)}{c_1 - c_0} \quad (8.14)$$

$$d_2 = \frac{1}{c_2 - c_1}\left(\frac{\text{decay\_error}(c_2) - \text{decay\_error}(c_0)}{c_2 - c_0} - \frac{\text{decay\_error}(c_1) - \text{decay\_error}(c_0)}{c_1 - c_0}\right) \quad (8.15)$$

Assuming this quadratic is a good approximation of the squared error, the minimum error was found by setting the first derivative to zero. Rearranging this derivative yielded an estimate of capacitance, $\hat{C}_t$:

$$\hat{C}_t = \frac{c_0 + c_1}{2} - \frac{d_1}{2d_2} \quad (8.16)$$

To implement capacitance estimation, an initial optimal capacitance, C_t^*, was determined for the square wave input by taking the natural log of the decay function, estimating its slope and calculating C_t^* as:

$$C_t^* = \frac{-1}{\text{slope}(R_t + R_l)} \quad (8.17)$$

Alternatively, the initial capacitance was set to 100 μF for the damped sinusoidal input. The capacitance step size, ΔC_t, was chosen as 5×10^{-6}, and values of convergence parameters, ε_1 and ε_2, were chosen as 0.75 and 0.005, respectively. decay_error(C_t) for C_t^* and ($C_t^* + \Delta C_t$) were then calculated. If decay_error(C_t^*) was greater than decay_error($C_t^* + \Delta C_t$), then c_i were defined as $c_0 = C_t^*$, $c_1 = C_t^* + \Delta C_t$, and $c_2 = C_t^* + 2\Delta C_t$. Otherwise, c_i were defined as $c_0 = C_t^* - \Delta C_t$, $c_1 = C_t^*$, and $c_2 = C_t^* + \Delta C_t$.

In the major loop of this algorithm, the minimum decay_error(C_t) for c_0, c_1, and c_2 was determined. The c_i value corresponding to this minimum was equated to C_t^*. $\hat{C}_t$ was then calculated from the equation above. If [decay_error(C^*_t) − decay_error($\hat{C}_t$)]/decay_error ($\hat{C}_t$) was less than or equal to ε_1 and ($C_t^* - \hat{C}_t$)/$\hat{C}_t$ was less than or equal to ε_2, the algorithm terminated. Otherwise, if decay_error(C_t^*) was greater than decay_error($\hat{C}_t$), then C_t^* was redefined as $C_t^* = \hat{C}_t$. The algorithm then looped back to calculation of c_i values. The capacitance estimation algorithm is illustrated in Figure 8.13.

8.5.3 Validation of Proposed Method

To validate the accuracy of the proposed method, resistance and capacitance were calculated using known resistor and capacitor values and either the low-current square wave or high-current defibrillator pulse. The resistors used were 24.2, 48.9, 74.6, 100.2, and 500.2 Ω. The capacitor used was 50.9 µF. The resistance estimates from the low- and high-current methods were compared to true values and estimates from the peak ratio method. The capacitance estimates were compared to the true value.

8.5.4 Swine Studies

To determine the linearity, with respect to current, of transthoracic resistance and capacitance, these parameters were measured in swine, using the low-current square wave input and high-current damped sinusoidal input. Five female Yorkshire swine (50.2 ± 3.3 kg), age 3–6 months, were fasted overnight and anesthetized with sodium pentobarbital (6–10 mg/kg bolus, followed by 6–10 mg/kg/hr infusion), after sedation with an intramuscular injection of ketamine (20 mg/kg), xylazine (2 mg/kg), and atropine (0.1 mg/kg). Each animal was mechanically ventilated using a respirator (Harvard Model 605A or North America Drager AV). Pancuronium bromide (1 mg/kg/hr) was injected intravenously every 30 minutes. Propranolol HCl (1 mg/kg bolus, followed by 0.5 mg/kg/hr) was infused intravenously during the course of the study. An ear vein was cannulated for administration of anesthetic and other agents. Body temperature and ECG were continuously monitored. The femoral artery was catheterized, and arterial blood gases were obtained every 30 minutes.

Each animal was subjected to three defibrillation shocks of 200 J (Hewlett Packard 43110 defibrillator), spaced 4 minutes apart. For three of the animals, each shock was synchronized to the R wave. Shocks were administered using a single pair of R2 electrodes (Model 610), adhered to the shaved thorax in an anterior–anterior position. Immediately preceding the first shock, transthoracic resistance and capacitance were measured using low current. After each shock, eight low current measurements were made every 30 seconds. High-current measurements were made with voltages obtained during each shock. Transthoracic resistance and capacitance were calculated using the high- and low-current methods described above. Transthoracic resistance was also calculated as the peak voltage/peak current ratio. The resulting resistance and capacitance values were analyzed. Comparisons were made using a two-sided t test that was either paired or assumed equal variance. A probability less than 0.05 was considered significant. The resistance values

8.5 APPLICATION OF THE ARX MODEL TO PREDICTION OF TRANSTHORACIC IMPEDENCE

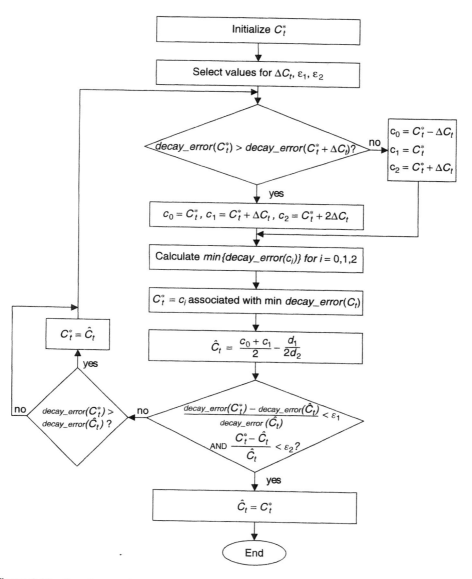

Figure 8.13 Capacitance estimation algorithm. ΔC_t = step size, ε_1 and ε_2 = convergence parameters. Based on [35].

obtained by the peak ratio and high-current methods were also compared using linear regression.

8.5.5 Swine Data Simulation

As a check for the validity of the transthoracic resistance and capacitance estimates, the scaled, digital voltage across the defibrillator, with baseline voltage subtracted, from each high-current study was converted to analog (National Instruments Lab PC+ board), and was output through a 50 Ω load resistor and a resistor–capacitor series combination possessing the estimated values. The input voltage and resistor–capacitor voltage were digitized, with a sampling frequency of 20 kHz. The voltage across the series combination was then plotted as a function of the current through the load resistor. This plot simulation was compared to a plot of the scaled, transthoracic voltage as a function as the transthoracic current during the study. Both plots were also compared to a plot of the transthoracic current times the estimated resistance versus the transthoracic current.

8.5.6 High-Current Electrode Studies

Finally, in order to isolate the electrode contribution to transthoracic impedance, the high-current impedance in 13 pairs of R2 electrodes (Model 610) was measured. Each electrode pair was positioned face-to-face, and after a 1 minute stabilization period, was subjected to at least one defibrillation shock of 200 J (Hewlett Packard 43110). In the first group of four electrode pairs, each pair was subjected to three defibrillation shocks, spaced 4 minutes apart. The remaining nine electrode pairs were divided into three groups of three electrodes each. For each of these groups, the three electrodes were also connected in series. In one group, the individual pairs were shocked once before being shocked once in series. In the other two groups, the individual pairs were shocked once after being shocked once in series. Electrode resistance and capacitance were calculated using the high-current method described above, with a voltage divider network of 2000:1 across the defibrillator and a voltage divider network of 20:1 across the electrodes. Comparisons were made using a two-sided t test that was either paired or assumed equal variance. A probability less than 0.05 was considered significant.

8.5.7 Results

During the validation experiment using the damped sinusoidal input, the five resistance values, as determined by the peak ratio method, were measured with an absolute error of $25 \pm 19\%$ (Table 8.1). This method was most accurate around values of 50 Ω. In contrast, the proposed method accurately estimated each resistance with an absolute error of $9 \pm 10\%$. Further, the proposed method also enabled capacitance measurement for the first

Table 8.1 Validation experiment: comparison of resistance estimates and capacitance estimates to true values. From [38]

True resistance (Ω)	Peak ratio resistance (Ω)	Low or high current (Ω)	True capacitance (μF)	Low-current capacitance (μF)	High-current capacitance (μF)
24.2	38.2	18	50.9	51.0	59.9
48.9	43.7	42	50.9	51.3	51.9
74.6	58.7	72	50.9	51.1	47.0
100.2	78.6	102	50.9	51.2	47.0
502	439.3	494	50.9	51.1	42.1
Absolute error %:	$25 \pm 19\%$	$9 \pm 10\%$		$0 \pm 0\%$	$10 \pm 7\%$

8.5 APPLICATION OF THE ARX MODEL TO PREDICTION OF TRANSTHORACIC IMPEDENCE

time, with mean absolute errors of 0 or 10 ± 7% for the low- and high-current measurements, respectively.

During the swine studies, the transthoracic resistance values obtained from the low-current method did not approximate those obtained from the high-current method. The mean low-current resistance at each time point ranged from 194 to 3321 Ω, whereas the mean high-current resistance at each time point ranged from 41 to 43 Ω. With each shock, the low-current resistance 30 sec after the shock decreased significantly from the resistance measurement immediately preceding the shock ($p \leq 0.002$ for the first shock, $p \leq 0.016$ for the second shock, $p \leq 0.015$ for the third shock). In contrast, the high-current resistance did not decrease significantly between shocks. The mean low-current resistance values for all five animals are plotted in Figure 8.14. The mean high-current resistance was 41 ± 5 Ω for the first shock ($n = 3$), 42 ± 7 Ω for the second shock ($n = 4$), and 43 ± 6 Ω for the third shock ($n = 4$). Due to a software error, the voltages during the first shock in one animal and the voltages during all three shocks of another animal were not recorded. To assess the accuracy of the peak ratio method, the peak voltage/peak current ratio for the peak voltage sample was also calculated.

Similarly, the transthoracic capacitance values obtained from the low- and high-current measurements were not equivalent. The mean values for each time point were an order of magnitude apart (21.5 to 95.3 μF vs. 341.6 to 410.6 μF, respectively). The low-current capacitance increased significantly only in response to the first shock ($p \leq 0.0001$), whereas the high-current capacitance did not change. The mean low-current capacitance values are plotted in Figure 8.14. With high current, the mean capacitance was 410.6 ± 52.4 μF for the first shock ($n = 3$), 381.3 ± 24.4 μF for the second shock ($n = 4$), and 341.6 ± 71.6 μF for the third shock ($n = 4$). Based on the initial mean low- or high-current measurement, the time constant, $R_t C_t$, decreased 76% from 71.1 to 16.8 msec as the current increased.

A typical high-current data pair is shown in Figure 8.15. Figure 8.15A contains a plot of the defibrillator and transthoracic voltages for swine 1003b, shock 1. Since the

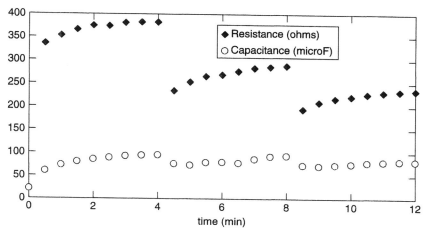

Figure 8.14 Mean transthoracic resistance and capacitance ($n = 5$) versus time for the low current square wave input. A 200 J shock was administered at $t = 0$, 4, and 8 min. For higher data resolution, the initial resistance at $t = 0$ min of 3321 Ω is omitted. From [38].

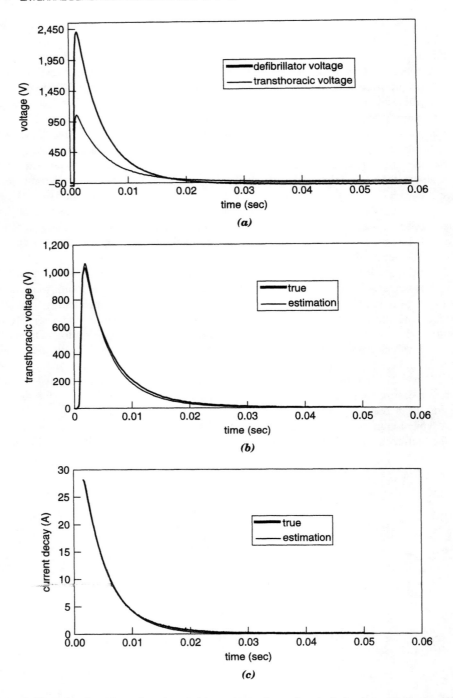

Figure 8.15 Estimation of transthoracic resistance and capacitance from typical voltage data (swine 1003, shock1). a) Defibrillator and transthoracic voltage. b) Estimated and true decimated, lowpass filtered, transthoracic voltage. Resistance = 35 Ω. c) Estimated and true transthoracic current decay. Capacitance = 355.2 μF. From [38].

8.5 APPLICATION OF THE ARX MODEL TO PREDICTION OF TRANSTHORACIC IMPEDENCE

transthoracic resistance was approximately equal to the load resistance, the transthoracic voltage peak was approximately half that of the defibrillator voltage peak. After these voltages were decimated and lowpass filtered, the transfer function between them was determined, in order to estimate the transthoracic resistance. In Figure 8.15B, the estimated transthoracic voltage is shown with the decimated, filtered transthoracic voltage. The estimated resistance was then used to model the decay current, in order to estimate the transthoracic capacitance. In Figure 8.15C, the estimated current decay is shown with the actual transthoracic current decay. For this data set, the estimated transthoracic resistance and capacitance were 35 Ω and 355.2 μF, respectively.

For all the high-current studies, each plot of transthoracic voltage versus current resulted in an elliptical plot. When the scaled defibrillator voltage was passed through the load resistor and estimated resistor–capacitor combination, the scaled transthoracic voltage was recovered. Typical plots of the observed voltage versus current and simulated voltage versus current are shown in Figure 8.16, for swine 1003, shock 1. The estimated transthoracic resistance and capacitance were 35 Ω and 355.2 μF, respectively. The discrete component values used were 35 Ω and 357 μF, respectively. Note that the observed transthoracic voltage cannot be recovered by multiplying the observed transthoracic current by the estimated resistance.

In the electrode studies, the mean high-current resistance did not change significantly with shock number, as it was 1 ± 0 Ω for the first, second, and third shocks. However, the mean capacitance from the first shock was significantly different from the means of the second and third shocks ($p \leq 0.047$ and $p \leq 0.048$, respectively). The mean high-current capacitance was 299.4 ± 29.5 μF for the first shock, 264.6 ± 30.7 μF for the second shock, and 246.13 ± 51.7 μF for the third shock. Similarly, in the electrode series studies,

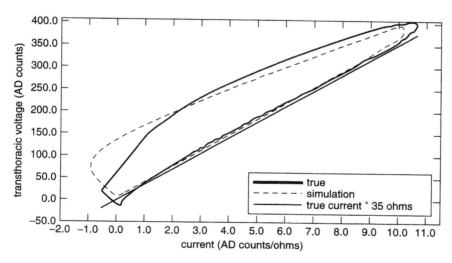

Figure 8.16 Typical swine transthoracic voltage versus current plot (swine 1003, shock 1). The simulation was obtained by passing the scaled, defibrillator voltage from a voltage divider network through a 50 Ω load resistor, 35 Ω resistor, and 357 μF capacitor. The estimated transthoracic resistance and capacitance are 35 Ω and 355.2 μF, respectively. The conversion factor for the observed data was 2.65 volts/AD counts. Without a voltage divider network, the conversion factor for the simulated data was 0.00122 volts/AD counts. The voltage for a transthoracic impedance of 35 Ω alone is also plotted (straight line). From [38].

the mean high-current resistance did not change significantly with single or series electrodes (0 ± 0 Ω versus 1 ± 1 Ω, respectively). Further, in the electrode series studies, the mean capacitance also did not change significantly (251.6 ± 26.8 μF versus 267.8 ± 14.9 μF, respectively).

8.5.8 Discussion

As shown in the validation experiment, the peak ratio method estimated accurate resistance values in the range of 50 Ω using the damped sinusoidal input; as resistance values deviated from this range, the peak ratio absolute error increased. This range of accuracy may not overlap with the range of Kerbers' original measured impedance of 78.1 ± 19.4 Ω in 19 patients [26]. As Geddes has stated, "in defibrillation studies many authors use the term *impedance* [Geddes' italics] to mean the ratio of peak voltage to peak current for the defibrillating current pulse, regardless of waveform. Although this practice violates the strictest definition of impedance, the ratio so obtained informs about the quality of the conducting pathway" [40]. In contrast, the proposed method accurately measured the large range of resistance values and enabled accurate measurement of the corresponding capacitance. Assuming a series R–C model, reactance is then just $1/(2\pi \cdot$ frequency $\cdot$ capacitance).

Transthoracic impedance does contain a substantial reactive element that can be modeled as a capacitance in series with the transthoracic resistance. In swine, this capacitance is on the order of 400 μF. This estimate was obtained by minimizing the sum of squared differences between the observed and estimated waveforms at each sample, assuming a constant capacitance. When the damped sinusoid waveform was used as an input, this capacitance caused the swine transthoracic voltage versus current plot to appear elliptical, as in earlier studies [27, 31]. The elliptical shape was recovered by applying the scaled defibrillator voltage to a load resistor and discrete resistor–capacitor combination.

Note that the simulated ellipse approximates the observed ellipse well at high currents and voltages, but is much larger than the observed ellipse at low currents and voltages. The better fit at high currents and voltages is due to minimization of the squared error in the capacitance estimation algorithm, assuming a constant capacitance. However, the lack of fit at low currents and voltages may represent the true nonlinearity of the capacitance, which has been shown to increase with current. Initially, before the current is applied, the capacitance may be zero, and may act as an open circuit. At low applied current, the capacitance may increase to approximately 60 μF, as was measured. As the current increases to its peak value, the capacitance may also increase to approximately 400 μF. However, the observed voltage at this peak current would also include a steady state voltage due to interaction between the defibrillator capacitor and the capacitor at peak current. As the current drops back to zero, the capacitance would also drop, eventually returning to zero. The apparent capacitance of zero at the beginning and end of the current pulse would explain the lack of steady state voltage due to capacitor interactions. Thus, the estimated capacitance of 400 μF is a "lumped average" of the true capacitance, which varies during the course of the applied defibrillation pulse.

As shown in the electrode studies, this high capacitance cannot be solely attributed to the electrode capacitance, which accounts for approximately 250 μF of the total transthoracic capacitance. Since the electrode capacitance component is lower than the 400 μF total capacitance, the true model of electrode, electrode–tissue interface, and tissue capacitance interactions is probably more complicated than a simple capacitance series combination. As evidenced by the electrode series studies, in which the high-current capacitance between single and series electrode pairs did not vary significantly, these capacitances in series do not follow the traditional circuit capacitance models.

Transthoracic impedance is a nonlinear function of current. In the swine studies, transthoracic resistance declined, whereas transthoracic capacitance increased, with increasing current. Together, the time constant, $R_t C_t$, decreased 76% as the current increased. Although the transmyocardial impedance was not measured, this nonlinearity is probably present in this parameter as well. While testing the utility of low energy shocks for estimating transmyocardial impedance, Leitch et al. observed a significant inverse relationship between impedance and peak current [41]. This nonlinearity has also been observed by Brewer et al. in implantable cardioverter-defibrillators with constant tilts as decreasing shock duration with increasing shock energy [42]. Due to this nonlinearity, transthoracic resistance and capacitance during defibrillation cannot be accurately estimated or predicted using low-current inputs.

Although the electrode and electrode–tissue interface certainly contribute, the dominant source of this nonlinear capacitance may be the myocardium itself. Previously, the impedance of the Purkinje fiber was modeled as a resistance in series with a parallel combination of membrane resistance and membrane capacitance [43], which can be alternatively expressed as an equivalent resistance–capacitance series combination. Similarly, the frequency response of dog ventricular myocardial fibers was modeled by Beeler and Reuter as a capacitance in parallel with a series combination of the membrane potential and membrane resistance. This parallel combination was then placed in series with a second resistance [44]. The cell membrane is known to exhibit rectification properties [45], thereby preventing current from flowing uniformly in both directions across the membrane. Additionally, the membrane undergoes a host of active processes that are triggered by the shock itself [45, 46, 47]. The time constant of the membrane response during the shock has been shown to differ from the time constant of the recovery of the membrane after the shock [48, 49]. More studies are required to characterize this capacitance in terms of its magnitude for a given population, and in other species such as man.

8.6 TRANSTHORACIC IMPEDANCE AS THE BASIS OF EXTERNAL DEFIBRILLATION WAVEFORM OPTIMIZATION

Transthoracic resistance and capacitance may determine the **defibrillation threshold,** the minimum shock for successful defibrillation. For a given defibrillator circuit and load resistor, the transthoracic current is uniquely determined by the transthoracic resistance and capacitance. In this section, work conducted by the author in 1996 is summarized [35, 36, 37, 38].

8.6.1 Application of an ARX Model to External Defibrillation Waveform Optimization

As an initial test for the hypothesis that transthoracic resistance and capacitance determine defibrillation threshold, the data from three sets of studies were combined:

1. The studies described above, in which swine transthoracic resistance and capacitance were estimated as $42 \pm 6\ \Omega$ and $374.8 \pm 55.8\ \mu F$, respectively

2. Heartstream studies in swine in which V_{50}, the total voltage required for 50% successful defibrillation, was determined [8]
3. Human studies from Kerber's group, in which the peak current and subsequent damped sinusoid defibrillation success rate were recorded for the first shocks in 100 ventricular fibrillation patients [7]

For reference purposes, Kerbers' data is reproduced in Table 8.2.

The peak damped sinusoid current for the Heartstream studies was determined from circuit analysis, based on reported and estimated values. The defibrillator components were reported as a 50 μF capacitor and 30 mH, 10 Ω inductor. A load resistor of 1 Ω was also assumed. As the 19 swine used in this study possessed a mean resistance of 42 ± 6 Ω, as determined using Geddes' method, it was assumed that the transthoracic resistance was 42 Ω. Further, it was also assumed, based on the authors' swine results above, that the transthoracic capacitance varied as 300, 350, or 400 μF.

However, V_{50} was not reported, as Heartstream could not estimate the damped sinusoid V_{50}. In order to estimate V_{50}, voltages higher than V_{50} were required in their threshold determination protocol. Unfortunately, the maximum capacitor voltage of 3200 V available in Heartstream's custom-built defibrillator did not consistently defibrillate. Therefore, damped sinusoid V_{50} was assumed to be 3200 V. Using circuit analysis, the peak current was calculated as 52, 51, or 50 A, for corresponding capacitances of 300, 350, or 400 μF, respectively. From Kerber's chart of success rates, this current range corresponds to a 50% success rate. If the assumed transthoracic capacitor were shorted out and neglected, the peak current would only be 45 A, which corresponds to an 83% success rate.

Similarly, the peak current for the monophasic truncated exponential was calculated. With a total duration of 12 msec, Heartstream calculated the monophasic truncated exponential V_{50} as 2185 ± 361 V. Using circuit analysis, the peak current was calculated as 59, 58, or 57 A, for corresponding capacitances of 300, 350, or 400 μF, respectively. This current range is not accurately estimated by Kerber's chart of success rates, which is based on damped sinusoid shocks. However, if these peak currents are scaled by the ratio of average truncated exponential to damped sinusoid currents (9.21/10.66 = 0.86), a peak current range of 51, 50, or 49 A is obtained. Again, this current range corresponds to a

Table 8.2 Peak Current Versus First Damped Sinusoid Shock Success. Reprinted from [7] with kind permission from Lippincott, Williams and Wilkins, Philadelphia, PA

Peak current (A)	Successful/ total shocks	% Successful
≤ 17	0/6	0
18–21	1/8	13
22–25	9/13	69
26–29	7/11	64
30–33	17/22	77
34–37	12/16	75
38–41	9/9	100
42–45	5/6	83
46–53	2/4	50
≥ 54	2/5	40

50% success rate. As before, if the assumed transthoracic capacitor were shorted out and neglected, the scaled peak current would only be 44 A, which corresponds to an 83% success rate.

Based on the peak current simulations described above, the hypothesis that transthoracic resistance and capacitance determine the defibrillation threshold is a reasonable one. The method based on an ARX model enables the peak current to be estimated, regardless of waveform shape. Using Kerbers' data, the success rate is then determined as a function of this peak current.

8.6.2 Clinical Studies in Dogs

More recently, Raymond Ideker's group at the University of Birmingham predicted the relative efficacy of shock waveforms for transthoracic defibrillation in dogs [50]. Ideker had previously seen early summaries of the author's work [38]. Rather than a series R-C circuit, Ideker's group used an equivalent parallel R–C circuit to model transthoracic impedance. To simplify calculations, they approximated the time constant (a function dependent on the total circuit resistance and capacitance) with Pspice (MicroSim Corporation, Irvine, CA), rather than individual estimates of resistance and capacitance. Using this time constant model, they were able to predict the relative efficacy of three monophasic waveforms (ascending ramp, descending ramp, and square wave), and the biphasic waveforms with the lowest defibrillation thresholds. In a manner similar to the comparisons in Section 8.5, comparisons between the monophasic waveforms were made using the waveform peak current, normalized to the defibrillation threshold of the square wave. The average measured and estimated peak current thresholds in 6 dogs (19 measurements each) are illustrated in Figure 8.17 for a time constant of 2.8 msec. A squared correlation coefficient of $r^2 = 0.88$ was observed ($p < 0.01$).

8.6.3 Potential Applications

Estimation of the resistance and capacitance comprising transthoracic impedance during external defibrillation has many applications. First, it may be used to determine the most

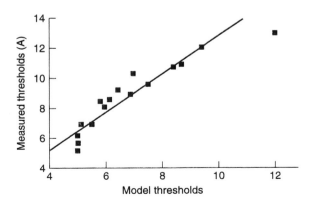

Figure 8.17 Experimental threshold plotted against the model thresholds for peak current for the monophasic waveform. Reprinted from [49] with permission from Mosby, Inc., St. Louis, MO.

advantageous waveforms for clinical and public access to transthoracic defibrillation. Second, in patients that are defibrillated more than once, the defibrillation voltage from the first shock could be used to model that patients' transthoracic resistance and capacitance. These values, in turn, could be used to calculate the minimum energy setting for the next shock and other waveform parameters for subsequent successful defibrillation. Third, this method could be extended to internal defibrillation and used to optimize subsequent shocks in patients who receive more than one shock. Finally, the proposed estimation method could be used during internal defibrillator surgery to automatically adjust the defibrillator energy setting after a single test pulse.

8.7 SUMMARY

In this chapter, we have discussed the application of linear system identification to external defibrillation. During external defibrillation, a strong current of short duration is administered across the thorax via paddles or electrode adhesive pads to convert rapid twitching of the ventricles, or ventricular fibrillation, to a slower rhythm that allows the heart to pump blood.

Ventricular fibrillation may be observed with an electrocardiogram, or ECG. An ECG is a body surface recording of the electrical activity generated by the heart, based on the differential voltage between two points on the body. During ventricular fibrillation, the conduction of cardiac excitation is transiently or permanently blocked. Conduction block may be caused by myocardial ischemia, congestive heart failure, hypothermia, or electric shock. In response, potential pacemaker cells in the ventricle that are normally silent become irritable and simultaneously rapidly discharge. The resulting rhythm is so chaotic that distinct, complete waves are indistinguishable. The ventricles no longer contract, but only twitch rapidly. Ventricular fibrillation is a type of sudden cardiac arrest, and requires immediate defibrillation to prevent sudden death.

Defibrillation refers to the application of a strong electric shock in an effort to convert fibrillation to a slower rhythm that allows the heart to pump blood. The shock may be applied externally through paddles or electrode adhesive pads positioned on the thorax using an external defibrillator, or internally through electrodes in contact with heart tissue using an internal defibrillator. Successful defibrillation stimulates cells by passing an adequate current intensity through the cells for an adequate period of time.

Depending on the components within the external defibrillator, a defibrillator shock may be delivered as a damped sinusoid, monophasic truncated exponential, or biphasic truncated exponential waveform. Although the damped sinusoid has been used since the early 1960s, more external defibrillators now use the biphasic truncated exponential waveform, with different waveform parameters, which was developed for automatic external defibrillators. A clinical need exists to determine an underlying mechanism for optimization of defibrillation waveforms, in order to efficiently allocate resources for this research.

An optimal waveform may be determined by understanding the transthoracic impedance. Because the defibrillation energy level is preset before shock administration, the transthoracic impedance uniquely determines the quantity of current that flows to the myocardium. Traditionally, impedance has been assumed to be purely resistive, and has been measured using the traditional peak ratio method. However, as early as 1976, studies demonstrated that impedance is not purely resistive, as plots of simultaneous voltage versus current during defibrillation discharge resulted in ellipses, rather than the expected straight line with approximate slope of 50 Ω.

Instead, transthoracic impedance was modeled as a resistance and capacitance in series. The ARX model discussed in Chapter 7 was used to estimate the resistive and reactive components of transthoracic impedance. This model was first validated by estimating known resistance and capacitance values in a series R–C circuit. It was then used to estimate transthoracic resistance and capacitance during external defibrillation in swine and during voltage discharges across electrode adhesive pads. This model enables the peak current of various waveform shapes to be calculated. The peak current is then used to determine defibrillation success.

Estimation of the resistance and capacitance comprising transthoracic impedance during external defibrillation has many applications. First, it may be used to determine the most advantageous waveforms for clinical and public access to transthoracic defibrillation. Second, in patients that are defibrillated more than once, the defibrillation voltage from the first shock could be used to model that patients' transthoracic resistance and capacitance. These values, in turn, could be used to calculate the minimum energy setting for the next shock and other waveform parameters for subsequent successful defibrillation. Third, this method could be extended to internal defibrillation and used to optimize subsequent shocks in patients who receive more than one shock. Finally, the proposed estimation method could be used during internal defibrillator surgery to automatically adjust the defibrillator energy setting after a single test pulse.

8.8 REFERENCES

[1]. Dubin, D. *Rapid Interpretation of EKG's*. 5th ed. Cover Publishing: Tampa, FL, 1996.
[2]. Gliner, B. E, Jorgenson, D. B., Poole, J. E., White, R. D., Kanz, K., Lyster, T. D., Leyde, K. W., Powers, D. J., Morgan, C. B., Kronmal, R. A., and Bardy, G. H. Treatment of out-of-hospital cardiac arrest with a low-energy impedance-compensating biphasic waveform automatic external defibrillator. *Biomed Instr & Tech, 32*, 631–644, 1998.
[3]. Prevost, J. L. and Batelli, F. Some effects of electric discharge on the heart of mammals. *C. R. Academy of Science, 129*, 1267, 1899.
[4]. Gurvich, N. L. and Yuniev, G. S. Restoration of heart rhythm during fibrillation by a condenser discharge. *American Review of Soviet Medicine, 4*, 252, 1947.
[5]. Lown, B., Neuman, J., Amarasingham, R., and Berkovits, B. V. Comparison of alternating current with direct current electroshock across the closed chest. *Am J Cardiol, 10*, 223–233, 1962.
[6]. Edmark, K. W. Simultaneous voltage and current waveforms generated during internal and external direct current defibrillation. *Surgical Forum, 51*, 326–333, 1963.
[7]. Kerber, R. E., Martins, J. B., Keinzle, M. G., Constantin L., Olshansky, B., Hopson, R., and Charbonnier, F. Energy, current, and success in defibrillation and cardioversion: Clinical studies using an automated impedance-based method of energy adjustment. *Circulation, 77*, 1038–1046, 1988.
[8]. Gliner, B. E., Lyster, T. E., Dillon, S. M., and Bardy, G. H. Transthoracic defibrillation of swine with monophasic and biphasic waveforms. *Circulation, 92*, 1634–43, 1995.
[9]. Bardy, G. H., Ivey, T. D., Allen, M. D., Johnson, G., Mehra, R., and Greene, H. L. A prospective, randomized evaluation of biphasic vs. monophasic waveform pulses on defibrillation efficacy in humans. *J Am Coll Cardiol, 14*, 728–733, 1989.

[10]. Diack, A. W., Welborn, W. S., Rullman, R. G., Walter, C. W., and Wayne, M. A. An automatic cardiac resuscitator for emergency treatment of cardiac arrest. *Med Instrum, 13,* 78–83, 1979.

[11]. Murphy, D. M. Rapid defibrillation: fire service to lead the way. *J Emerg Med Serv, 12,* 67, 1987.

[12]. Emergency Cardiac Care Committee and Subcommittees, American Heart Association. Guidelines for cardiopulmonary resuscitation and emergency cardiac care. I. Introduction. *JAMA, 268,* 2172–2183, 1992.

[13]. Gillum, R. F. Sudden coronary death in the United States. *Circulation, 79,* 756–765, 1989.

[14]. Eisenberg, M. S., Horwood, B. T., Cummins, R. O., Reynolds-Haertle, R., and Hearne, T. R. Cardiac arrest and resuscitation: A tale of 29 cities. *Ann Emerg Med, 19,* 179–186, 1990.

[15]. Becker, L. B., Ostrander, M. P., Barrett, J., and Knodos, G. T. Outcome of CPR in a large metropolitan area—where are the survivors? *Ann Emerg Med, 20,* 355–361, 1991.

[16]. Lombardi, G., Gallagher, E. J., and Gennis, P. Outcome of out of hospital cardiac arrest in New York City: The Pre-Hospital Arrest Survival Evaluation (PHASE) Study. *JAMA, 271,* 678–683, 1994.

[17]. White, R. D., Vukov, L. F., and Bugliosi, T. F. Early defibrillation by police: initial experience with measurement of critical time intervals and patient outcome. *Ann Emerg Med, 23,* 1099–1113, 1994.

[18]. Cobb, L. A., Fahrenbruch, C. E., Walsh, T. R., Copass, M. K., Olsufka, M., Breskin, M., and Hallstrom, A. P. Influence of cardiopulmonary resuscitation prior to defibrillation in patients with out-of-hospital ventricular fibrillation. *JAMA, 281,* 1182–1188, 1999.

[19]. Ornato, J. P., Craren, E. J., Gonzalez, E. R., Garnett A. R., McClung, B. K., and Newman, M. M., Cost-effectiveness of defibrillation by emergency medical technicians. *Am J Emerg Med, 6,* 108–112, 1988.

[20]. Goldman, L. Cost awareness in medicine. In *Harrison's Textbook of Internal Medicine,* Isselbacher, K. J., Braunwald, E., and Wilson, J. D., eds., 13th ed. McGraw-Hill: New York, 1994, pp. 38–42.

[21]. Wolbrink, A., and Borrillo, D. Airline use of automatic external defibrillators: shocking developments. *Aviat Space Env Med, 70,* 87–88, 1999.

[22]. Pereira, J. For the fearful: Portable defibrillators. *Wall Street Journal,* Sept. 1, 1999. p. B1.

[23]. IMS Health, *Hospital Supply Index.* IMS America: Plymouth Meeting, PA, 1997.

[24]. Geddes, L. A., Tacker, W. A., Jr., Cabler, P., Kidder, H., and Gothard, R. The impedance of electrodes used for ventricular defibrillation. *Med Instr, 9,* 177–178, 1975.

[25]. Geddes, L. A., Tacker, W. A., Jr., Schoenlein, W., Minton, M., Grubbs, S., and Wilcox, P. The prediction of the impedance of the thorax to defibrillation current. *Med Instr, 10,* 159–162, 1976.

[26]. Kerber, R. E., Kouba, C., Martins, J., Kelly, K., Low, R., Hoyt, R., Ferguson, D., Bailey, L., Bennett, P., and Charbonnier, F. Advance prediction of transthoracic impedance in human defibrillation and cardioversion: Importance of impedance in determining the success of low-energy shocks. *Circulation, 70,* 303–308, 1984.

[27]. Machin, J. Thoracic impedance of human subjects. *Med Biol Engr Comput, 16,* 169–178, 1978.

[28]. ———, DF39—Automatic external defibrillators and remote-control defibrillators. *Association for the Advancement of Medical Instrumentation,* 1993.

[29]. Hewlett-Packard. *43100A Defibrillator/Monitor with Recorder Manual,* 10th ed. Hewlett-Packard: McMinnville, OR, 1990, pp. 2–6.

[30]. Hewlett-Packard, *CodeMaster Defibrillator/Monitors—XL+ (M1722A/B), XL (M1723A/B) Service Manual,* 3rd ed. Hewlett-Packard: McMinnville, OR, 1993, pp. 5–20.

[31]. Dahl, C. F., Ewy, G. A., Ewy, M. D., and Thomas, E. D. Transthoracic impedance to direct current discharge: Effect of repeated countershocks. *Med Instr, 10,* 151–154, 1976.

[32]. Tang, A. S., Yabe, S., Wharton, J. M., Dolker, M., Smith, W. M., and Ideker, R. E. Ventricular defibrillation using biphasic waveforms: The importance of phasic duration. *J Am Coll Cardiol, 13,* 207–214, 1989.

[33]. Lawrence, J. H., Brin, K. P., Halperin, H. R., Platia, E. V., Tsitlik, J. E., Levine, J. H., and Guarnieri, T. The characterization of human transmyocardial impedance during implantation of the automatic internal cardioverter defibrillator. *PACE, 9,* 745–755, 1986.

[34]. KenKnight, B. H., Eyuboglu, B. M., and Ideker, R. E. Impedance to defibrillation countershock: Does an optimal impedance exist?, *PACE, 18,* 2068–2087, 1995.

[35]. Baura, G. D. *Method and Apparatus for Electrode and Transthoracic Impedance Estimation.* U.S. Patent 6,016,445. January 18, 2000.

[36]. Baura, G. D. *Method and Apparatus for High Current Electrode, Transthoracic and Transmyocardial Impedance Estimation.* U.S. Patent 6,058,325. May 2, 2000.

[37]. Baura, G. D. *Method and Apparatus for High Current Electrode, Transthoracic and Transmyocardial Impedance Estimation.* U.S. Patent 6,253,103. June 26, 2001.

[38]. Baura, G. D. Reevaluation of impedance to defibrillation countershock. Manuscript submitted to *IEEE Trans Biomed,* August, 1996.

[39]. Reklaitis, G. V., Ravindran, A., and Ragsdell, K. M. *Engineering Optimization: Methods and Applications.* Wiley: New York, 1983. pp. 48–52.

[40]. Geddes, L. A. Electrodes for transchest and ICD defibrillation and multifunctional electrodes. In *Defibrillation of the Heart: ICDs, AEDs, and Manual.* Edited by W. Tacker, Jr. Mosby–Year Book: St. Louis, 1994, p. 104.

[41]. Leitch, J. W., Yee, R., Klein, G. J., and Jones, D. L. Utility of low energy test shocks for estimation of cardiac and electrode impedance with implantable defibrillators. *PACE, 13,* 410–416, 1990.

[42]. Brewer, J. E., Tvedt, M. A., Adams, T. P., and Kroll, M. W. Low voltage shocks have a significantly high tilt of the internal electric field than do high voltage shocks. *PACE, 18*[pt, II], 214–220, 1995.

[43]. Fozzard, H. A. Membrane capacity of the cardiac Purkinje fiber. *J Physiol, 182,* 255–267, 1966.

[44]. Beeler, G. W. and Reuter, H. Voltage clamp experiments on ventricular myocardial fibers. *J Physiol, 207,* 165–190, 1970.

[45]. Cohen, I. S., Falk, R. T., and Kline, R. P. Voltage-clamp studies on the canine Purkinje strand. *Proc R Soc Lond B Biol Sci, 217,* 215–236, 1983.

[46]. Luo, C. H. and Rudy, Y. A dynamic model of the cardiac ventricular action potential. I. Simulations of ionic currents and concentration changes. *Circ Res, 74,* 1071–1096, 1994.

[47]. DiFrancesco, D. and Nobel, D. A model of cardiac electrical activity incorporating ionic pumps and concentration changes. *Philos Trans R Soc B Biol Sci, 307,* 353–398, 1985.

[48]. Zhou, X., Rollins, D. L., Smith, W.M. Responses of the transmembrane potential of myocardial cells during a shock. *J Cardiovasc Electrophysiol, 6,* 252–263, 1995.

[49]. Knisley, S. B., Smith, W. M., and Ideker, R. E. Effect of field stimulation on cellular repolarization in rabbit myocardium. Implications for reentry induction. *Circ Res, 70,* 707–715, 1992.

[50]. White, J. B., Walcott, G. P., Wayland, J. L., Jr., Smith, W. M., and Ideker, R. E. Predicting the relative efficacy of shock waveforms for transthoraic defibrillation in dogs. *Ann Emerg Med, 34,* 309–320, 1999.

[51]. Gregory, T. K., and Stevenson, J. W. Medical Thermometer. *U.S. Patent 5,632,555.* May 27, 1997.

[52]. Welch Allyn, Inc. Skaneateles Falls, NY. San Diego facility test data, 1999.

192 EXTERNAL DEFIBRILLATION WAVEFORM OPTIMIZATION

Further Reading

Electrocardiography: Guyton, A. C. and Hall, J. E. Textbook of Human Physiology. 10th edition. Saunders: Philadelphia, 2000.

Defibrillation: Tacker, W. A., Jr., ed. *Defibrillation of the Heart: ICDs, AEDs, and Manual.* Mosby, St. Louis, 1994.

8.9 DIGITAL THERMOMETRY EXERCISES

Engineer D at Startup Company has been asked to upgrade an oral digital thermometer for faster measurement. A competitive product, the Welch Allyn SureTemp thermometer, measures oral temperature within 4 seconds (Figure 8.18). The Welch Allyn thermometer heats the temperature probe, in order to minimize the time interval between the rise of the initial probe temperature to steady state temperature within the mouth. A predictive algorithm then enables an estimate of steady state temperature to be made within 4 seconds [51]. Engineer D believes that he can upgrade Startup Company's older technology that requires about 38 seconds for the probe temperature to reach steady state (three time constants) in the mouth. Rather than initially heating the probe, he wishes to use linear sys-

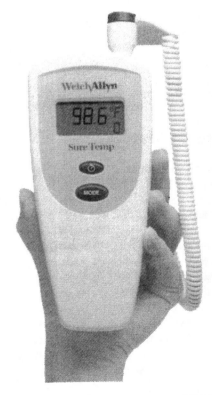

Figure 8.18 Welch Allyn Sure Temp Digital Thermometer. Courtesy of Welch Allyn, Inc. Skaneateles Falls, NY.

tem identification to predict at 10 seconds the steady state temperature at 38 seconds. We recreate his analysis below.

8.9.1 Clinical Data

The temperature file in Figure 8.19 was recorded by Engineer D from an employee at Startup Company. Three hundred seconds of data were recorded with a sampling frequency of 1 Hz and resolution of 12 bits/sample; the data were saved in units of °F. A similar temperature file from the Sure Temp thermometer is also illustrated in Figure 8.19 for comparison. These data were generously provided by Welch Allyn, Inc. [52].

Please move the *temperature.dat* file from ftp://ftp.ieee.org/uploads/press/baura to your *Matlab\work* directory. Input the data using "load temperature.dat -ascii."

8.9.2 Useful Matlab Functions

The following functions from the Signal Processing Toolbox are useful for completing this exercise: arx, idsim, and present.

8.9.3 Exercises

1. Subtract the baseline value of temperature before further processing. Why is this necessary?

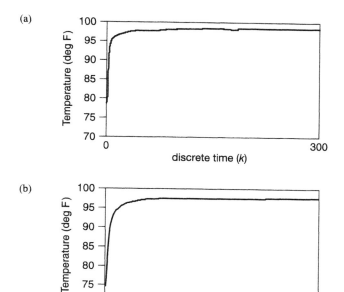

Figure 8.19 Temperature plots. a) Welch Allyn Sure Temp with initial temperature of 78.7 °F. b) Older thermometer with initial temperature of 74.5°F. The sampling frequency = 1 Hz. From [52].

Table 8.3 ARX Model choices

N	M	D
3	2	1
5	3	1
10	5	1

2. Create the input signal for an ARX model. Should the input be an impulse or step function? (Hint: How long is heat transferred orally to the probe?)
3. a) Determine the optimum values of N, M, and D in an ARX model from the choices in Table 8.3. b) What measures are used to choose the optimum model order? c) How do you explain the results?
4. a) Check your calculations by simulating $y(t) = 1 - e^{-0.8t}$ for $0 \leq t < 1$ in sample increments of 0.00333 (300 samples total). Identify this waveform with $N = 1$, $M = 1$, and $D = 1$. b) How do the values obtained for the measures of validity with this simulation differ from the values obtained in Step 3?
5. a) Identify the first 10 samples of the temperature data with $N = 3$, $M = 2$, and $D = 1$. b) Repeat the identification for the first 20 samples of the temperature data. c) How accurate are the steady state predictions based on 10–20 samples of early data? d) How can these results be explained?
6. Name two reasons why linear prediction of steady state oral temperature is not recommended when based on a few data points.

9

NONLINEAR SYSTEM IDENTIFICATION

In Chapter 7, we assumed that the relationship between a single input, $u(k)$, and single output, $y(k)$, was linear and time-invariant. With these assumptions, the system operator was an impulse response, $h(k)$. In this chapter, we wish to relax these constraints. First, rather than studying a single input, single output system, let us study the more generalized **multiple input, u(k), multiple output, y(k), (MIMO) system.** Second, let us allow the system operator to be nonlinear but time-invariant (Figure 9.1). With these assumptions in place, we may describe the system operator as an **artificial neural network** (ANN).

As its name implies, an artificial neural network refers to a mathematical model of human brain processing. Indeed, in the 1940s, physiologists and electrical engineers worked together toward this goal. Over time, it was discovered that these models did not simulate human neuron processing. However, because these models are very useful for the major applications of **function approximation** (modeling and prediction), **pattern classification,** and **combinatorial optimization** (minimization of a performance function in which the solution involves multiple steps), engineers continue to investigate these models.

In this chapter, we discuss some of the basic neural network architectures for function approximation and pattern classification. These architectures may be classified by their **learning algorithms,** or type of parameter identification. The two main classes of learning algorithms are **supervised** and **unsupervised learning.** With supervised learning, the network is trained by presentation of input and output pairs. With unsupervised learning, the network automatically classifies input data according to the distribution of input data and their relations. The application of combinatorial optimization, which includes the **Hopfield network,** is beyond the range of our discussion. The use of neural networks, often with fuzzy logic, in **artificial intelligence** is also beyond the range of our discussion. References for these topics are provided at the end of the chapter.

9.1 HISTORICAL REVIEW

A biological **central** (brain) **neuron** consists of a **soma,** the cell body with its nucleus, and numerous projections. The majority of these projections are **dendrites** that receive in-

196 NONLINEAR SYSTEM IDENTIFICATION

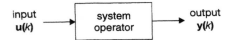

Figure 9.1 Nonlinear time-invariant, multiple input, multiple output system.

puts from hundreds of other neurons. Typically, only one **axon** projects from the soma and communicates to the dendrite of another neuron through a **synapse**. When the cell is excited above its **resting potential** (voltage) to a **threshold** (voltage), an **action potential** (see Section 8.1.1) is transmitted across the axon to its the **presynaptic terminal**. Within the presynaptic terminal of the axon, the action potential causes proteins called **neurotransmitters** to be released. These drift across the synaptic junction and initiate depolarization of the **postsynaptic membrane** of the corresponding dendrite (Figure 9.2).

9.1.1 McCullough–Pitts Model

In 1943, Warren McCullough and Walter Pitts presented the first mathematical model of a single idealized neuron [1]. In this simple **McCullough–Pitts model**, each input, u_i, is weighted by some value, w_i, that represents the synaptic strength. The weighted inputs are summed, and if a **threshold**, θ, is reached, a response, x, is generated. The response is modulated by a nonlinear **activation function**, $g(x)$, to obtain the output, y, of the neuron. In this model, the threshold and weights are assumed to have reached steady state.

An illustration of this process is given in Figure 9.3. The expression for the output is

$$y = g\left(\sum_{i=1}^{M} u_i w_i - \theta\right) \tag{9.1}$$

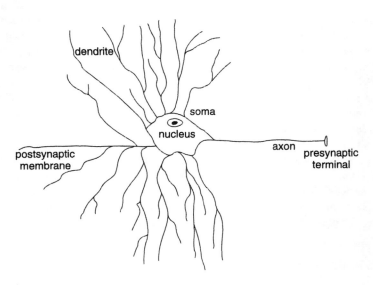

Figure 9.2 Schematic of typical central neuron, with labeling of one of the numerous postsynaptic membranes and dendrites.

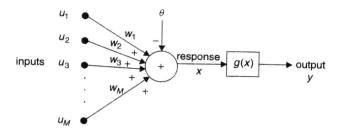

Figure 9.3 McCullough–Pitts model.

This simple open-loop system attracted the interest of many researchers, who developed next-generation systems that incorporated learning and adaptation.

9.1.2 Rosenblatt Perceptron

Frank Rosenblatt's **perceptron** model, which was first described in 1958 [2], incorporated supervised learning into a model of the neuron. Rosenblatt used the hard-limited nonlinearity of a step function,

$$g(x) = \begin{cases} 1, & x > 0 \\ 0, & x \leq 0 \end{cases} \tag{9.2}$$

to condition the inputs. He also set the threshold to 1 and added a bias weight, w_0, of unspecified value θ. The perceptron was trained to perform a pattern recognition task.

A deficiency of the early perceptron was that it failed to pass a benchmark test called **linear separation**. Linear separation is the differentiation between two linearly separable sets of patterns. The perceptron could not be used to perform an **exclusive-or** (XOR) function (Figure 9.4). That is, the four possible combinations of two binary inputs could not be classified, or linearly separated, for the XOR function. It was severely criticized by Minsky and Papert [3], who believed that claims about the perceptron were exaggerated. Eventually, more advanced perceptrons were designed that could solve the XOR problem.

During training of the perceptron, a stimulus excites its inputs and produces an output. The sample, k, represents one iteration in the learning process during which a particular pair of inputs and output are presented to the network. This output, $\hat{y}(k)$, is compared to the desired output, $y(k)$, and their difference, the error signal, $\varepsilon(k)$, is used to adjust the value of

$$\mathbf{w}(k+1) = \mathbf{w}(k) + \rho\{y(k) - g[\mathbf{w}(k)^T\mathbf{u}(k)]\}\mathbf{u}(k) \tag{9.3}$$

where ρ = **learning rate**,

$$\mathbf{u}(k) = [u_1(k)\ u_2(k)\ \ldots\ u_M(k)]^T \tag{9.4}$$

$$\mathbf{w}(k) = [w_1(k)\ w_2(k)\ \ldots\ w_M(k)]^T \tag{9.5}$$

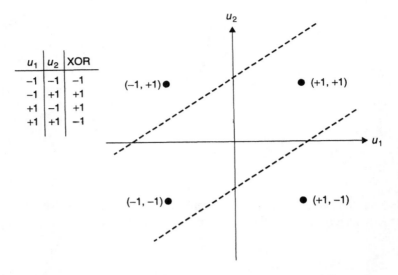

Figure 9.4 Separation of the XOR function is not possible with a single line.

As in Chapter 3, **u**(k) represents the input vector of order M and **w**(k) represents the weight vector. With this perceptron learning rule, it can be proven that the combination of desired outputs, weights, and inputs converge to a size larger than some margin, ϑ:

$$y(k)\sum_{i=1}^{M} w_i(k)u_i(k) > \vartheta \qquad (9.6)$$

9.1.3 Adaline and Madeline

Bernard Widrow, whose adaptive filtering algorithm was discussed in Chapter 3, also developed neural network models. In the late 1950s, Widrow developed a single neuron model called **ADALINE** (adaptive linear neuron) and a multiple neuron model called **MADALINE** (many ADALINE) [4]. As shown in Figure 9.5, the function used was a *sgn* function, where

$$sgn(x) = \begin{cases} +1, & x \geq 0 \\ -1, & x < 0 \end{cases} \qquad (9.7)$$

In ADALINE, the error signal is the difference between the analog response, $x(k)$, and desired output, rather than the difference between the output and desired output. The weights were adapted using the least mean squares algorithm discussed in Chapter 3. Note that although the output of ADALINE is nonlinear, the learning algorithm is linear. The first commercial application of neural networks was an implementation based on ADALINE developed by Robert Lucky to adaptively equalize telephone channels [5].

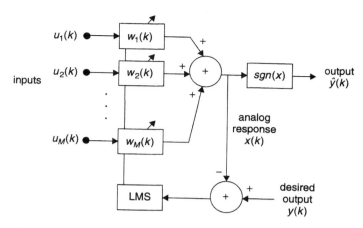

Figure 9.5 Adaptive threshold element (ADALINE). LMS = least mean squares algorithm.

9.2 SUPERVISED MULTILAYER NETWORKS

All the architectures described above are limited in their ability to "learn" a nonlinear function representing the relationship between input and output. Even in ADALINE, which has been the most successfully implemented, the nonlinearity is not part of the learning algorithm. This limitation was resolved by choosing a continuous activation function, and minimizing mean squared error using a method similar to the algorithms in Chapters 3 and 7. This solution was first described by Paul Werbos [6] in 1974, and was rediscovered independently by David Rumelhart [7] and David Parker [8]. Because the system error is propagated back from the outputs of the network to the inputs, it is referred to as **back propagation.**

The network architecture that utilizes this algorithm is a **multilayer feedforward network.** Feedforward networks easily model nonlinear systems, possess a high degree of generalization, and are suitable for parallel processing. However, these networks are slow to train, as the parameter estimates converge slowly. They also cannot be easily analyzed for network behavior. In other words, if the network does not generalize well, it cannot be easily modified, but must be retrained.

9.2.1 Multilayer Feedforward Networks

A multilayer feedforward network contains one layer of inputs, $\mathbf{u}(k)$, one or more hidden layers of neurons, $\mathbf{x}(k)$, and one output layer of neurons, $\hat{\mathbf{y}}(k)$. Each weight between the input and hidden layers is represented by $w_{lm}(k)$, a weight connecting hidden neuron $x_l(k)$ to input $u_m(k)$. Similarly, each weight between the hidden and output layers is represented by $W_{nl}(k)$, a weight connecting output $\hat{y}_n(k)$ to hidden neuron $x_l(k)$. The weights are fully connected between layers. The desired outputs at iteration k are represented by $\mathbf{y}(k)$. A network with one hidden layer is illustrated in Figure 9.6. Please note that common nomenclature for the total number of layers is based on the sum of the hidden and output layers, and excludes the input layer. Therefore, the network in Figure 9.6 contains two layers.

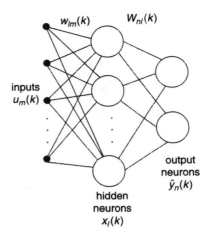

Figure 9.6 Multilayer feedforward network with two layers. An activation function, $g(x)$, and summation are contained within each neuron.

The processing at each hidden neuron may be represented as

$$x_l(k) = g[f_l(k)] = g\left[\sum_{m=1}^{M} w_{lm}(k)u_m(k)\right] \qquad (9.8)$$

The processing at each output may be represented as

$$\hat{y}_n(k) = g[f_n(k)] = g\left[\sum_{l=1}^{L} W_{nl}(k)x_l(k)\right] = g\left\{\sum_{l=1}^{L} W_{nl}(k)g\left[\sum_{m=1}^{M} w_{lm}(k)u_m(k)\right]\right\} \qquad (9.9)$$

The thresholds have been omitted in these definitions, but could easily be added to the summation terms.

9.2.2 Back Propagation

The error, $\varepsilon(k)$, in this network may be described as

$$\varepsilon(k) = \mathbf{y}(k) - \hat{\mathbf{y}}(k) \qquad (9.10)$$

In a manner similar to our discussion in Chapter 7, let us define a squared error cost function,

$$\xi[\mathbf{w}(k)] = \frac{1}{2}\sum_{n=1}^{N}[y_n(k) - \hat{y}_n(k)]^2 \qquad (9.11)$$

where $\mathbf{w}(k)$ contains the weights from both layers, w_{lm} and W_{nl}.
Substitution of Eq. (9.9) yields

$$\xi[\mathbf{w}(k)] = \frac{1}{2}\sum_{n=1}^{N}\left\{y_n(k) - g\left[\sum_{l=1}^{L}W_{nl}(k)g\left[\sum_{m=1}^{M}w_{lm}(k)u_m(k)\right]\right]\right\}^2 \qquad (9.12)$$

Since Eq. (9.12) is clearly a continuous differentiable function of every weight, we may use a **gradient descent method** to learn appropriate weights (recall Section 3.2). With this method, we change each weight by an amount proportional to the gradient of the cost function at the present location. Let us calculate the partial derivatives of the cost function to obtain our learning rules.

For the hidden-to-output connections, the gradient descent method results in

$$\Delta W_{nl}(k) = \frac{\partial \xi[\mathbf{w}(k)]}{\partial W_{nl}(k)} = -\sum_{n=1}^{N}[y_n(k) - \hat{y}_n(k)]g'[f_n(k)]x_l(k) \qquad (9.13)$$

Similarly, for the input-to-hidden connections, the gradient descent method results in

$$\Delta w_{lm}(k) = \frac{\partial \xi[\mathbf{w}(k)]}{\partial w_{lm}(k)} = \frac{\partial \xi[\mathbf{w}(k)]}{\partial x_l(k)} \cdot \frac{\partial x_l(k)}{\partial w_{lm}(k)} \qquad (9.14)$$

$$\Delta w_{lm}(k) = -\sum_{n=1}^{N}\{[y_n(k) - \hat{y}_n(k)]g'[f_n(k)]W_{nl}(k)\} \cdot g'[f_l(k)]u_m(k) \qquad (9.15)$$

We may then construct our learning rules as

$$W_{nl}(k+1) = W_{nl}(k) + \rho \Delta W_{nl}(k) \qquad (9.16)$$

$$w_{lm}(k+1) = w_{lm}(k) + \rho \Delta w_{lm}(k) \qquad (9.17)$$

where ρ is the learning rate.

Eqs. (9.13), (9.15)–(9.17) constitute the back propagation algorithm. These equations back propagate the error within the system until it is minimized. To start, the weights are initialized to random values. During each iteration, k, of training, an input–output pair, or pattern, is presented to the network. As the weights are updated, the cost function decreases, adapting to the local gradient. If the patterns are presented in random order, the path through the control space becomes stochastic, allowing wider exploration of the control surface. Note that the update rules are local, requiring minimal storage. To simplify the nomenclature, with the exceptions of Sections 9.2.5 and 9.5.1, we omit using subscripts to differentiate the various input–output pairs, $\{\mathbf{x}(k), \mathbf{y}(k)\}$, where P is the total number of input–output pairs.

9.2.3 Activation Function

In order to use the back propagation algorithm, the activation function must be differentiable. Further, we would like it to saturate at both extremes as $0 \rightarrow 1$ or $-1 \rightarrow +1$. The following functions meet these criteria:

$$g(x) = \frac{1}{1 + \exp(-2\beta x)} \qquad (9.18)$$

$$g(x) = \frac{1}{2}\tanh \beta x \qquad (9.19)$$

where tanh = hyperbolic tangent.

Typically, the **steepness parameter,** β, is set to 1 or 0.5, for Eqs. (9.18) and (9.19). The derivatives of these functions are readily expressed in terms of the functions themselves. The derivative of Eq. (9.18) is

$$g'(x) = 2\beta g[1 - g(x)] \tag{9.20}$$

The derivative of Eq. (9.19) is

$$g'(x) = \beta[1 - g^2(x)] \tag{9.21}$$

9.2.4 Necessary Number of Hidden Layers

Up to this point, we have considered networks with one hidden layer. Our analysis may be easily generalized to more than one hidden layer. However, we need to know how many hidden layers are necessary for a given desired accuracy.

In 1988, George Cybenko proved that at most two hidden layers are required, with arbitary accuracy obtained given enough units per layer [9]. Cybenko [10] and Hornik, et al. [11] also proved that one hidden layer is enough to approximate any continuous function. A less rigorous proof for two hidden layers was also presented by Lapedes and Farber. They argued that any "reasonable" function can be represented by a linear combination of localized "bumps." A bump refers to a multidimensional parabolic function with a global maximum that is nonzero only in a small region of its domain. By demonstrating that each bump can be constructed with two hidden layers, they proved that at most two hidden layers are required to represent a "reasonable" function [12].

The choice of the number of hidden neurons is discussed in Section 9.5.

9.2.5 Selection of Input Variables

Performance of the feedforward network depends heavily on the selected input variables. The two approaches to input determination are **extraction of input variables** and **selection of input variables.**

With extraction, the original set of input variables is optimized. A common method for extraction is **principal component analysis,** also known as the **Karhunen–Loeve transformation** [13]. This optimal transformation enables decorrelation of the input variables.

To utilize this technique, the input correlation matrix, $\underline{R}_i(k)$, of the inputs is first estimated as

$$\underline{R}_i(k) = \frac{1}{P} \sum_{i=1}^{P} \mathbf{u}_i(k)\mathbf{u}_i^T(k) \tag{9.23}$$

Let $\lambda_i(k)$ and $\mathbf{q}_i(k)$, $1 \leq i \leq M$, denote the eigenvalues and eigenvectors of the matrix, $\underline{R}_i(k)$, respectively, which are arranged in descending order of the eigenvalues such that $\lambda_i(k) \geq \ldots \geq \lambda_M(k)$. We then construct the transformation matrix, $\underline{T}(k)$, whose rows are the eigenvectors, as

$$\underline{T}(k) = \begin{bmatrix} \mathbf{q}_i^T(k) \\ \vdots \\ \mathbf{q}_M^T(k) \end{bmatrix} \tag{9.24}$$

The Karhunen–Loeve transformation of each training pattern, $\mathbf{u}_i^*(k)$, is then just

$$\mathbf{u}_i^*(k) = \underline{\mathbf{T}}(k)\,\mathbf{u}_i(k),\ 1 \le i \le P \qquad (9.25)$$

With a few equations, we have transformed the input vectors in the direction of their eigenvectors. We have decorrelated the input vectors, minimized the mean squared error between the transformed and original inputs, and minimized the total entropy of the inputs.

In contrast, with selection, the second approach to determination of input variables, only the relevant input variables are utilized. It is desirable that the selected inputs be orthogonal to each other. Accordingly, Shigeo Abe's group deleted input variables that were expressed by the linear combination of other input variables [14]. Later, Abe's group analyzed the sensitivity of a trained network, that is, the change in output when some input was changed. They then deleted input variables that were redundant [15].

9.2.6 Other Supervised Networks

A multilayer feedforward network assumes that the data moves from input to output. In a similar network, the **recurrent network,** data moves in both directions, giving rise to feedback (Figure 9.7a). In another network, the **radial basis function network,** the weights in the feedforward network are replaced by Gaussian functions at each hidden neuron (Figure 9.7b). This summation at the output neuron is equivalent to a linear output function. References for both types of networks are provided at the end of the chapter.

9.2.7 Feedforward Network Example

As an example of a feedforward network, let us discuss the ability to recognize handwritten zip codes from the U.S. mail. A network to solve this problem was developed by Le Cun, et al. of AT&T Bell Laboratories [16]. The example database used for training and testing consisted of 9298 isolated numerals digitized from handwritten zip codes. Typical examples are shown in Figure 9.8. Note the large variety of sizes, writing styles, instruments, and writing quality. Many of the digits would be difficult for a human to classify. 7291 sample digits were used in training; 2007 digits were used in testing. Each digit was first normalized to fill an area consisting of 40 × 60 black and white pixels. These patterns were then reduced to 16 × 16 pixel images using a linear transformation that mapped the grey levels of the image into a range of $\{-1, +1\}$.

The network used possessed three hidden layers—J1, J2, and J3—and an output layer (Figure 9.9). J1, the first hidden layer, consisted of 768 neurons. These neurons were arranged as 12 feature detectors, each composed of 64 neurons. Each group of 64 neurons was arranged in an 8 × 8 square; each neuron received information only from a 5 × 5 contiguous square of pixels of the original input. All 64 neurons within a feature detector possessed the same 25 weight values. The location of each 5 × 5 square shifted by 2 pixels between neighbors in the hidden layer. These extra arrangements enabled each feature detector, composed of 64 neurons, to detect one feature with high resolution.

Normally, all 256 inputs would be fully connected to all 768 hidden neurons, requiring 196,608 weights. Using special weighting rules, the number of connections were reduced

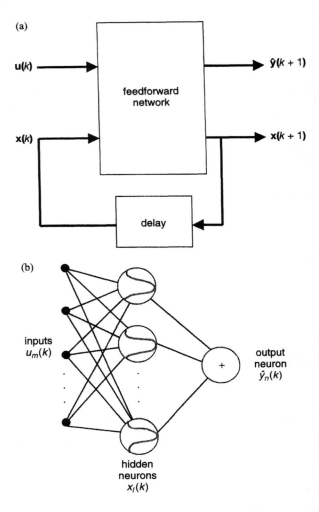

Figure 9.7 Other supervised networks. (a) Recurrent network. (b) radial basis function network.

to 19,968. As a result, only 768 thresholds and 300 weights remained as free parameters in the first hidden layer during training.

The second hidden layer, J2, similarly consisted of 192 hidden neurons, arranged as 12 feature detectors. Each feature detector was composed of 16 (4 × 4) neurons each. Each neuron received information only from groups of 25 neurons, arranged as 5 × 5 receptive fields in the previous layer. Using special weighting rules, the number of connections was reduced from 38,592 for full connection to 2592. As a result, only 192 thresholds and 200 weights remained as free parameters in the second hidden layer.

The third hidden layer, J3, consisted of 30 neurons, receiving information from all 192 neurons of the second layer. With full connections, J3 contained 5760 weights and 30 thresholds. The 10 output neurons, representing 10 digits, were fully connected to J3, and contained 300 weights and 10 thresholds.

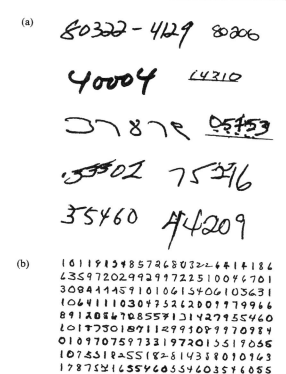

Figure 9.8 (a) Examples of handwritten zip codes and (b) normalized digits from the training/test database. Reprinted from [16] with kind permission from MIT Press, Cambridge, MA.

This network was trained using back propagation. The training set was presented to the network 23 times in random order, and assumed that each digit occurred with the same probability in the training set. The weights and thresholds were initialized with random values. The outputs were continuous values within the range $\{-1, +1\}$, rather than the typical ± 1 extreme values used in classification. This prevented the weights from growing indefinitely during training.

After training, only 10 digits were misclassified (0.14%). However 102 mistakes (5%) were reported during testing. By rejecting some of the more illegible test patterns (12.1%), the misclassification during testing was decreased to 1%. Simpler networks with fewer feature mapping levels were also evaluated, but produced inferior results.

9.3 UNSUPERVISED NEURAL NETWORKS: KOHONEN NETWORK

Unlike supervised neural networks, unsupervised networks possess the ability to self-organize and independently estimate their own parameters. In the next two sections, we consider two architectures: the **Kohonen network** and the **adaptive resonance theory network**.

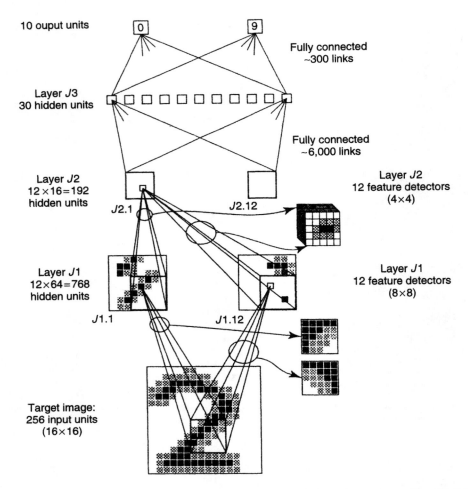

Figure 9.9 Architecture of multilayer feedforward network for handwritten character recognition. Reprinted from [16] with kind permission from MIT Press, Cambridge, MA.

9.3.1 General Concepts

The Kohonen network possesses a similar structure to a single-layer feedforward network. Each input, $u_m(k)$, is fully connected to each output neuron, $\hat{y}_n(k)$. As before, weights are labeled $w_{nm}(k)$, signifying the connection of input m to output n. However, Teuvo Kohonen proposed that the output neurons be arranged in two dimensions. Clusters of nearby neurons around a central neuron, $\hat{y}_c(k)$, are termed neighborhoods, $N_c(k)$. For example, the neurons in Figure 9.10 are arranged as a 5 × 5 dimensional array. A neighborhood may be defined as a 3 × 3 array subset array around a central array.

Rather than using back propagation, the Kohonen network relies on **competitive learning** to determine the value of its weights. At each iteration, the neuron, $\hat{y}_c(k)$, which minimizes the following **Euclidean distance** relationship, is found:

9.3 UNSUPERVISED NEURAL NETWORKS: KOHONEN NETWORK

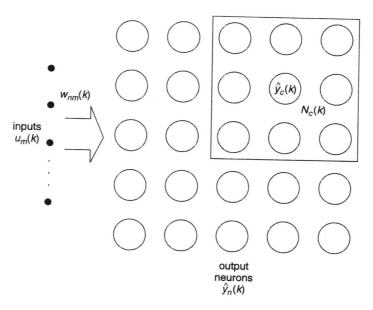

Figure 9.10 Kohonen network with 25 output neurons arranged in a 5 × 5 array. To simplify the illustration, the fully connected weights are represented as an open arrow. The network possesses the same weight connections as a one-layer feedforward network. The neighborhood, $N_c(k)$, is a 3 × 3 array.

$$\|\mathbf{u}(k) - \mathbf{w_j}(k)\| = \sqrt{\sum_{i=1}^{m}[u_i(k) - w_{ji}(k)]^2} \quad (9.26)$$

where

$$\mathbf{w_j}(k) = [w_{j1}(k)\ w_{j2}(k) \ldots w_{jm}(k)]^T \quad (9.27)$$

The weights associated with the localized neighborhood of neurons around the central neuron are then updated according to the gradient descent method. This self-organization is intended to mimic localization of brain function. For example, brain utilization of glucose during external activity has been mapped using positron emission tomography, and has been shown to be localized to specific brain regions.

In general, the training time of the Kohonen network is shorter than that of a multilayer network. However, the generalization ability of the Kohonen network is lower because the class boundaries are not determined precisely without desired outputs.

9.3.2 Competitive Learning Algorithm

The cost function used in this algorithm is a variation of the Euclidean distance:

$$\xi[\mathbf{w_j}(k)] = [\mathbf{u}(k) - \mathbf{w_j}(k)]^2 \quad (9.28)$$

As discussed above for mimicry of brain function, we wish to minimize the distance between inputs and weights. Weight updates are based on the gradient descent method, in which we set the gradient equal to zero:

$$\Delta w_{nm}(k) = \frac{\partial \xi[\mathbf{w_j}(k)]}{\xi w_{nm}(k)} = -2[\mathbf{u}(k) - \mathbf{w_j}(k)] = 0 \qquad (9.29)$$

$$\Delta w_{nm}(k) = \mathbf{u}(k) - \mathbf{w_j}(k) = 0 \qquad (9.30)$$

Therefore, for a neighborhood of neurons, $N_c(k)$, weights are updated as

$$w_{nm}(k+1) = w_{nm}(k) + \rho(k)\Delta w_{nm}(k) \qquad (9.31)$$

$$w_{nm}(k+1) = w_{nm}(k) + \rho(k)[\mathbf{u}(k) - \mathbf{w}(k)], \quad \hat{y}_n(k) \text{ within } N_c(k) \qquad (9.32)$$

where $\rho(k)$ is the **adaptive learning rate.** Weights associated with neurons not in the neighborhood remain the same:

$$w_{nm}(k+1) = w_{nm}(k), \quad \hat{y}_n(k) \text{ not in } N_c(k) \qquad (9.33)$$

Initially, the weights should be set to random different values. As before, input vectors should be presented in random order and reused. For approximately the first 1000 iterations, $\rho(k)$ should start with a value close to unity, thereafter decreasing monotonically. The learning rate may decrease linearly, exponentially, or inversely proportionally to k. After these iterations, $\rho(k)$ should be set to small values of ~ 0.1.

The neighborhood should be chosen carefully. If it is too small, the self-organization will not be global, but will result in small subsets within a larger grouping. To avoid this phenomenon, a fairly large neighborhood (more than half the diameter of the network) that shrinks with time should be initially used. A rule of thumb for good statistical accuracy is to use a number of iterations that is at least 500 times the number of neurons.

9.3.3 Kohonen Network Example

Let us compare the performance of a multilayer feedforward network to that of a Kohonen network in a license plate recognition system. Takatoo, et al. extracted four license plate digits from video images of moving automobiles. They then extracted 12 features from each digit, such as the number of holes, curvature at certain locations, and distance from the left-hand side.

The 12 features were fed into a two-layer feedforward network. The number of hidden neurons varied. Ten output neurons represented the digits 0 to 9. In representing the desired output vector, the output neuron representing the true digit was set to 0.99. All other output neurons in this vector were set to 0.01 (Figure 9.11). 200 training and 1430 test data were used that had been gathered from moving vehicles. In the training data, 20 data were included for each digit. In the testing data, 143 data were included for each digit. The network was trained until a convergence criterion was met. It was required that each absolute value of the difference between the desired and measured outputs at each neuron be ≤ 0.01 for all training data.

9.3 UNSUPERVISED NEURAL NETWORKS: KOHONEN NETWORK

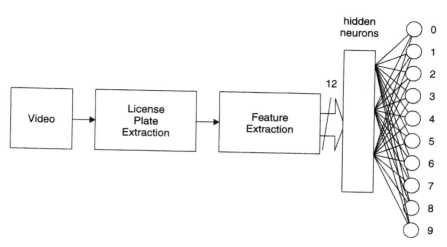

Figure 9.11 License plate recognition system.

The network was trained for hidden layers containing 4, 6, 8, 10, or 12 hidden neurons. Four hidden neurons did not result in a high recognition rate with test data. However, the other hidden layers resulted in recognition rates of 97 to 98% [17].

For comparison purposes, Shigeo Abe repeated training using a Kohonen network with 12 inputs and 100 outputs, arranged in a 10 × 10 array. The firing neurons for the training data after training are shown in Figure 9.12. Note that the 12-dimensional input data were

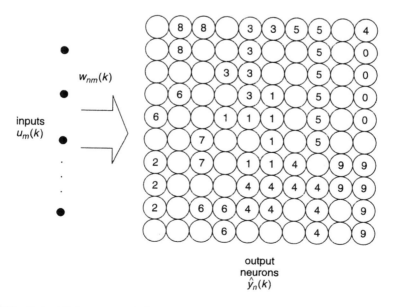

Figure 9.12 Trained Kohonen network for license plate digit classification system. Labeled neurons fire for the associated digit. Unlabeled neurons did not fired for any digit during training. Based on [18].

compressed into two-dimensional data. A neuron that is blank represents a neuron that did not fire for any training data. The recognition rate for this network was "comparable" to that of the multilayer feedforward network (the actual statistics were not stated) [18].

9.4 UNSUPERVISED NETWORKS: ADAPTIVE RESONANCE THEORY NETWORK

One limitation of the competitive learning algorithm in the Kohonen network is that the stability of the formed categories is never guaranteed. If a finite fixed sequence of patterns is continually presented, the weight involved in a particular pattern may change endlessly. Although these changes may be presented by reducing the learning rate gradually to zero, such a measure will freeze the learned categories, causing the network to lose its **plasticity**, or ability to react to new data.

In response to this **stability–plasticity dilemma**, Stephen Grossberg and Gail Carpenter developed adaptive resonance theory (ART) networks. At least four versions of ART exist: ART1, ART2, ART3, and ARTMAP. ARTMAP is a supervised network, whereas the other networks are unsupervised. ART3 uses a distributed search procedure, whereas ART1 and ART2 utilize competitive learning. ART1 uses binary inputs, whereas ART2 uses continuous value inputs. To simplify the discussion, we focus on an abridged model of ART1, given by Hertz, Korgh, and Palmer [19].

9.4.1 General Concepts

Adaptive resonance refers to the ability of the input and a stored category to "resonate" when they are sufficiently similar. When an input pattern is not sufficiently similar to any existing category, a new category is formed, using a previously uncommitted output neuron. If no uncommitted neurons are available, then a novel input leads to no response.

As with other networks, the ART1 algorithm is based on updating weight vectors. Before discussing the network implementation, let us focus on input vectors, $\mathbf{u}(k)$, and binary weight vectors, $\mathbf{w_n}(k)$, both of which contain M binary members, or bits. We assume that N output neurons or categories, $\hat{\mathbf{y}}(k)$, are present, and that these neurons can be enabled or disabled. Early assumptions of brain behavior are imitated with this network. When an input is presented, a category is found that resembles the input. A matching process then occurs. If the input should not be in this category, the selected category is rendered inactive. If the input is accepted into the category, "learning" proceeds, with past inputs being matched with the current input.

For this algorithm, we begin with $\mathbf{w_n}(k) = 1$ for all n, representing an uncommitted state. When a new input pattern is presented, the following steps occur:

1. All output neurons are enabled.
2. The winning output neuron, $\hat{y}_c(k)$, is determined. This winning neuron is associated with the largest value of the product of a normalized weight vector, $\overline{\mathbf{w}}_n(k)$, times input vector:

$$[\overline{\mathbf{w}}_n(k) \cdot \mathbf{u}(k)] = \left[\frac{\mathbf{w}_n(k)}{e + \sum_{l=1}^{N} w_{nl}(k)} \right]^T \cdot \mathbf{u}(k) \qquad (9.34)$$

where e is a small number and $w_{nl}(k)$ is the lth component of $\mathbf{w_n}(k)$. An uncommitted unit wins if no better choice exists. The winning weight vector is designated $\mathbf{w_c}(k)$. During this step, a category is chosen.

3. The ratio, r, which is the fraction of bits in $\mathbf{u}(k)$ that are also in $\mathbf{w_c}(k)$, is computed:

$$r = \left[\frac{\mathbf{w_n}(k)}{\sum_{l=1}^{N} u_l(k)} \right]^T \cdot \mathbf{u}(k) \qquad (9.35)$$

4. If $r \geq$ a **vigilance parameter,** ν, (set between 0 and 1) resonance occurs. In other words, a match has been found. Continue with Step 5. Otherwise, the category vector, $\mathbf{w_c}(k)$, is rejected. As a result, output neuron $\hat{y}_c(k)$ is disabled; continue with Step 2.

5. The winning weight vector $\mathbf{w_c}(k)$ is adjusted by deleting any bits in it that are not also in $\mathbf{u}(k)$. This is a logical AND operation which represents the "learning" process.

The algorithm terminates in one of three ways. If a matching category is found, it is adjusted in Step 5 and the category $\hat{y}_c(k)$ is output. If no match is found, an uncommitted weight vector is selected and made equal to the input in Step 5; the category $\hat{y}_c(k)$ is output. Finally, if no matches or uncommitted vectors exist, all units are disabled, leading to no output.

This network possesses plasticity until all output neurons have been used. It possesses stability because all weight changes cease after a finite number of presentations of a fixed set of input patterns. This algorithm is based on biologically reasonable assumptions, rather than our previously discussed cost function derivations.

9.4.2 ART1 Network Implementation

As stated above, this discussion is based on an abridged model of ART1 [19], which is illustrated in Figure 9.13. The abridged network possesses two layers with hidden neurons,

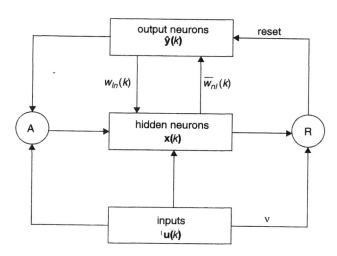

Figure 9.13 The ART1 network. Based on [19].

$\mathbf{x}(k)$, and output neurons, $\mathbf{y}(k)$. The layers are fully connected in both directions. The forward weights, $\overline{w}_{nl}(k)$, are normalized copies of the backward weights, $w_{ln}(k)$. The backward weights, inputs, hidden neurons, and output neurons are all binary, as are A and R. R is the reset signal, which can disable output neurons. A is an auxiliary unit that is turned on according to the status of inputs and outputs. The hidden neurons are equal to the inputs only if no outputs are on; otherwise, a logical AND operation occurs.

ART1 networks are not easy to adjust, and are very sensitive to noise in the input data. Incorrect classification of patterns may occur when data are presented in certain sequences. This order dependency can be traced to the coding of the category templates, which are memoryless and grant more importance to ones than to zeros [20].

9.5 MODEL VALIDATION

As we have discussed before in Chapter 7 for linear models, we wish to choose the best nonlinear model, given various models for the same set of input–output pairs (for a supervised network) or set of inputs (for an unsupervised network). Model validation involves analyzing the results of parameter identification and model plausibility in order to select an "optimum" model. One parameter identification validity measure for linear models, **residual statistics,** is a measure of system noise that requires time-based data. Since the input data for neural networks may be presented in random order during training, residual statistics are not applicable to neural networks. We discuss parameter identification validity measures for the multilayer feedforward, Kohonen, and ART1 networks.

9.5.1 Multilayer Feedforward Network Parameter Identification

Assuming our multilayer feedforward network contains two layers, various models are based on differing numbers of hidden neurons. The number of hidden neurons is a critical choice because it directly affects the computation time spent training the network. Bernard Widrow has suggested as a rule of thumb that the number of hidden neurons be equal to 10 times the number inputs, M [21]. As a more analytic approach, we may utilize validity measures similar to those utilized with linear models: the **correlation matrix** and **goodness of fit.**

The correlation matrix of the hidden neurons, $\underline{\mathbf{R}}_\mathbf{h}(k)$, is first estimated as

$$\underline{\mathbf{R}}_\mathbf{h}(k) = \frac{1}{P} \sum_{i=1}^{P} \mathbf{x}_i(k)\mathbf{x}_i^T(k) \tag{9.36}$$

where P is the number of input–output pairs and i is the subscript designating a particular pair. The necessary number of hidden neurons is determined by calculating the rank of this matrix. If a hidden neuron is expressed by a linear combination of other neurons, then this hidden neuron is redundant [22].

The second validity measure, goodness of fit, may be determined using a more generalized version of Akaike's final prediction error criterion (FPE)—the **information criterion,** AIC [23]. FPE was discussed in Chapter 7 and is applicable to linear models only. The more flexible AIC may be used to assess nonlinear models. Because a more complicated model with more parameters may better fit experimental data, the number of model parameters is weighed against the number of data points and mean squared error:

$$AIC = NN \ln \sum_{i=1}^{P} \frac{1}{\sigma^2(k)} [\mathbf{y}_i(k) - \hat{\mathbf{y}}_i(k)]^2 + 2p \qquad (9.37)$$

where P is the number of input–output pairs, i is the subscript designating a particular training pair, $\sigma^2(k)$ is the system variance, NN is the number of data points, and p is the number of parameters. In terms of this criterion, the best model is that which yields the lowest value of AIC.

9.5.2 Kohonen Network Parameter Identification

For the Kohonen network, we may also use the validity measure of goodness of fit. However, instead of the covariance matrix, we monitor network parameter identification through **topology preservation**. That is, we monitor the quality of training by quantifying the accuracy of various maps in preserving the network relations of the input space.

A parameter that quantifies the lack of topology preservation is topographic error, ε_t. At iteration k, we look for the nearest weight vector, $\mathbf{w}_{n1}(k)$, and second nearest weight vector, $\mathbf{w}_{n2}(k)$. We assume that some of the points in space between $\mathbf{u}(k)$ and $\mathbf{w}_{n2}(k)$ are mapped to $\mathbf{w}_{n1}(k)$, whereas the rest are mapped to $\mathbf{w}_{n2}(k)$. If corresponding neurons $\hat{\mathbf{y}}_{n1}(k)$ and $\hat{\mathbf{y}}_{n2}(k)$ are adjacent, then the mapping is locally continuous. If they are nonadjacent, then a local discontinuity, or local topographic error, has been discovered. These errors are summed for all P input vectors and normalized as:

$$\varepsilon_t = \frac{1}{P} \sum_{i=1}^{P} \varkappa[\mathbf{u}_i(k)] \qquad (9.38)$$

where

$$\varkappa[\mathbf{u}_i(k)] = \begin{cases} 1, & \text{two nearest neighbors are nonadjacent} \\ 0, & \text{otherwise} \end{cases} \qquad (9.39)$$

The degree of topology preservation is inversely proportional to the topological error. In terms of this criterion, the best model is that which yields the lowest value of ε_t [24].

9.5.3 ART1 Network Parameter Identification

As with the feedforward and Kohonen networks, the Akaike information criterion may be used to assess various orders of ART1 networks. An ART1 network possesses the **number of templates** property. That is, it has been proven that the number of output neurons, or templates, required is smaller than the number of total input patterns for certain values of network parameters [25]. Therefore, the AIC is sufficient for assessing model validity.

9.5.4 Model Plausibility

After the model fit has been analyzed, the model plausibility should be examined. First, the model should flexible, that is, capable of describing most of the different system dynamics that can be expected in the application in question. Both the number of parameters and the way they enter the model are important. Second, the model should be as simple as possible, avoiding unneeded complexity.

9.6 SUMMARY

In this chapter, we have reviewed the basic concepts underlying nonlinear system identification, assuming that the system operator of a time-invariant multiple input, multiple output system is an artificial neural network. As its name implies, an artificial neural network refers to a mathematical model of human brain processing. Indeed, in the 1940s, physiologist and electrical engineers worked together towards this goal. Over time, it was discovered that these models did not simulate human neuron processing. However, because these models are very useful for the major applications of function approximation, pattern classification, and combinatorial optimization, engineers continue to investigate these models. The basic neural network architectures for function approximation and pattern classification may be classified by their learning algorithms. With supervised learning, the network is trained by presentation of input and output pairs. With unsupervised learning, the network automatically classifies input data according to the distribution of input data and their relations.

In 1943, Warren McCullough and Walter Pitts presented the first mathematical model of a single idealized neuron in the brain. This model was expanded by Frank Rosenblatt to incorporate supervised learning. The resulting perceptron was trained to perform a pattern recognition task. However, the early perceptron could not be trained to perform an exclusive-or (XOR) function. That is, the four possible combinations of two binary inputs could not be classified, or linearly separated, for the XOR function. It was severely criticized by Minsky and Papert, who believed that claims about the perceptron were exaggerated. Eventually, more advanced perceptrons were designed that could solve the XOR problem.

The early supervised learning architectures were limited in their ability to "learn" a nonlinear function representing the relationship between input and output. This limitation was resolved by choosing a continuous activation function, and minimizing mean squared error using a method similar to the algorithms in Chapters 3 and 7. Because the system error is propagated back from the outputs of the network to the inputs, it is referred to as back propagation.

The network architecture that utilizes this algorithm is a multilayer feedforward network. Feedforward networks easily model nonlinear systems, possess a high degree of generalization, and are suitable for parallel processing. However, these networks are slow to train, as the parameter estimates converge slowly. They also cannot be easily analyzed for network behavior. In other words, if the network does not generalize well, it cannot be easily modified, but must be retrained.

A multilayer feedforward network assumes that the data moves from input to output. In a similar network, the recurrent network, data moves in both directions, giving rise to feedback. In another network, the radial basis function network, the weights in the feedforward network are replaced by Gaussian functions at each hidden neuron and a summation at the single output neuron.

Unlike supervised neural networks, unsupervised networks possess the ability to self-organize and independently estimate their own parameters. The Kohonen network possesses a similar structure to a single-layer feedforward network. However, the output neurons are now arranged in two dimensions. Rather than using back propagation, the Kohonen network relies on competitive learning to determine the value of its weights. At each iteration, the output central neuron that minimizes the Euclidean distance relationship is found.

The weights associated with the neighborhood of neurons around the central neuron are then updated according to the gradient descent method. This self-organization is intended to mimic localization of brain function. In general, the training time of the Kohonen network is shorter than that of a multilayer network. However, the generalization ability of the Kohonen network is lower because the class boundaries are not determined precisely without desired outputs.

One limitation of the competitive learning algorithm in the Kohonen network is that the stability of the formed categories is never guaranteed. Adaptive resonance theory networks were developed to solve this problem. Adaptive resonance refers to the ability of the input and a stored category to "resonate" when they are sufficiently similar. When an input pattern is not sufficiently similar to any existing category, a new category is formed, using a previously uncommitted output neuron. If no uncommitted neurons are available, then a novel input leads to no response.

As we have discussed before in Chapter 7 for linear models, we wish to choose the best nonlinear model, given various models for the same set of input–output pairs (for a supervised network) or set of inputs (for an unsupervised network). For both supervised and unsupervised networks, we may evaluate goodness of fit using a more generalized version of Akaike's final prediction error criterion—the information criterion, AIC. Different validation measures are also utilized for the multilayer feedforward, Kohonen, and ART networks to evaluate particular features in each network.

9.7 REFERENCES

[1]. McCulloch, W. S. and Pitts, W. A logical calculus of the ideas immanent in nervous activity. *Bull Math Biophys, 5,* 115–133, 1943.

[2]. Rosenblatt, F. The perceptron: A probabilitic model for information storage and organization in the brain. *Psychol Rev, 65,* 386–408, 1958.

[3]. Minsky, M. L. and Papert, S. A. *Perceptrons: An Introduction to Computational Geometry.* MIT Press: Cambridge, MA, 1969.

[4]. Widrow, B. Generalization and Information Storage in Networks of Adaline "Neurons." In *Self Organizing Systems,* M. C. Yovits, G. T. Jacobi, and G. D. Goldstein, eds. Spartan Books: Washington, DC, 1962, pp. 435–461.

[5]. Lucky, R. W. Automatic equalization for digital communication. *Bell Syst Tech J, 44,* 547–588, 1965.

[6]. Werbos, P. *Beyond Regressions: New Tools for Prediction and Analysis in the Behavioral Sciences.* Ph.D. Thesis, Harvard University: Boston, MA, 1974.

[7]. Rumelhart, D. E., Hinton, G. E., and Williams, R. J. Learning representations by back-propagating errors. *Nature, 323,* 533–536, 1986.

[8]. Parker, D.B. *Learning Logic.* Technical Report TR-47. Center for Computational Research in Economics and Management Science, Massachusetts Institute of Technology: Cambridge, MA, 1985.

[9]. Cybenko, G. *Continuous Valued Neural Networks with two Hidden Layers are Sufficient.* Technical report. Department of Computer Science, Tufts, University: Medford, MA, 1988.

[10]. Cybenko, G. Approximation by superpositions of a sigmoidal function. *Mathematics of Control, Signals, and Systems, 2,* 303–314, 1989.

[11]. Hornik, K, Stinchcombe, M., and White, H. Multilayer feedforward networks are universal approximators. *Neural Networks. 2,* 359–366, 1989.

[12]. Lapedes, A. and Farber, R. How neural nets work. In *Neural Information Processing Systems,* D. Z. Anderson, ed. American Institute of Physics: New York, 1988, pp. 442–456.

[13]. Malki, H. A. and Moghaddamjoo, A. Using the Karhunen–Loe've transformation in the backpropagation training algorithm. *IEEE Trans Neur Net, 2,* 162–165, 1991.

[14]. Kayama, M., Abe, S., Takenaga, H., and Morooka, Y. Constructing optimal neural networks by linear regression analysis. In *Proceedings of Neuro-Nimes '90.* Nimes, France, November, 1990. pp. 363–376.

[15]. Takenaga, H., Abe, S., Takatoo, M., Kayama, M., Kitamura, T., and Okuyama, Y. Input layer optimization of neural networks by sensitivity analysis and its application to recognition of numerals. *Electr Eng Japan, 111,* 130–138, 1991.

[16]. Le Cun, Y., Boser, B., Denker, J. S., Henderson, D., Howard, R. E., Hubbard, W., and Jackel, L. D. Backpropagation applied to handwritten zip code recognition. *Neural Comp, 1,* 541–551, 1989.

[17]. Takatoo, M., Kanasaki, M., Mishima, T., Shibata, T., and Ota, H. Gray scale image processing technology applied to vehicle license recognition system. In *Proceedings International Workshop on Industrial Applications of Machine Vision and Machine Intelligence.* Tokyo, Japan. February, 1987. pp. 76–79.

[18]. Abe, Shigeo. *Neural Networks and Fuzzy Systems: Theory and Applications.* Kluwer Academic: Boston, 1997. p. 98.

[19]. Hertz, J., Krogh, A., and Palmer, R. G. *Introduction to the Theory of Neural Computation.* Addison-Wesley: Reading, MA, 1991. pp. 228–232.

[20]. Leuba, A. and Koen, B. V. Eliminating order dependency of classification in ART1 networks. In *Proceedings SPIE: Applications and Science of Artificial Neural Networks.* Orlando, FL. April, 1995. Vol. 2492, pp. 779–787.

[21]. Widrow, B. ADALINE and MADALINE—1963, Plenary Speech. In *Proceedings IEEE First International Conference on Neural Networks.* San Diego, CA. 1987. pp. 143–158.

[22]. Hu, Y. H., Xue, Q., and Tompkins, W. J. Structural simplification of a feed-forward multi-layer perceptron artificial neural network. In *Proceedings International Conference on Acoustics, Speech, and Signal Processing.* Toronto, Canada. May, 1991. Vol. 2, pp. 1061–1064.

[23]. Akaike, H. A new look at the statistical model identification. *IEEE Trans Auto Cont, AC-19,* 716–723, 1974.

[24]. Kiviluoto, K. Topology preservation in self-organizing maps. *Proceedings IEEE International Conference on Neural Networks.* Washington, DC June, 1996. Vol. 1, pp. 294–299.

[25]. Georgiopoulos, M., Huang, J., and Heileman, G. L. A survey of learning results in ART architectures. In *Proceedings SPIE: Applications and Science of Artificial Neural Networks.* Orlando, FL. April, 1995. Vol. 2492, pp. 416–424.

Further Reading

Combinatorial Optimization: Abe, S. *Neural Networks and Fuzzy Systems: Theory and Applications.* Kluwer Academic: Boston, 1997, pp. 7–43.

Artificial Intelligence: Chen, C. H., ed. *Fuzzy Logic and Neural Network Handbook.* McGraw-Hill, New York: 1996.

Supervised Networks:

Multilayer Feedforward: Werbos, P. J. *The Roots of Backpropagation: From Ordered Derivatives to Neural Networks and Political Forecasting.* Wiley: New York, 1994.

Recurrent: Parlos, A. G., Chong, K. T., and Atiiya, A. F. Application of the recurrent multilayer perceptron in modeling complex process dynamics. *IEEE Trans Neur Net, 5,* 255–266, 1994.

Radial Basis Function: Powell, M. J. D., Radial basis functions for multivariable interpolation: a review. In *Algorithms for Approximation,* ed. J. C. Mason and M. G. Cox. Oxford University Press: Oxford, 1987. pp. 143–167.

Unsupervised Networks:

Kohonen: Kohonen, T. *Self-Organizing Maps,* 2nd ed. Springer-Verlag: New York, 1997.

Adaptive Resonance Theory: Carpenter, G. A ., and Grossberg, S., eds. *Pattern Recognition by Self-Organizing Neural Networks.* MIT Press: Cambridge, MA, 1991.

9.8 RECOMMENDED EXERCISES

Supervised Networks: See Looney, C. G. *Pattern Recognition Using Neural Networks: Theory and Algorithms for Engineers and Scientists.* Oxford University Press: NY, 1997. Exercises 3.1–3.4.

Unsupervised Networks: See Zurado, J. M. *Introduction to Artificial Neural Systems.* West: St. Paul, MN, 1992. Problems P7.16–P7.21.

10

IMPROVED SCREENING FOR CERVICAL CANCER

In this chapter, we discuss an industrial application of artificial neural networks: the identification of abnormal cervical cells that are consistent with cervical cancer. Abnormal cells are identified from a **Papanicolaou (Pap) smear,** the procedure during which cervical cells are collected during a pelvic exam and preserved on a glass slide for classification. Neural network identification has been used to supplement human screening of these abnormal cells, in order to reduce the rate of false-negative smears.

10.1 PHYSIOLOGY

The **cervix** is one of the internal organs of the female reproductive tract. It is the lower portion of the **uterus,** or womb, that is a narrow muscular canal connecting the body of the uterus to the **vagina,** or birth canal (Figure 10.1). The cervix secretes copious amounts of mucus at certain times during the reproductive cycle. Abnormalities of the cervix, such as low-grade infection or inflammation, or abnormal hormonal stimulation of the cervix can lead to a viscous mucous plug that prevents fertilization. It has been theorized that stretching of the cervix by the fetus' head initiates labor.

10.1.1 Cervical Cells

The cervix is lined with **epithelial** cells. The epithelia are interconnected sheets of cells that work in concert to secrete mucus. The outer surface of the cervix is lined by epithelial flat cells called **squamous** cells. The canal of the cervix is lined with epithelial tall cells called **columnar** cells. These two cell types meet at the **squamo–columnar junction,** which may be located on the outside of the cervix or within the cervical canal. This junction is also called the **tranformation zone** because it is the site where tall columnar cells are constantly transformed into flat squamous cells (Figure 10.1). Within the transformation zone, abnormal growth, or **dysplasia,** may develop. Cervical dysplasia, which is limited to the cervical lining, is a precursor to invasive cervical cancer.

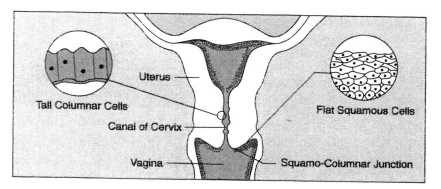

Figure 10.1 Cervical Structures. Courtesy of Paul Indman, MD, www.gynalternatives.com, Los Gatos, CA.

10.1.2 Cervical Cancer

Cancer is a multifactorial disease characterized by cell growth without the normal restraints that regulate normal cells. Most types of cancer cells, or **carcinomas,** form a lump or mass called a **tumor.** Cells from a cancerous, or **malignant,** tumor may break away and travel to other parts of the body, where they can continue to grow. This spreading process is referred to as **metastasis.**

Cervical cancer is one of the most common malignancies in women, accounting for 15,700 new cases and 4900 deaths in the United States each year. Worldwide, it is second only to breast cancer as the most common malignancy in terms of incidence and mortality. More than 471,000 new cases are diagnosed each year, predominantly among the economically disadvantaged, in both developing and industrialized nations [1].

When isolated cancer cells ($\leq$ 3 mm in depth, $\leq$ 7 mm wide) are detected only in the cervix, a **simple hysterectomy** (surgical removal of the uterus only) is virtually 100% curative of patients. As the cancer spreads, a **radical hysterectomy** (surgical removal of the uterus and adjacent tissues) and/or **radiation therapy** (treatment with high-energy rays such as x-rays) is warranted.

With radiation therapy alone, local control can be achieved in more than 95% of small ($\leq$ 4 cm) tumors, with cure rates exceeding 85%. Even with larger tumors (7–10 cm), local control can be achieved in 50% of the cases. Radiation therapy is particularly effective in treating invasive cervical carcinomas, as compared to other cancers of the epithelium, because cervical carcinomas rarely have metastasized at diagnosis. Further, these carcinomas tend to follow an orderly predictable pattern in spreading to other locations. Moreover, it is possible to deliver a high dose of radiation, with acceptably low risk of normal tissue complications, because the hollow uterus and vagina can accept a radioactive source while shielding other tissues from the radiation. In order to increase the control rates, researchers are currently investigating the combination of chemotherapy using the drug cisplatin and radiation therapy [2].

10.1.3 The Role of Human Papillomavirus in Cervical Cancer

Cervical cancer has been linked to the **human papillomavirus,** (HPV), which commonly manifests as warts. More than 200 types of HPV exist, all of which infect the squamous epithelia of the skin and mucuous membranes. Less than 40 types infect the genital tract; these genital viruses are transmitted through sexual intercourse. Genital infection with HPV is extremely common, and is estimated to affect 10 to 15% of the sexually active population, especially between the ages of 18 to 28. Most genital infections are subclinical, with only 1% of the population having evidence of genital warts and 4% of women possessing abnormal cervical cells. Most HPV infections are transient; a woman may be infected by more than one viral type in her lifetime [3, 4].

An international study of over 1000 cervical cancers from 22 counties recently reported that 93% were associated with HPV. HPV 16 was found in 50% of the cases. HPV 18, 31, 33, 35, and 45 were found in 29% of the cases [5]. The protein products of the HPV E6 and E7 genes block the actions of critical cell regulatory proteins to suppress cell proliferation. Unregulated, cell proliferation leads to dysplasias, most of which are likely to regress. However, a small number of dysplasia cases, due to HPV types found in cervical cancer, are characterized by cells with significantly atypical nuclei. These cells are considered the true precursor of cervical cancer. Although it is not known what other factors cause this precursor to progress to full carcinogenesis, one theory is that persistent HPV infection causes a second bout of dysplasia, which then leads to malignancy [4].

The presence of HPV infection may be determined through various molecular screening tests for HPV **deoxyribose nucleic acid** (DNA). One screening test, the HPV Hybrid Capture II (HC II), is manufactured by Digene. HCII detects HPV DNA efficiently from cervical cell samples, requiring only 1000 DNA copies for identification. The test can be used in a clinical laboratory setting; results may be obtained in less than 24 hours. More commonly, physicians examine the vagina and cervix under magnification, after application of 5% acetic acid. This examination is known as **colposcopy.** For unknown reasons, many HPV-associated lesions demonstrate whitening after this application. A biopsy of a suspected lesion is then taken for further analysis [6].

10.2 PAP SMEAR

As stated above, the Pap smear is the current standard screening method for detection of cervical cancer. During a Pap smear, a sample of mucus and cells is scraped from the transformation zone of the cervix, and is spread in a relatively even monolayer on a glass slide for classification.

10.2.1 Test Development

Classification of cervical cell samples was first introduced as a cervical cancer detection method by Babes in 1926. During the same time frame, George Papanicolaou noted that abnormal cells were present in the vaginal pool in the presence of early cervical cancer. Both published their observations in 1928. In 1943, Papanicolaou and Traut published the first book on the diagnosis of cervical **neoplasia** (new cell growth) by vaginal pool smear. Later, Papanicolaou, with other investigators, developed a classification system based on the degree of abnormality of the cells on the smear. Since its introduction in the United States in 1947, the Pap smear has been credited with decreasing the incidence of cervical cancer from 44 per 100,0000 to 5 to 8 per 100,000 women. It has reduced the mortality rate of cervical cancer by greater than 70%.

10.2.2 Cell Classification

The Pap smear classification system was revised in 1988 by the National Cancer Institute in Bethesda, Maryland. Reclassification was conducted for consistency with the natural history of HPV and cervical dysplasia, which is also known as **cervical intraepithelial neoplasia** (CIN). **The Bethesda system** (TBS) classifies cervical cells and evaluates the specimen for adequacy, or quality. In addition to appropriate labeling, a slide should contain adequate numbers of well-preserved and well-visualized squamous epithelial cells and an adequate endocervical/transformation zone component. Determination of specimen quality is a major contribution because retrospective reviews of smears from women with cervical cancer have shown that many smears were unsatisfactory. The main two cellular diagnoses are **benign** cellular changes and epithelial abnormalities. Benign, or noncancerous, cellular changes may be due to infection, reactive changes, or relative cellular changes. Relative cellular changes are associated with inflammation, atrophy with inflammation, or an intrauterine contraceptive device (Figure 10.2).

Classification of epithelial abnormalities is based on the progression of cervical dysplasia. Normally, the cervical lining is composed of organized layers of uniformly shaped cells, with the bottom layer containing round cells. As these cells mature, they rise to the surface and flatten out to become flat squamous cells. During mild cervical dysplasia, which is designated as CIN I, this growth process is disrupted with a few abnormal cells. During moderate dysplasia, or CIN II, the abnormal cells are distributed in about half the thickness of the lining. If the abnormal growth progresses to severe dysplasia or **carcinoma-in-situ** (known as CIN III), the entire thickness becomes disordered, but the abnormal cells have not spread below the lining. Once the abnormal growth invades the tissue, it becomes invasive cancer (Figure 10.3).

Epithelial cell abnormalities can be classified into six categories. **Atypical squamous cells of undetermined significance** (ASCUS) classification refers to unusual cells that are not abnormal enough to be classified as dysplasia (Figure 10.4). This classification

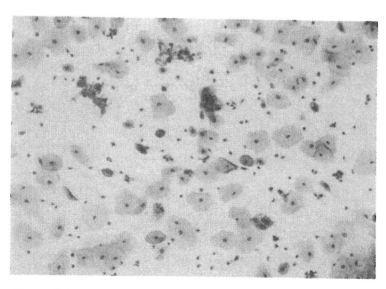

Figure 10.2 Benign Pap smear. Courtesy of TriPath Imaging, Burlington, NC.

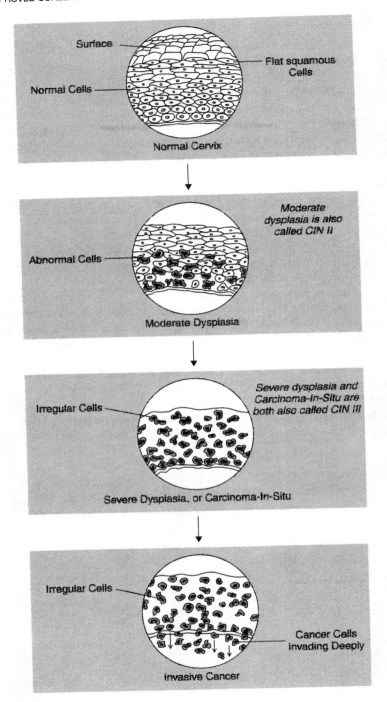

Figure 10.3 Progression of cervical dysplasia from normal to invasive cancer. Courtesy of Paul Indman, MD, *www.gynalternatives.com*, Los Gatos, CA.

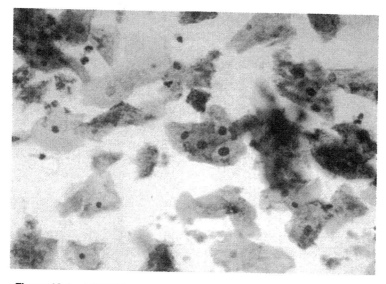

Figure 10.4 ASCUS Pap smear. Courtesy of TriPath Imaging, Burlington, NC.

usually results in a second Pap smear for verification. Approximately 5 to 40% of women with this cytology possess dysplasia. **Low-grade squamous intraepithelial lesion (LSIL)** classification is associated with hollow cells, called **koilocytes,** that possess atypical nuclei and/or mild dysplasia (Figure 10.5). The abnormalities are confined to significantly superficial and intermediate cells. LSIL Pap smears possess a lower risk of cervical cancer detection than do ASCUS smears. Detection of LSIL cells usually results in colposcopy for further examination.

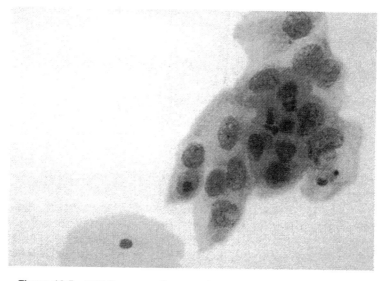

Figure 10.5 LSIL Pap smear. Courtesy of TriPath Imaging, Burlington, NC.

High-grade squamous intraepithelial lesion (HSIL) classification is associated with moderate or severe dysplasia or carcinoma-in-situ. The cells are usually deficient in **nucleoli** (the dense region of the nucleus that contains DNA), are granular or reticular, and are often found in patterns of lines (Figure 10.6). HSIL detection results in a colposcopy and an appropriately direct cervical biopsy. 76 to 88% of women with an HSIL Pap smear have moderately or severe dysplasia detected at colposcopy. When the smear is suggestive of cancer, cancer is detected 78 to 95% of the time.

Squamous cell carcinoma classification refers to a malignant invasive tumor of squamous cells. These cells occur singly or in aggregates; their nuclei contain coarse granular clumps (Figure 10.7). **Adenocarcinoma** classification refers to a malignant invasive tumor composed of endocervical, endometrial (from the lining of the uterus), or extrauterine (outside of the uterus) cells (Figure 10.8).

The last epithelial category, **atypical glandular cells of undetermined signficance** (AGUS), is the most difficult to diagnose. This classification is associated with cellular changes in glandular cells exceeding those expected in a benign reactive or reparative reaction, yet are not abnormal enough to be clearly neoplasic (Figure 10.9). 20 to 80% of women with this diagnosis have significant dysplasia, adenocarcinoma, or squamous carcinoma. Women with this classification should have colposcopy of the cervix and vagina, as well as an **endocervical curettage** (scraping of the inner uterine lining), performed [7].

10.2.3 Limitations

The sensitivity of the Pap smear is limited by **sampling error,** which occurs when abnormal cells are not placed on the smear. It is also limited by **detection error,** wherein possibly a few dozen abnormal cells are not identified among hundreds of thousands of normal cells present in a well-taken smear. This sensitivity has been estimated as 40 to 80% for HSIL cells, and is also poor for AGUS cells. Specificity is a problem, as screening laboratories are overburdened by borderline and mildly abnormal smears [8].

10.3 PROBLEM SIGNIFICANCE

The incidence of cervical cancer has increased since 1978, with more than 55,000 cases reported in the United States in 1996. It is estimated that 12,800 new cases of invasive cervical cancer will be diagnosed in 1998, and that 4600 women will die from this disease [9].

10.3.1 Clinical Significance

Two issues prevent eradication of cervical cancer. First, underserved women, who possess a higher risk for cervical cancer, do not obtain regular Pap smears. Second, false-negatives may occur in Pap smears, especially when they are conducted irregularly. The specificity, or true-negative rate, of conventional Pap smears was recently assessed by the Duke University Center for Health Policy Research, as 51%. In contrast, the sensitivity, or true-positive rate, was assessed as 98%. About 67% of false-negatives result from sampling error, whereas 33% result from detection error [10].

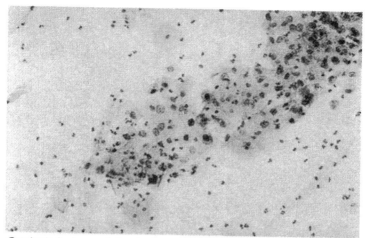

Cytologic features of dysplasia are seen on Pap smear. Some of the cells at the center show increased nuclear/cytoplasmic ratio, with darker and more irregular nuclei than the normal squamous cells with large amount of cytoplasm and small, pyknotic nuclei.

(a)

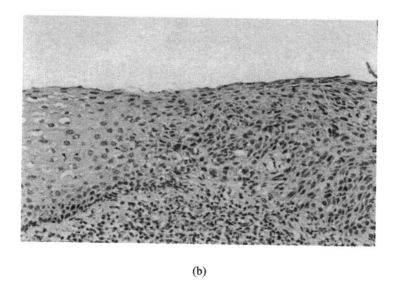

(b)

Figure 10.6 (a) HSIL Pap smear and (b) cervical biopsy. The biopsy contains normal cells on the left and dysplastic cells on the right. The smear and biopsy were obtained from different patients. Images from WebPath (*http://medstat.med.utah.edu/WebPath/*webpath.html), courtesy of Edward C. Klatt, MD.

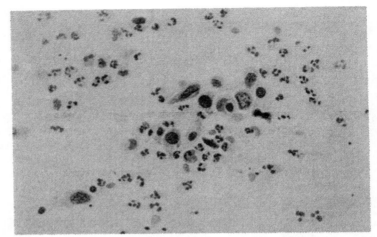

The cytologic features of invasive squamous cell carcinoma are demonstrated on a Pap smear. Note the cells with hyperchromatic, pleomorphic nuclei and increased nuclear/cytoplasmic ratio. The background has inflammatory cells.

(a)

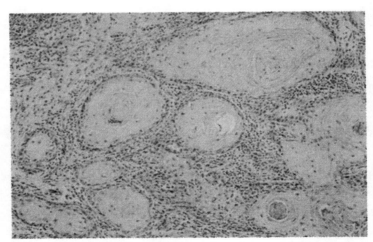

A biopsy reveals invasive squamous cell carcinoma of the cervix.

(b)

Figure 10.7 a) Squamous cell carcinoma Pap smear and b) cervical biopsy. The smear and biopsy were obtained from different patients. Images from WebPath (*http://medstat.med.utah.edu/WebPath/*webpath.html), courtesy of Edward C. Klatt, MD.

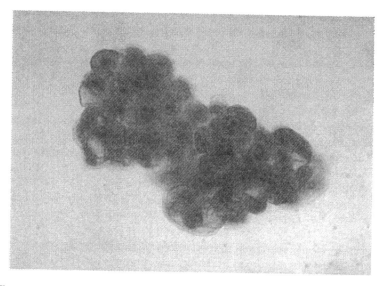

Figure 10.8 Adenocarcinoma Pap smear. Courtesy of TriPath Imaging, Burlington, NC.

10.3.2 Market Forces

Three approaches have been taken to reduce the rate of Pap smear false-negatives. **Thin-layer cytology** aims at reducing the sampling error. The ThinPrep 2000, manufactured by Cytyc Corporation, filters cervical cell samples and transfers them to slides with fewer artifacts. Two computer rescreening technologies aim at reducing the detection error. PAP-NET, manufactured by Neuromedical Systems, uses artificial neural networks to identify

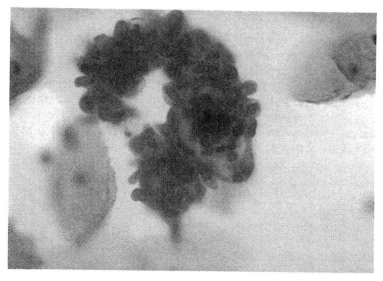

Figure 10.9 AGUS Pap smear. Courtesy of TriPath Imaging, Burlington, NC.

cells or clusters of cells that require review by a cytotechnologist. Similarly, Autopap 300 QC, manufactured by Neopath, uses fuzzy logic to identify abnormal cell clusters for further review.

These three technologies were recently assessed for cost effectiveness in enhancing Pap smear sensitivity. Based on laboratory surveys, it was assumed that the ThinPrep adds $9.75 per slide in 1996 dollars compared to conventional Pap smear testing. It was also assumed that PAPNET adds $10 and AutoPap adds $1.50 per slide. A hypothetical program served as a cohort of 20 to 65 year old women who began screening at the same age and are representative of the U.S. population. The true-negative rate was assumed to be 95% for all technologies; the true-positive rate for each technology was taken from the literature. Based on this model, a Pap smear with 10% rescreen cost $75.75 per slide, whereas the other technologies cost between $81 to $86 per slide. When used with triennial screening, each technology produced more dollars per year of life saved ($16,000 to $381,000 per year of life saved) than conventional Pap testing every two years ($9000 per year of life saved) [11].

In late 1999, AutoCyte, a thin-layer cytology competitor of Cytyc Corporation, merged with Neopath and acquired the intellectual property of Neuromedical Systems. The new company is called TriPath Imaging. The AutoCyte system was approved by the Food and Drug Administration (FDA) in June, 1999; the Cytyc system was approved in May, 1996 [12]. TriPath Imaging is in the process of obtaining FDA approval for a combined AutoCyte/AutoPap system.

10.4 SEMIAUTOMATION OF CERVICAL CANCER SCREENING

In this section, the ThinPrep and AutoPap systems are described in detail. The PAPNET system is described in Section 10.5.

10.4.1 ThinPrep 2000 Processor

For a typical Pap smear, the sample brush or swab containing cervical cells is smeared across a glass slide. To prepare a cervical sample for ThinPrep processing, the sample brush is rinsed in a vial containing a buffered alcohol preservative solution. The vial is then input to the ThinPrep 2000 system. Under microprocessor control, blood, mucous, and nondiagnostic debris are removed from the sample, enabling the sample to be mixed. The sample is then drawn through a filter using a series of negative pressure pulses so that a thin, even layer of diagnostic cellular material may be collected. The cellular material is transferred to a glass slide using computer-controlled mechanical positioning and positive air pressure. Finally, the slide is ejected into a cell fixative bath, ready for staining and evaluation.

In a clinical trial of 6747 women from three screening centers and three hospitals, a conventional Pap test was performed, after which a ThinPrep slide was prepared. Each unlabeled slide was read independently by a different cytotechnologist, and classified according to the Bethesda System. For the three screening centers, which were chosen because of their higher rates of LSILs and HSILs, 65% more diagnoses of LSIL and higher (HSIL, squamous carcinoma, adenocarcinoma) were made on the ThinPrep slides ($p < 0.001$). The difference in diagnoses at the hospitals were insignificant. With the ThinPrep method, there was an 11% increase in all six sites in the number of satisfactory specimens ($p < 0.001$) [13]. Based on this study, the ThinPrep 2000 system is capable of reducing sampling error and the rate of false-negative classifications. In May, 2000, an upgraded system, the ThinPrep 3000, received FDA approval. This instrument is capable of processing 80 slides simultaneously (Figure 10.10).

10.4.2 AutoPap Systems

The AutoPap 300 QC and Primary Screening Systems were manufactured by NeoPath until 1999, when NeoPath merged with AutoCyte. NeoPath was founded by University of Washington professor Alan Nelson in 1989. Classification technology in these systems is based on fuzzy logic, most likely because the application of artificial neural networks to cell classification was patented by competitor Neuromedical Systems in 1990 [14]. Fuzzy logic models, which are rule-based, are described in Chapter 11. The discrimination rules for different cell types, which are implemented in the AutoPap systems, are illustrated in Table 10.1. Each rule should be interpreted as the intersection of six conditions to describe a particular cell type. The rules were developed by noted cervical cytologist Stanley Patten, who had classified cervical cells since the 1960s [15].

The primary, unassisted screening performance of the AutoPap Primary Screening System (Figure 10.11) was assessed through a clinical trial of 25,124 slides at five commercial laboratory sites. Each slide was classified by a cytotechnologist and by the AutoPap system. The classification of ASCUS+ (ASCUS and above), LSIL, LSIL+ (LSIL and above), and HSIL+ (HSIL, squamous carcinoma, adenocarcinoma) slides was compared. In all four categories, the sensitivity was increased using AutoPap (86 vs. 79%, 91 vs. 84%, 92 vs. 86%, and 97 vs. 93%, respectively). These increases in sensitivity were all significant ($p < 0.05$) [16].

Figure 10.10 ThinPrep 3000 Processor. Courtesy of Cytec Corporation, Boxborough, MA.

Table 10.1 AutoPap discrimination rules for different cervical cell types. Reprinted from [15] with permission from John Wiley & Sons.

	Chromatin particles	Chromatin distribution	Cell shape	Cyto texture	Cell borders	Cell arrangement
Intermediate squamous	Fine	Even	Polygonal	Homogeneous	T	Isolated
Endometrial	Fine	Even	Round	Finely vacuolated	F	Isolated/cluster
Squamous metaplastic (P)	Fine	Even	Round/oval	Homogeneous	F	Isolated
Squamous metaplastic (I)	Fine	Even	Round/oval	Primitive	T	Isolated
Squamous metaplastic (M)	Fine	Even	Small polygonal	Homogeneous	T	Isolated
Atypical endometrial hyperplasia	Fine	Even	Round	Vacuolated	F	Cluster
Endometrial Adenoca	Fine	Uneven	Round	Vacuolated	F	Isolated/cluster
Endocervical	Fine	Even	Columnar	Granular	T	Isolated/sheets
Atrophic squamous	Fine	Even	Round/oval	Homogeneous	T	Isolated
Squamous atypical repair	Fine	Even	Round/oval	Homogeneous	T	Sheets
ASCUS	Fine/oval	Even	Polyg/round	Homogeneous	T	Isolated
Metaplastic dysplasia	Fine/clumped	Even	Round/oval	Homogeneous	T	Isolated
Intermediate CIS	Fine/coarse	Even	None	Primitive	F	Isolated/syncytial
Large cell CIS	Fine	Even	None	Primitive	T	Isolated/syncytial
Nonkeratinizing sq CA	Fine/coarse	Uneven	None	Primitive	F	Isolated/syncynal
Keratinizing dysplasia	Opaque	Even	Pleomorphic	Homogeneous	T	Isolated
Keratinizing sq carcinoma	Opaque	Even/uneven	Pleomorphic	Homogeneous	T	Isolated
Small cell CIS	Fine/coarse	Even	Oval	Primitive	F	Syncytial
Small cell sq carcinnina	Fine/coarse	Uneven	Oval	Primitive	F	Syncytial
Atypical immature squamous metaplastic type	Fine	Even	Round	Homogeneous	T	Isolated
Endocervical atypta	Fine	Even	Columnar/oval	Homogeneous	T	Sheets
Adenocarcinoma in situ	Coarse	Even	Columnar	Granular	T	Isolated/sheets
Endocarvical adenoca	Fine	Uneven	Columnar	Granular	F	Cluster

P: primitive; I: immature; M: mature; Adenoca: adenocarcinoma; ASCUS: atypical squamous cells of undetermined significance; CIS: carcinoma in situ; sq: squamous; CA: carcinoma; T: true; F: false.

Figure 10.11 AutoPap Primary Screening System. Courtesy of TriPath Imaging, Burlington, NC.

10.5 CERVICAL CANCER SCREENING USING NEURAL NETWORKS

An artificial neural network was used to train the cervical cell classifier in the PAPNET system.

10.5.1 Training Data

According to Neuromedical patents, the PAPNET Neural Network Processor (NNP) contains one two-layer feedforward network, utilizing back propagation [17]. Each input is based on the pixels of a scanned pap smear image within an approximate area of 2×2 cm^2, containing 50,000 to 300,000 cells. Inputs were chosen based on identification of abnormal cells or cell clusters that were subsets of larger scanned images. A seriously abnormal smear may contain only a few dozen abnormal cells scattered among hundreds of thousands of normal cells.

The hidden layer consists of nodes that number approximately 25% the number of input pixels. The activation function used is the hyperbolic tangent function [14]. The output layer consists of a single neuron. An output of 0.9 (maximum output = 1) signifies that a premalignant or malignant cell has been detected. The threshold for detection of a premalignant or malignant cell is 0.65. During training of an engineering prototype, an unspecified number of 24×24 pixel images were input to the network to determine the weights. The misclassification rate after training was not reported [17].

10.5.2 PAPNET System

In contrast to the AutoPap system, which is capable of primary, unassisted screening, the PAPNET system prescreens images from a Pap smear, and then displays the 128 images from the smear that are most likely to be premalignant or malignant. Pap smears are sent

232 IMPROVED SCREENING FOR CERVICAL CANCER

Figure 10.12 PAPNET System. Reprinted from [18] with permission by the International Academy of Cytology, Chicago, IL.

to a central scanning center, where a scanning station (Figure 10.12) is used to prescreen images from each smear. During prescreening, a conventional image processor is used to isolate potential premalignant or malignant cells for further review. In an early version of PAPNET, this isolation of abnormal cells was implemented using neural networks [14, 19]. However, this isolation is now implemented using conventional image processing techniques.

According to patents, for isolation of target cells, this "primary classifier" performs an **erosion** of the image, which refers to isolation of all objects within the image that are greater than or equal to the smallest size of a pathological cell nucleus. During erosion, binary pixels are inverted from "1" to "0," based on various mathematical manipulations. The remaining objects are then **dilated** by adding pixels. As with erosion, binary pixels during dilation are inverted based on mathematical manipulations. Dilation is stopped before the objects begin to touch. Based on an engineering prototype, for every 1000 objects in a typical benign Pap smear, no more than 15 objects pass the erosion/dilation screen.

The primary classifier then subjects the remaining objects to an **integrated optical density** (IOD) screen. IOD refers to the sum of the pixel grey values for each object. Because premalignant cells tend to possess large, dark nuclei, the IOD threshold is set to filter out objects below the threshold displayed by a truly premalignant or malignant cell. Based on an engineering prototype, for every 15 objects that pass the erosion/dilation screen, only 10 will pass the IOD screen. The 10 objects may include other objects such as cell clumps, debris, clumps of leucocytes, and mucus. The remaining 10 objects are passed to the neural network for further classification [14].

In actual implementation, the neural network, or "secondary classifier" is presented with similar, but not identical images as those output by the primary classifier. Each secondary classifier input possesses the same central cell as its corresponding primary classifier output. However, the secondary classifier processes images of cell clusters, rather than images of single cells. After classification, the 64 images that most likely to be pre-malignant or malignant are recorded on digital tape, each as a single cell and as part of a cluster of cells. The digital tape corresponding to each Pap smear is sent back to a cytology lab for further review by a cytotechnologist. The cytotechnologist reviews all 128 images and classifies the Pap smear. This process is illustrated in Figure 10.13.

10.5.3 Data Analysis

The primary, assisted screening performance of the PAPNET system was assessed through a clinical trial of 21,700 slides at five commercial laboratory sites. Each slide was classified by the combination of PAPNET system and cytotechnologist and also by an independent cytologist. Because the trial was performed in Europe, classification was performed according to the criteria recommended by the British Society for Clinical Cytology: inadequate specimen, benign, mild dysplasia, severe dysplasia, invasive carcinoma, glandular neoplasia, moderate dysplasia, and borderline nuclear changes.

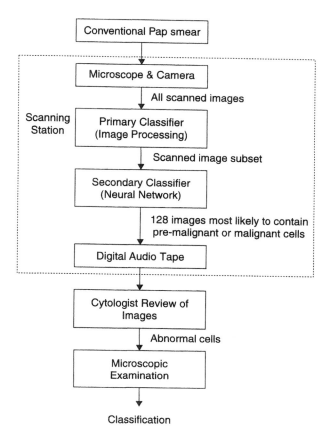

Figure 10.13 PAPNET system data processing.

The overall agreement between conventional and PAPNET-assisted screening was 89.8%. The sensitivity was similar for correctly identified abnormal smears on conventional (83%) and PAPNET-assisted screening (82%). PAPNET-assisted screening showed significantly better specificity (77%, $p < 0.001$) than conventional screening (42%) for identification of benign smears. The total mean time for screening and report for conventional screening amounted to 10.4 minutes per smear, whereas PAPNET-assisted screening required only 3.9 minutes [20].

10.5.4 System Comparison

Ideally, cervical cancer screening would be an excellent application for comparing the effectiveness of fuzzy models versus neural networks during classification. However, only the fuzzy AutoPap system was assessed for its ability to screen for various Pap smears in an unassisted manner. In contrast, the neural PAPNET system could only assist a cytotechnologist during this screen. The inability of the PAPNET system to screen in an unassisted manner should not reflect a limitation in neural network technology. Rather, this limited performance may reflect the PAPNET implementation.

As stated in Section 9.2.5, performance of a feedforward network depends heavily on the selected input variables. For some reason, the PAPNET designers chose to input analog pixels of an image rather than specific, relevant parameters of each image. If selection or extraction of input variables is not performed, the resulting neural network will be limited in its utility. Further, the PAPNET designers seemed unsure how to perform primary classification effectively, as the methodology for this classification changed from patent to patent [14, 17, 21]. For these reasons, the neural network may have been only able to assist, rather than perform, the screening process. It is therefore not surprising that the sensitivity of screening did not change with PAPNET assistance.

AutoPap survived in the marketplace because of superior technology. TriPath Imaging, which purchased both the AutoPap and PAPNET systems, probably acquired PAPNET for its market share, rather for than its technology.

10.6 SUMMARY

In this chapter, we discuss an industrial application of artificial neural networks: the identification of abnormal cervical cells that are consistent with cervical cancer. Abnormal cells are identified from a Papananicolaou (Pap) smear, the procedure during which cervical cells are collected during a pelvic exam and preserved on a glass slide for classification. Neural network identification has been used to supplement human screening of these abnormal cells, in order to reduce the rate of false-negative smears.

Cervical cancer is one of the most common malignancies in women, accounting for 15,700 new cases and 4900 deaths in the United States each year. Worldwide, it is second only to breast cancer as the most common malignancy in terms of incidence and mortality. More than 471,000 new cases are diagnosed each year, predominantly among the economically disadvantaged, in both developing and industrialized nations. The Pap smear is the current standard screening method for detection of cervical cancer. During a Pap smear, a sample of mucus and cells is scraped from the transformation zone of the cervix, and is spread in a relatively even monolayer on a glass slide for classification.

The Pap smear classification system was revised in 1988 by the National Cancer Institute in Bethesda, Maryland. Reclassification was conducted for consistency with the natural history of HPV and cervical dysplasia, which is also known as cervical intraepithelial neoplasia. The Bethesda system classifies cervical cells and evaluates the specimen for adequacy, or quality. The main two cellular diagnoses are benign cellular changes and epithelial abnormalities. Benign, or noncancerous, cellular changes may be due to infection, reactive changes, or relative cellular changes. Relative cellular changes are associated with inflammation, atrophy with inflammation, or an intrauterine contraceptive device.

Classification of epithelial abnormalities is based on the progression of cervical dysplasia. Normally, the cervical lining is composed of organized layers of uniformly shaped cells, with the bottom layer containing round cells. As these cells mature, they rise to the surface and flatten out to become flat squamous cells. During mild cervical dysplasia, this growth process is disrupted with a few abnormal cells. During moderate dysplasia, the abnormal cells are distributed in about half the thickness of the lining. If the abnormal growth progresses to severe dysplasia or carcinoma-in-situ, the entire thickness becomes disordered, but the abnormal cells have not spread below the lining. Once the abnormal growth invades the tissue, it becomes invasive cancer. Epithelial cell abnormalities can be classified into six categories: atypical squamous cells of undetermined significance (ASCUS), low-grade squamous intraepithelial lesion (LSIL), high-grade squamous intraepithelial lesion (HSIL, squamous cell carcinoma, adenocarcinoma), and atypical glandular cells of undetermined signficance (AGUS).

The sensitivity of the Pap smear is limited by sampling error, which occurs when abnormal cells are not placed on the smear. It is also limited by detection error, when a few abnormal cells are not identified among the more numerous normal cells present in a well-taken smear. PAPNET, manufactured by Neuromedical Systems, attempts to reduce the detection error. PAPNET uses artificial neural networks to identify cells or clusters of cells that require review by a cytotechnologist.

The primary, assisted screening performance of the PAPNET system was assessed through a clinical trial of 21,700 slides at five commercial laboratory sites. Each slide was classified by the combination of PAPNET system and cytotechnologist and also by an independent cytologist. Because the trial was performed in Europe, classification was performed according to the criteria recommended by the British Society for Clinical Cytology: inadequate specimen, benign, mild dysplasia, severe dysplasia, invasive carcinoma, glandular neoplasia, moderate dysplasia, and borderline nuclear changes.

The overall agreement between conventional and PAPNET-assisted screening was 89.8%. The sensitivity was similar for correctly identified abnormal smears on conventional (83%) and PAPNET-assisted screening (82%). PAPNET-assisted screening showed significantly better specificity (77%, $p < 0.001$) than conventional screening (42%) for identification of benign smears. The total mean time for screening and report for conventional screening amounted to 10.4 minutes per smear, whereas PAPNET-assisted screening required only 3.9 minutes.

10.7 REFERENCES

[1]. National Institutes of Health. *Cervical Cancer*. NIH Consensus Development Statement. April 1–3, 1996, *14*, 1–38, 1996.

[2]. Eifel, P. J. Chemoradiation for carcinoma of the cervix: Advances and opportunities. *Radiat Res, 154,* 229–236, 2000.

[3]. Carr, J. and Gyorfi, T. Human papillomavirus: Epidemiology, transmission, and pathogenesis. *Clin Lab Med, 20,* 235–255, 2000.

[4]. McLachlin, C. M. Human papillomavirus in cervical neoplasia: Role, risk factors, and implications. *Clin Lab Med, 20,* 257–270, 2000.

[5]. Wick, M. J. Diagnosis of human papillomavirus gynecologic infections. *Clin Lab Med, 20,* 271–287, 2000.

[6]. Bosch, F. X., Manos, M. M., Munoz, N., Sherman, M., Jansen, A. M., Peto, J., Schiffman, M. H., Moreno, V., Kurman, R., and Shah, K. V. Prevalence of human papillomavirus in cervical cancer: A worldwide perspective. International biological study on cervical cancer (IBSCC) Study Group. *J Natl Cancer Inst, 87,* 796–802, 1995.

[7]. Cox, J. T. Evaluation of abnormal cervical cytology. *Clin Lab Med, 20,* 303–343, 2000.

[8]. Cuzick, J., Sasieni, P., Davies, P., Adams, J., Normand, C., Frater, A., van Ballegooijen, M., and van den Akker-van Marle, E. A systematic review of the role of human papilloma virus (HPV) testing within a cervical screening programme: Summary and conclusions. *J Cancer, 83,* 561–565, 2000.

[9]. American Cancer Society. *Cervical Cancer—Overview.* www.cancer.org. November, 2000.

[10]. Agency for Health Care Policy and Research. *Evaluation of Cervical Cytology.* www.ahcpr.gov/clinic/cervsumm.htm. November, 2000.

[11]. Brown, A. D., and Garber, A. M. Cost-effectiveness of 3 methods to enhance the sensitivity of Papanicolaou testing. *JAMA, 281,* 347–353, 1999.

[12]. Food and Drug Administration Center for Devices and Radiological Health (CDRH), *CDRH Premarket Approval Database.* www.accessdata.fda.gov/scripts/cdrh/cfdocs/cfPMA/pma.cfm. November, 2000.

[13]. Lee, K. R., Ashfaq, R., Birdsong, G. G., Corkill, M. E., McIntosh, K. M., and Inhorn, S. L. Comparison of conventional Papanicolaou smears and a fluid-based, thin-layer system for cervical cancer screening. *Obstet Gynecol, 90,* 278–284, 1997.

[14]. Rutenberg, M. R. Neural network based automated cytological specimen classification system and method. *U.S. Patent 4,965,725.* October 23, 1990.

[15]. Lee, J. S. J., and Nelson, A. C. Stanley F. Patten, Jr., M.D., Ph.D. and the development of an automated Papanicolaou smear screening system. *Cancer (Cancer Cytopathology), 81,* 332–326, 1997.

[16]. Wilbur, D. C., Prey, M. U., Miller, W. M., Pawlick, G. F., Colgan, T. J., and Taylor, D. D. Detection of high grade squamous intraepithelial lesions and tumors using the AutoPap system: Results of a primary screening clinical trial. *Cancer (Cancer Cytopathology), 87,* 354–258, 1999.

[17]. Rutenberg, M. R. and Hall, T. L. Automated cytological specimen classification system and method. *U.S. Patent 5,740,270.* April 14, 1998.

[18]. Mango, L. J. Neuromedical Systems, Inc. *Acta Cytol, 40,* 53–59, 1996.

[19]. Mango, L. J. Computer-assisted cervical cancer screening using neural networks. *Can Lett, 77,* 155–162, 1994.

[20]. PRISMATIC Project Management Team, Assessment of automated primary screening on PAPNET of cervical smears in the PRISMATIC trial. *Lancet, 353,* 1381–1385, 1999.

[21]. Boon, M. E., Kok, L. P., Mango, L. J., Rutenberg, A., and Rutenberg, M. R. Automated specimen classification system and method. *U.S. Patent 5,544,650.* August 13, 1996.

[22]. Martin, J. F. Continuous cardiac output derived from arterial pressure waveform using pattern recognition. *U.S. Patent 5,797,395.* August 25, 1998.

[23]. Martin, J. F., Volfson, L. B., Kirzon-Zolin, V. V., and Schukin, V. G. Application of pattern recognition and image classification techniques to determine continuous cardiac output from the arterial pressure weaveform. *IEEE Trans BME, 41,* 913–920, 1994.

[24]. Sugimachi, M. National Cardiovascular Center Research Institute, Osaka, Japan. Research data, 1992.

Further Reading

The Bethesda System Pap Smear Classifications: Kurman, R. J. and Solomon, D. *The Bethesda System For Reporting Cervical/Vaginal Cytologic Diagnoses: Definitions, Criteria, and Explanatory Notes for Terminology and Specimen Adequacy.* Springer-Verlag: New York, 1994.

Image Processing: Russ, J. C. *The Image Processing Handbook,* 2nd ed. CRC Press: Boca Raton, FL, 1995.

10.8 CARDIAC OUTPUT EXERCISES

Engineer E at Startup Company recently read *U.S. Patent 5,797,395.* The method described in this patent uses a pattern recognition methodology known as typical shape function (TFA) analysis to classify blood pressure waveforms and determine cardiac output (CO) [22]. According to the accompanying journal article, "for over 200,000 individual heart beats, covering a wide range of hemodynamic conditions, the mean error, in calculated CO compared to ultrasonic flow probe determined CO, was 2.8% with a standard deviation of 9.8%" [23]. Engineer E recently studied neural networks, and believes the methodology in this patent could be further improved by substituting a feedforward neural network for TFA. We recreate his analysis below.

10.8.1 Clinical Data

The radial data file in Figure 2.10 was recorded from a patient at Startup Company. Ten seconds of data were recorded with a sampling frequency of 200 Hz and cutoff of 1 MHz, and saved in units of mmHg. These data were generously provided by Dr. Masaru Sugimachi [24].

Please move the *radial.dat* file from ftp://ftp.ieee,org/uploads/press/baura to your *Matlab\bin* directory. Input the data using "load radial.dat -ascii."

10.8.2 Useful Matlab Functions

The following functions from the Neural Network Toolbox are useful for completing this exercise: newff, init, train, and sim.

10.8.3 Exercises

1. Determine five waveform features that could be input to the neural network. How would you check for orthogonality?
2. Construct a feedforward neural network to classify these features. Assume you have 1000 training data. (a) How many hidden layers are required? (b) How many

hidden neurons per layer are required? (c) How many output neurons are required? (d) What activation function would you choose?
3. (a) What typical range of cardiac outputs would you expect to see in a hospital population? (b) How many training sets are required for classification of this range? Explain.
4. (a) Do you believe a neural network can classify cardiac output from blood pressure? Explain. (b) How would you explain the results reported in the journal article?

11

FUZZY MODELS

In Chapter 7, we assumed that the relationship between a single input, u(k), and single output, y(k), was a linear and time-invariant impulse response, h(k). In Chapter 9, we relaxed these constraints to a more generalized multiple input, **u**(k), multiple output, **y**(k), system with a nonlinear but time-invariant system operator that could be described as an artificial neural network. In this chapter, let us further relax our constraints so that our system operator is nonlinear, time-invariant, and sufficiently complex that it may not easily be described by mathematical equations or numeric representations. For example, the movement of a syringe piston in a syringe pump may be affected by so many factors that this movement is not readily describable by numeric values. Given such a system, the system operator may be described using **fuzzy logic**.

As its name implies, fuzzy logic is the logic underlying modes of reasoning that are approximate rather than exact. Conventional approaches to knowledge representation are based on **bivalent** (two states: true/false) logic. However, these approaches are unable to deal with the issues of uncertainty and precision. In contrast, fuzzy logic is derived from the fact that most modes of human reasoning, especially common sense, are approximate in nature. Models based on fuzzy logic are very useful for the major applications of control, **pattern recognition,** and **function approximation** (modeling and prediction).

In this chapter, we discuss the components of the ubiquitous **linguistic fuzzy model,** which is the original fuzzy model. Control and pattern recognition examples based on this model structure are discussed. The other model structures—the **fuzzy relational model** and **Takagi–Sugeno model**—are beyond the range of our discussion. As stated in Chapter 9, the use of fuzzy logic, often with neural networks, in **artificial intelligence** is also beyond the range of our discussion. References for these topics are provided at the end of the chapter.

11.1 HISTORICAL REVIEW

In 1965, Lotfi Zadeh at the University of California at Berkeley introduced the concept of **fuzzy sets** as a generalization of conventional set theory [1]. Fuzzy sets are mathematical models of the vagueness present in our natural language when we describe phenomena that do not possess sharply defined boundaries. These sets are rooted in the notions of similarity between situations, and are formally based on the concept of multivalued logic.

For example, consider a conventional **crisp** set that contains the set of numbers **z** between 3 and 5. We represent this set as

$$\mathbf{z} = \{r \in \Re \mid 3 \leq r \leq 5\} \tag{11.1}$$

where r represents a real number and $\Re$ represents the set of real numbers. Equivalently, we may describe **z** by its **membership,** or characteristic, **function,** $\mu(\mathbf{z})$:

$$\mu(\mathbf{z}) : \Re \to \{0, 1\} \tag{11.2}$$

where

$$\mu(\mathbf{z}) = \begin{cases} 1, & 3 \leq r \leq 5 \\ 0, & \text{otherwise} \end{cases} \tag{11.3}$$

This membership function is illustrated in Figure 11.1(a). Every real number, r, either is or is not a member of **z**.

Alternatively, consider a fuzzy set that contains the set of numbers **u** *that are close to* 4. A unique or crisp membership function no longer exists for this set. Instead, we may construct a membership function that possesses the properties of **normality** [$\mu(4) = 1$], **monotonicity** (values closer to 4 have memberships closer to 1), and **symmetry.** Two membership functions that meet these criteria are illustrated in Figure 11.1(b) and (c).

Zadeh's 1973 journal article describing the application of fuzzy logic to control [2] inspired Ebrahim Mamdani, then at Queen Mary College in London, to create the first fuzzy controller [3]. Mamdani's dissertation was concerned with the application of learning techniques to process control. Before utilizing fuzzy logic, Mamdani had incorporated Bayesian learning and rule-based approaches into controllers. Reading and understanding Zadeh's article and implementing a working controller required only one week of work [4].

The model Mamdani created is now known as the linguistic fuzzy model, or the **Mamdani model.** In this model, crisp inputs are transformed into fuzzy inputs using **fuzzification. Rule base inference** is then used to map the fuzzy inputs into fuzzy outputs. The fuzzy outputs are transformed into crisp outputs by **defuzzification.** The **knowledge base** provides the fuzzy control rule base for fuzzy inference and data for fuzzification and defuzzification membership functions. The linguistic fuzzy model is illustrated in Figure 11.2.

After this work was published, fuzzy logic research in the 1970s and early 1980s was primarily conducted in Japan and Europe. Two other models were developed in 1985: the fuzzy relational model by Witold Pedrycz [5], and the Takagi–Sugeno model by Hideyuiki Takagi and Michio Sugeno [6]. During this period, the applications of fuzzy control were centered on industrial systems. In 1987, the first consumer product (a shower head) was manufactured by Matsushita. This shower head quickly led to other consumer products, including washing machines, vacuum cleaners, camcorders, cameras, and air conditioners. According to Takagi, who conducts his research at the Central Research Laboratories at Matsushita, the proliferation of fuzzy logic devices can be traced to the Japanese consumer.

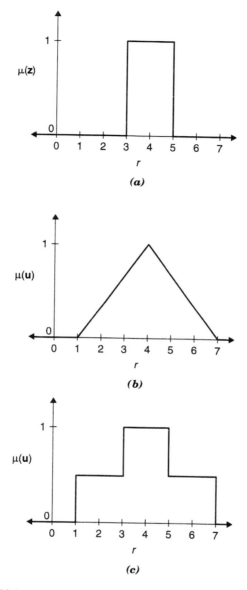

Figure 11.1 Membership functions. (a) Crisp set; (b and c) fuzzy sets.

He believes "the Japanese consumer strongly requires more intelligent and more sensitive appliances with finer capabilities. Manufacturers address this by increasing the number of sensors and the amount of information available to the device" [7].

Eventually, the enormous success of mainly Japanese commercial applications, at least partly dependent on fuzzy technologies, led to a surge of curiosity in the United States about the utility of fuzzy logic for scientific and engineering applications. IEEE began sponsoring an annual conference on fuzzy systems in 1992. One year later, Jim Bezdek at

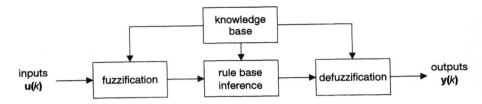

Figure 11.2 Linguistic fuzzy model.

the University of Western Florida became the founding editor of the *IEEE Transactions on Fuzzy Systems.*

11.2 FUZZIFICATION

Given a fuzzy system with multiple crisp inputs, **u**(k), let us convert these crisp inputs to fuzzy inputs, **U**(k). Each fuzzy input, $U_{mi}(k)$ is an ordered pair containing a **linguistic label,** $label_i$, and **degree of membership,** $\mu_{mi}[u_m(k)]$:

$$U_{mi}(k) = \{label_i, \mu_{mi}[u_m(k)]\} \tag{11.4}$$

Here, k refers to the sample and m refers to the input number. The degree of membership may also be referred to as the **possibility** (as opposed to statistical probability) with which a crisp input belongs to a membership function. The conversion from crisp to fuzzy input is known as fuzzification. Each crisp input is converted to its fuzzy equivalent using a family of membership functions, which may function as a **fuzzy partition.**

11.2.1 Membership Functions

A set of membership functions is typically represented with five to nine trapezoids. Note that a triangular function is a subset of a trapezoid (Figure 11.3). The value of a membership function for a given input is termed the degree of membership. The maximum degree of membership is termed the **height.** If the height equals 1, then a primary fuzzy set is called a **normalized** fuzzy set. The **support** of a primary fuzzy set $U_m(k)$ in the universal set $\mathcal{U}$ is a crisp set that contains all the elements of $u_m(k)$ that possess nonzero degrees of membership. The **core** of a primary fuzzy set $U_m(k)$ in the universal set $\mathcal{U}$ is a crisp set that contains all the elements of $u_m(k)$ that possess degrees of membership equal to one.

The structure of the membership functions may be determined empirically or analytically. Often, empirical determination of membership functions is sufficient for development of a solid fuzzy model. Analytical construction of the membership functions is accomplished through optimization techniques such as neural networks or least squares estimation that are beyond the range of our discussion.

11.2.2 Fuzzy Partitions

A family of primary fuzzy sets or membership functions may be described as a fuzzy partition if three constraints are met. First, the family of primary fuzzy sets must fully de-

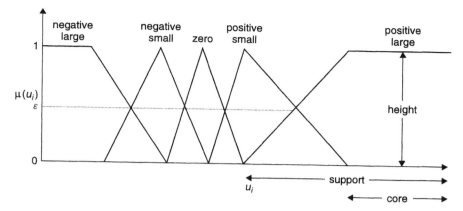

Figure 11.3 Group of membership functions that qualifies as a fuzzy partition. Definitions are illustrated for the positive large membership function.

scribe the universal set $\mathcal{U}$. In other words, each element of this universal set must be assigned to at least one membership function with a nonzero degree of membership. The **level of coverage** is the minimum value, ε. Because each membership function is associated with a linguistic label, this property assures that any piece of information defined in $\mathcal{U}$ is properly represented in terms of a generic linguistic label (Figure 11.3). Second, each membership function must be unimodal. In this way, each membership region possesses a clear semantic meaning. Third, according to Pedrycz, all degrees of membership at any point of $\mathcal{U}$ must sum up to 1 [5].

When these constraints are satisfied, a family of primary fuzzy sets may be considered a fuzzy partition. A fuzzy partition becomes more **specific** as the total number, or **cardinality**, of membership functions and primary fuzzy sets, P, increases. The **scope of perception** allows elements of $\mathcal{U}$ with the highest grades of membership to be grouped together. These sets also exhibit **robustness**, that is, input information imprecision is well tolerated. With overlapping membership functions, decisions are distributed over more than one input class, making the system more robust. Therefore, when an input is degraded by noise, the resulting output is minimally affected.

11.2.3 Example

Let us discuss the fuzzy controller within a washing machine developed by Hitachi Corporation [8]. Because the original description was given in Japanese, we rely on the discussion by Schwartz et al. to describe a simplified version of this controller [9].

The inputs for control are the amount of clothes, $u_1(k)$, and quality of clothes, $u_2(k)$. The amount of clothes is determined by the force of rotation caused by the clothes when the motor has been turned off after initial rotation with a small amount of water. Amount of clothes is partitioned as "small," $\mu_{11}[u_1(k)]$; "normal," $\mu_{12}[u_1(k)]$, and "large," $\mu_{13}[u_1(k)]$. Quality of clothes is determined by force of rotation caused by the clothes when the motor has been turned off after a second rotation with a larger amount of water. Quality of clothes is partitioned as "soft," $\mu_{21}[u_2(k)]$; "rather soft," $\mu_{22}[u_2(k)]$; "rather hard," $\mu_{23}[u_2(k)]$; and "hard," $\mu_{24}[u_2(k)]$. The membership functions for each linguistic variable are given in Figure 11.4.

244 FUZZY MODELS

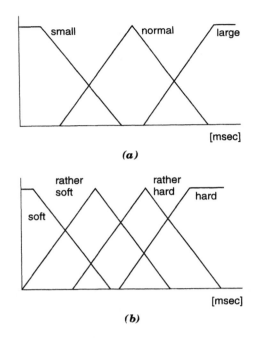

Figure 11.4 Input membership functions. (a) Amount of clothes. (b) Quality of clothes. From [9].

Let us assume that the value of $u_1(k)$ results in two fuzzy inputs:

$$U_{11}(k) = \{\text{normal}, \mu_{11}[u_1(k)]\} \tag{11.5}$$

$$U_{12}(k) = \{\text{large}, \mu_{12}[u_1(k)]\} \tag{11.6}$$

To simplify the problem, let us assume that the value of $u_2(k)$ results in only one fuzzy input (although using the membership functions in Figure 11.4 could result in three fuzzy inputs):

$$U_{21}(k) = \{\text{rather hard}, \mu_{21}[u_2(k)]\} \tag{11.7}$$

11.3 RULE BASE INFERENCE

Once the fuzzy inputs have been determined, the input labels are used to derive the output labels for various combinations, based on a given set of rules. The various combinations of fuzzy input degrees of membership are also combined using fuzzy logic to determine the output degrees of membership. This process is known as rule base inference. As with the fuzzy inputs, each fuzzy output, $Y_{ni}(k)$, consists of an ordered pair containing a linguistic label and degree of membership:

$$Y_{ni}(k) = \{\text{label}_i, \mu_{ni}[y_n(k)]\} \tag{11.8}$$

11.3.1 Rules

A rule, or **fuzzy conditional statement,** is composed of an **antecedent** and a **consequent.** An antecedent contains several preconditions; a consequent contains one or more output actions. The antecedents and consequents utilize linguistic labels. An example of a rule is "IF amount of clothes is small and quality of clothes is soft, THEN churning strength is weak and washing time is short." The maximum number of rules is equal to the product of the cardinalities of the fuzzy partitions or membership functions. Rules may be determined by empirical modeling of the system. Alternatively, rules may be determined using a neural network. For efficient implementation, rules are often arranged in tables.

11.3.2 Fuzzy Logic

The mathematical operators used to combine the fuzzy input degrees of membership are called fuzzy set intersections. We consider three common fuzzy set intersections, $[\mu_{ij}(k) \cap \mu_{mj}(k)]$. The first is the **Zadeh intersection,** which results in taking the minimum degree of membership:

$$[\mu_{ij}(k) \cap \mu_{mj}(k)]_Z = [\mu_{ij}(k) \wedge \mu_{mj}(k)] = \min\,[\mu_{ij}(k), \mu_{mj}(k)] \qquad (11.9)$$

The second is the **product intersection,** which results in taking the product of the two degrees of membership:

$$[\mu_{ij}(k) \cap \mu_{mj}(k)]_P = [\mu_{ij}(k) \cdot \mu_{mj}(k)] \qquad (11.10)$$

The third is the **Lukasiewicz intersection,** which results in taking the maximum of the two degrees of membership:

$$[\mu_{ij}(k) \cap \mu_{mj}(k)]_L = [\mu_{ij}(k) \vee \mu_{mj}(k)] = \max\,[\mu_{ij}(k), \mu_{mj}(k)] \qquad (11.11)$$

In the first fuzzy controller, Mamdani used the Zadeh intersection. These intersections easily generalize with three or more inputs.

11.3.3 Example

Continuing with our washing machine example, the twelve basic rules for control are illustrated in Table 11.1. The control outputs are "churning strength," which is listed as the top output for each input combination, and "washing time," which is listed as the bottom output for each input combination. For our assumed inputs in Eqs. (11.5)–(11.7), we use the Zadeh intersection to determine the output degrees of membership. The output membership functions are shown in Figure 11.5. As shown in Table 11.1, the inputs "large amount" and "rather hard quality" result in the outputs "strong churning" and "long washing time." Now, applying Zadeh intersection, as shown in Figure 11.6, these outputs possess degrees of membership equal to that of "rather hard quality." Similarly, the inputs "normal amount" and "rather hard quality" result in the outputs "normal churning" and "normal washing time," with degrees of membership equal to normal amount.

246 FUZZY MODELS

Table 11.1 Washing machine rules. From [9]

		Amount of clothes		
		Small	Normal	Large
Quality of clothes	Soft	weak short	rather weak short	normal normal
	Rather soft	rather weak soft	normal normal	normal normal
	Rather hard	rather weak short	normal normal	strong long
	Hard	rather weak short	normal normal	strong long

11.4 DEFUZZIFICATION

After the fuzzy outputs have been determined, they are transformed to crisp outputs. This process is known as defuzzification. Defuzzification procedures are used to select an adequate decision among those deemed adequate by the output possibility distribution. Because one or more outputs may have been determined by rule base inference, these out-

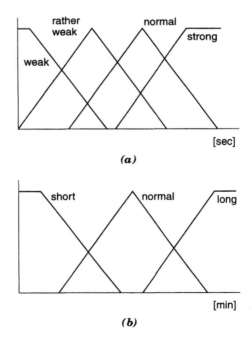

Figure 11.5 Output membership functions. (a) Strength of churning. (b) Length of washing. From [9].

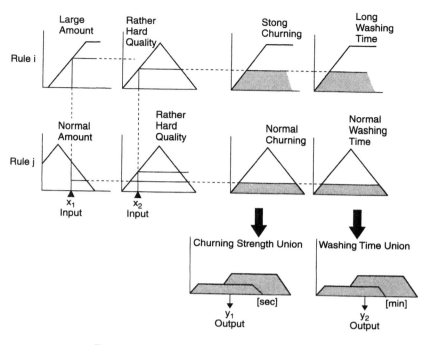

Figure 11.6 Rule base inference example. From [9].

put sets are combined. According to the **disjunctive interpretation** of a fuzzy relation, which was posed by Zadeh, Mamdani, and Assilian, the outputs must be combined by union to approximate the compatibility relation. From the union of the outputs, the crisp output is then determined by taking the **centroid** (center of area), maximum, or mean of maximum (Figure 11.7). The use of the Zadeh intersection for rule base inference and the centroid for defuzzification is known as the **Mamdani max-min** (union, Zadeh intersection) **inference.**

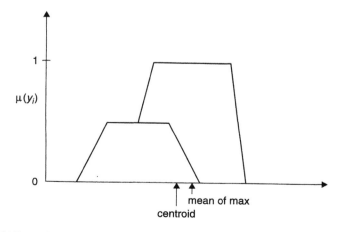

Figure 11.7 Defuzzification strategies. The max strategy does not result in a unique value.

11.4.1 Centroid

The centroid method generates the center of gravity of the possibility distribution of an output. It is a popular method because it is simple to implement. Assuming that q discrete crisp values exist for an output, $y_{ni}(k)$, then the output is calculated as

$$y_{ni}(k) = \frac{\sum_{s=1}^{q} \mu_{ns}[y_s(k)] y_s(k)}{\sum_{s=1}^{q} \mu_{ns}[y_s(k)]} \qquad (11.12)$$

where $y_s(k)$ are the discrete points in the relevant domain.

11.4.2 Maximum

The maximum method merely generates the one or more crisp values at which the possibility distribution of the output reaches its maximum value.

11.4.3 Mean of Maximum

The mean of maximum method generates the crisp value that corresponds to the mean value of all outputs whose membership functions reach the maximum. It is expressed as

$$y_{ni}(k) = \frac{\sum_{s=1}^{t} w_s(k)}{t} \qquad (11.13)$$

where $w_s(k)$ is the support value at which the membership function reaches the maximum value and t is the number of such support values.

11.4.4 Example

We calculate the outputs of our washing machine controller by determining the union of each set of fuzzy outputs and using the centroid method to obtain crisp output values. The results are shown in Figure 11.6. The union of strong and normal churning results in crisp output $y_1(k)$. That is, Eq. (11.12) is used with these output membership functions to calculate the center of gravity, $y_1(k)$. Similarly, the union of long and normal washing time results in crisp output $y_2(k)$.

11.5 KNOWLEDGE BASE

The knowledge base comprises knowledge of the application domain and the modeling goals. From the knowledge base, the input and output fuzzy partitions (or membership functions) and rule base are established. Aspects that are taken into account in construction of the membership functions and rule base are characteristics of human control behavior, development of process skills, individual differences between process operators, task factors affecting performance, and organization of the operator's behavior. The membership functions are selected to serve as meanings for the linguistic labels in the inference rules.

Normally, for a system designed exclusively on the basis of intuitive knowledge, fine tuning of the membership functions and rule base amounts to observing the results and then intuitively adjusting membership functions that were derived from intuition in the first place. For simple fuzzy models, such as those with two inputs, this trial and error method is adequate. However, with larger systems, the trial and error method is less feasible. Instead, neural networks may be utilized for efficient fine tuning.

11.5.1 Neural Network Tuning

The multilayer feedforward network discussed in Chapter 9 may be used to determine rule base inference. Recall that the parameters, or weights, of this nonlinear model are determined by presenting input/output pairs to the network, with the weights adjusted in order to minimize the mean squared error between the observed and true outputs. Assuming fixed input and output membership functions, these networks can learn and extrapolate complex relationships between the antecedents and consequents for rules containing single and conjunctive antecedent clauses. Further, multiple compatible rules can be stored in a single network structure, providing a natural mechanism for conflict resolution.

The utilized network is a three-layer network with two hidden layers and an output layer. The inputs and outputs are degrees of membership of the antecedents and consequents, respectively. Hidden layer one may not be fully connected to hidden layer two to ease the training burden. However, hidden layers one and two are fully connected to the input layer and the output layer, respectively (Figure 11.8). The known antecedent/consequent rules are used to train the network. After convergence, new antecedents may be fed into the network to generate corresponding new consequents [10].

For example, let us construct a network for the input and output fuzzy partitions in Figure 11.9. In this example, each input fuzzy partition contains 13 crisp values. Note that

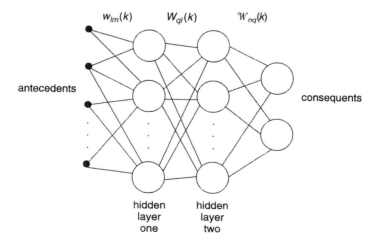

Figure 11.8 Multilayer feedforward network for tuning rule base.

250 FUZZY MODELS

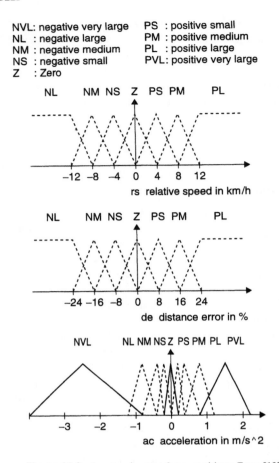

Figure 11.9 Input and output fuzzy partitions. From [13].

the inputs are the vertices of the membership functions. Therefore, there are 26 inputs to the neural network. Let us choose 20 neurons for each hidden layer. Because there are 62 crisp values in the output fuzzy partition, there are also 62 output neurons.

We may train the network using the subset of the rules we know from experience, i.e., a fraction of the rules presented in Table 11.2. For example, the rule "IF relative speed is negative small and distance error is zero, THEN acceleration is zero" translates into the input (listed top to bottom)

$$\{0\ 0\ 0\ 0.5\ 1\ 0.5\ 0\ 0\ 0\ 0\ 0\ 0\quad 0\ 0\ 0\ 0\ 0\ 0.5\ 1\ 0.5\ 0\ 0\ 0\ 0\ 0\ 0\}$$

and output

$$\{00000000000000000000000000000000\ 0.5\ 1\ 0.5\ 00000000000000000000\}.$$

After converting the rules in Table 11.2 to input/output pairs, the resulting training sets are presented to the multilayer feedforward network. The weights are initially randomized

Table 11.2 Distance controller rules. From [13]

		de						
		too close → danger!				too far		
rs		NL	NM	NS	Z	PS	PM	PL
rs < 0 → danger	NL	NVL	NVL	NVL	NL	NM	NS	NS
	NM	NVL	NL	NM	NS	Z	Z	Z
	NS	NL	NM	NS	Z	Z	Z	Z
	Z	NM	NS	Z	Z	Z	PS	PS
re > 0	PS	NS	Z	Z	Z	Z	PM	PL
	PM	NS	Z	Z	PS	PM	PL	PVL
	PL	NS	Z	Z	PS	PL	PVL	PVL

values that converge over time as the performance funciton of mean squared error is minimized.

After convergence, we may then convert the antecedents not used in training to numerical inputs, and present them to the network with converged weights. The resulting outputs can then be converted to their corresponding consequents. Thus, by using this method, we can derive the remaining system rules.

11.6 MODEL VALIDATION

Model validation is not as straightforward with a fuzzy model as with a traditional mathematical model. It is possible, however, to analyze the coverage of the input space by the rules. For an incomplete rule base, additional rules may be constructed from prior knowledge. The antecedents of these rules may be created from unused combinations of membership functions in the initial model. The identification data usually cover only a fraction of the complete product space of the model parameters. Therefore, the antecedents of the obtained rules include only those combinations of the linguistic terms that were identified from the data. It is possible that regions not covered by any rules are entered during simulation or prediction. This situation can be detected by observing the output degree of membership in the rule base. If no rule is activated above a specified threshold, an additional rule may be added to the rule base. The antecedent of this rule is given by the combination of linguistic terms that give the highest output degree of membership for the given data point [11].

11.7 FUZZY CONTROL

The most widely developed application for fuzzy models is fuzzy control. In conventional control theory, control of a moderately complex system is based on mathematical modeling of the system and use of the system output to generate a corrective action on a system input, thus attempting minimization of the error in an output. Such a conventional controller is often designed to operate in a narrow performance band. However, it is often impossible to derive a mathematical law of system behavior from physical principles because these principles are too complex. Fuzzy set theory provides an alternative methodology for modeling a complex, nonlinear system that operates over a wider performance band. Additionally, fuzzy control systems tend to require less development time than their conventional counterparts, and are therefore more cost-effective.

A fuzzy controller may be based on the linguistic fuzzy model in Figure 11.2. For control, the crisp outputs are fed back as inputs to the system. The crisp inputs are system outputs (Figure 11.10). The simplest controller utilizes two crisp inputs and one crisp output. Often, these inputs are an error parameter and change in error parameter. In practice, fuzzy control devices based on triangular and trapezoidal functions seem to perform as well as those based on some other form of membership function. Additionally, these functions are easier to implement [8]. As a rule of thumb, a minimum of five membership functions per input is required for smooth control.

Various approaches for analyzing the stability of a fuzzy controller have been proposed. These approaches requires several limiting assumptions, such as symmetric membership functions or systems with an order less than two [12]. Mamdani has argued that stability analysis is not a necessary requirement for acceptance by industry of a well-designed control system, and that prototype testing is more important than statistical analysis [4].

11.7.1 Fuzzy Control Example

Fuzzy control was applied to intelligent cruise control by Muller and Nocker at Daimler-Benz [13]. An intelligent cruise controller, called an **autonomous intelligent cruise control (AICC) system,** maintains speed and distance in a passenger car from preceding vehicles. Ordinary cruise control systems are of limited utility because increasing traffic density rarely enables driving at a preselected speed.

The test vehicle under investigation was a Merced-Benz 300E equipped with a five-beam infrared distance sensor and other sensors to measure the speed, acceleration, and steering angle of the vehicle. Driver actions were also monitored by measuring the actions on the pedals and switches. Every 100 ms, each beam of the infrared sensor delivered the distance to the detected target, the relative speed between the host vehicle and target, and information about the target size. The first input to the controller was the percentage dis-

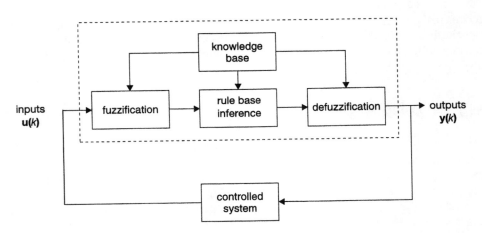

Figure 11.10 Fuzzy controller.

tance error, $de(k)$, using a reference distance of half the speed of the host vehicle in meters. The second input was the relative speed, $rs(k)$, between the two vehicles. The control output was the nominal value for acceleration, $ac(k)$, for the drive-by-wire host vehicle actuator system, which was limited to the range $\{-2.5, +1.5\}$ m/s^2. The fuzzy partitions for inputs and output are shown in Figure 11.9.

The rule set shown in Table 11.2 reflected the knowledge of an experienced driver and took into account special features of the sensor data. For example, if the target vehicle is driving faster than the host vehicle [$rs(k) > 0$], an experienced driver would not brake hard but would maintain speed or brake only slightly. To make the controller insensitive to noise, outputs of zero were located close to the operating point of $de(k)$ and $re(k)$ approximately equal to zero. Based on fine tuning, it was determined that the min–max inference did not result in as smooth control as the combination of product intersection and bounded sum union. Further, it was determined that the definition of reference distance needed refinement and acceleration behavior needed to be adapted to the driving situation.

An advanced AICC controller, composed of two fuzzy blocks, is shown in Figure 11.11. Note that the reference distance was modified to a function of the driver behavior, weather, and speed. The updated acceleration controller also utilized the steering angle as an input. Each of the two fuzzy blocks was refined into two fuzzy modules, each possessing at most three inputs. The advanced controller contained 200 rules.

Measurements on a highway are given in Figure 11.12. The target vehicle was driven at a speed of ~ 88 km/hr. After 17 seconds, the host vehicle was accelerated at a rate of

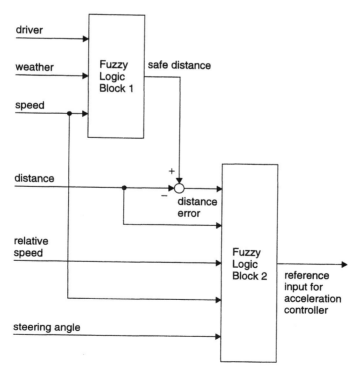

Figure 11.11 Block diagram of advanced distance controller. From [13].

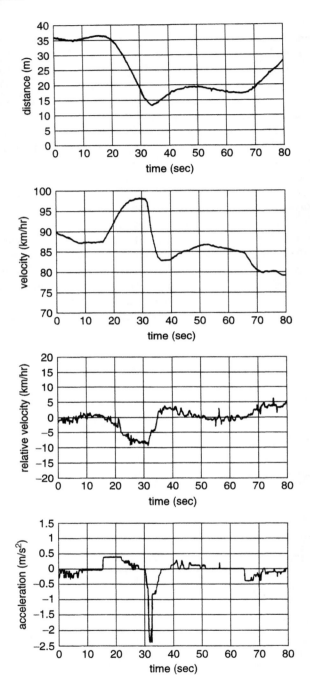

Figure 11.12 Measurements of real test drives. From [13].

0.4 m/s², closing in on the target vehicle. This caused the controller to apply the brake for a short time (last diagram, 30–33 sec). After a small overshoot of ~ 4 m, the reference distance was again established (top diagram at 40 sec).

11.8 FUZZY PATTERN RECOGNITION

In a similar manner, pattern recognition may be achieved by empirical tuning of the linguistic fuzzy model. Alternatively, fuzzy logic may be combined with the self-organization concepts discussed in Chapter 9 to achieve pattern recognition. We focus on the specific pattern recognition problem of **classification.**

11.8.1 Empirical Classification

The membership functions and rule base inference of the linguistic fuzzy model may be empirically tuned using expert knowledge to achieve pattern recognition. That is, a fuzzy model may be constructed that classifies specific patterns. For example, this model was recently applied to bacterial classification on the basis of **pulsed-field gel electrophoresis (PFGE)** [14]. Specific PFGE patterns from DNA appear to correlate with *E. coli* O158:H7 strains associated with the most severe human illness. During the PFGA procedure, DNA is digested by restriction enzymes into a number of large fragments measured in kilobase pairs (kb) that are secured to an agar block/rectangular gel slab. Under electrical stimulation, DNA fragments of different sizes advance particular distances through the gel slab. The resulting linear array of fragments is characteristic of different species of bacteria.

For example, the PFGE images of *E. coli* O158:H7 and non-*E. coli* O158:H7 isolates are shown in Figure 11.13. The labeled lanes A and B contain marker bands of known size. Primary lane measurements such as locations, intensities, and widths of bands with a certain range along the intensity profile were computed directly from 240 scanned images. By observation and study by microbiologists with expert knowledge, features were determined, such as a major band at 300 kb for *E. coli* O158:H7 patterns. These features are given in Table 11.3. One family of membership functions developed for the feature of intensity of the band at 300 kb is shown in Figure 11.14.

Based on these features, rules were generated. One rule was "IF there is a narrow and bright band at approximately 300 kb and there is a bright band at 450 kb, THEN the pattern is very likely *E. coli* O158:H7." Three groups of rules were used to determine the confidence level of identification of *E. coli* O158:H7. These are listed in Table 11.4. The rules and membership functions were trained using 120 *E. coli* O158:H7 and 120 non-*E. coli* O158:H7 patterns. After training, 95% of *E. coli* O158:H7 and 90.8% of non-*E. coli* O158:H7 training patterns were correctly identified.

11.8.2 Nearest Prototype Classification

Rather than using empirical tuning, a fuzzy model may be developed using the concept of **similarity.** As we have seen in Chapter 9, data may be classified by unsupervised learning based on minimization of a measure of similarity. The usual choice for measuring similarity is the square of the **Euclidean distance** between an input pattern and a template pattern. Kohonen's self organizing map (Section 9.3) is just one of many clustering algo-

256 FUZZY MODELS

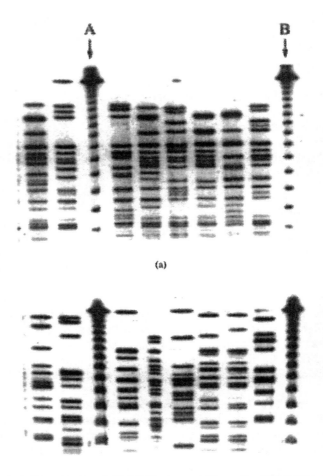

Figure 11.13 PFGE images of (a) *E. coli* O158:H7 isolates and (b) non-*E. coli* O158:H7 isolates. From [14].

Table 11.3 Features (variables) used for identification of *E. coli* O158:H7. From [14]

Variable names	Description
B3Loc	300 kb band's location w.r.t. markers
B3RI	300 kb band's relative height
B3Wid	300 kb band's relative width
B45Loc	450 kb band's location w.r.t. markers
B45RI	450 kb band's relative height
UpB45RI	upper split 450 kb band's relative
LoB45RI	lower split 450 kb band's relative
Center	center of split 450 kb bands
EcoliConfd	confidence value of *E. coli* O157:7

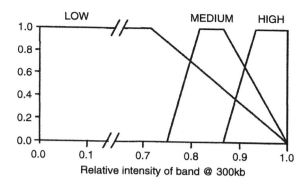

Figure 11.14 Membership functions for variable representing intensity of the band at 300 kb, used to model the variable brightness in this band. From [14].

rithms based on Euclidean distance. By iterating with this measure, similar **clusters** of data are formed. The data in the resulting clusters may be relabeled. Cluster validation is then performed using a well-defined performance objective, such as minimum apparent error rate, to validate the new fuzzy labels [15].

11.9 SUMMARY

In this chapter, we have reviewed the basic concepts underlying nonlinear system identification, assuming that the system operator of a time-invariant multiple input, multiple output system is sufficiently complex that it may not easily be described by mathematical equations. Given such a system, the system operator may be described using fuzzy logic.

As its name implies, fuzzy logic is the logic underlying modes of reasoning that are approximate rather than exact. Conventional approaches to knowledge representation are based on bivalent logic. However, these approaches are unable to deal with the issues of uncertainty and precision. In contrast, fuzzy logic is derived from the fact that most modes of human reasoning, especially common sense, are approximate in nature. Models based on fuzzy logic are very useful for the major applications of control, pattern recognition, and function approximation.

Lotfi Zadeh's 1973 journal article describing the application of fuzzy logic to control inspired Ebrahim Mamdani, then at Queen Mary College in London, to create the first fuzzy controller. The model Mamdani created is now known as the linguistic fuzzy model, or the Mamdani model. In this model, crisp inputs are transformed into fuzzy inputs using fuzzification. Rule base inference is then used to map the fuzzy inputs into fuzzy outputs. The fuzzy outputs are transformed into crisp outputs by defuzzification. The knowledge base provides the fuzzy control rule base for fuzzy inference and data for fuzzification and defuzzification membership functions.

Given a fuzzy system with multiple crisp inputs, $\mathbf{u}(k)$, we convert these crisp inputs to fuzzy inputs, $\mathbf{U}(k)$. Each fuzzy input, $U_{mi}(k)$ is an ordered pair containing a linguistic label, $label_i$, and degree of membership, $\mu_{mi}[u_m(k)]$. Here, k refers to the sample. The degree of membership may also be referred to as the possibility with which a crisp input belongs to a membership function. The conversion from crisp to fuzzy input is known as fuzzifi-

Table 11.4 Three groups of rules for identification of *E. coli* O158:H7. From [14]

Rules of Group I—General Variations within *E. coli* O157:H7

B3Loc	B3RI	B45Loc	B45RI	UpB45RI	LoB45RI	Center	EcoliConfd
Exact	High	Exact	High				High
Exact	High	Exact	Medium				Medium
Exact	High	SWExact	Medium				Medium
Exact	High	SWExact	High				Medium
SWExact	High	Exact	High				High
SWExact	High	Exact	Medium				Medium
SWExact	High	SWExact	Medium				Medium
SWExact	High	SWExact	High				Medium
Exact	Medium	Exact	High				High
Exact	Medium	Exact	Medium				Medium
Exact	Medium	SWExact	Medium				Medium
Exact	Medium	SWExact	High				Medium
SWExact	Medium	Exact	High				Medium
SWExact	Medium	Exact	Medium				Medium
SWExact	Medium	SWExact	Medium				Medium
SWExact	Medium	SWExact	High				Medium
Exact	High	Very Low		High	High	Exact	High
Exact	High	Very Low		High	Medium	Exact	Medium
Exact	High	Very Low		Medium	High	Exact	Medium
Exact	High	Very Low		Medium	Medium	Exact	Medium
Exact	High	Very Low		High	High	SWExact	High
Exact	High	Very Low		High	Medium	SWExact	Medium
Exact	High	Very Low		Medium	High	SWExact	Medium
Exact	High	Very Low		Medium	Medium	SWExact	Medium
SWExact	High	Very Low		High	High	Exact	Medium
SWExact	High	Very Low		High	Medium	Exact	Medium
SWExact	High	Very Low		Medium	High	Exact	Medium
SWExact	High	Very Low		Medium	Medium	Exact	Medium
SWExact	High	Very Low		High	High	SWExact	Medium
SWExact	High	Very Low		High	Medium	SWExact	Medium
SWExact	High	Very Low		Medium	High	SWExact	Medium
SWExact	High	Very Low		Medium	Medium	SWExact	Medium
SWExact	Medium	Very Low		High	High	Exact	Medium
SWExact	Medium	Very Low		High	Medium	Exact	Medium
SWExact	Medium	Very Low		Medium	High	Exact	Medium
SWExact	Medium	Very Low		Medium	Medium	Exact	Medium
SWExact	Medium	Very Low		High	High	SWExact	Medium
SWExact	Medium	Very Low		High	Medium	SWExact	Medium
SWExact	Medium	Very Low		Medium	High	SWExact	Medium
SWExact	Medium	Very Low		Medium	Medium	SWExact	Medium
Exact	Medium	Very Low		High	High	Exact	Medium
Exact	Medium	Very Low		High	Medium	Exact	Medium
Exact	Medium	Very Low		Medium	High	Exact	Medium
Exact	Medium	Very Low		Medium	Medium	Exact	Medium
Exact	Medium	Very Low		High	High	SWExact	Medium
Exact	Medium	Very Low		High	Medium	SWExact	Medium
Exact	Medium	Very Low		Medium	High	SWExact	Medium
Exact	Medium	Very Low		Medium	Medium	SWExact	Medium

Table 11.4 *Continued*

Rules of Group 2

B3Loc	B3RI	B45Wid	B45Loc	B45RI	LoB45RI	Center	EcoliConfd
Very Exact	High	Very Small	Exact	Medium			High
Very Exact	Medium	Very Small	Exact	Medium			High
Very Exact	High	Very Small			Medium	Exact	High
Very Exact	Medium	Very Small			Medium	Exact	High

Rules of Group 3

B3RI	B3Wid	B45RI	UpB45RI	LoB45RI	Center	EcoliConfd
Low						Small
	Large					Small
		Low	Low			Small
		Low		Low		Small
Medium		Medium				Small
		Low			Not Exact	Small

cation. Each crisp input is converted to its fuzzy equivalent using a family of membership functions, which may function as a fuzzy partition.

Once the fuzzy inputs have been determined, the input labels are used to derive the output labels for various combinations, based on a given set of rules. A rule, or fuzzy conditional statement, is composed of an antecedent and consequent. An antecedent contains several preconditions; a consequent contains one or more output actions. The antecedents and consequents utilize linguistic labels. The various combinations of fuzzy input degrees of membership are also combined using fuzzy logic to determine the output degrees of membership. The mathematical operators used to combine the fuzzy input degrees of membership are called fuzzy set intersections. This process is known as rule base inference. As with the fuzzy inputs, each fuzzy output, $Y_{ni}(k)$, consists of an ordered pair containing a linguistic label and degree of membership.

After the fuzzy outputs have been determined, they are transformed into crisp outputs. This process is known as defuzzification. Defuzzification procedures are used to select an adequate decision among those deemed adequate by the output possibility distribution. Because one or more outputs may have been determined by rule base inference, these output sets are combined. According to the disjunctive interpretation of a fuzzy relation, which was posed by Zadeh, Mamdani, and Assilian, the outputs must be combined by union to approximate the compatibility relation. From the union of the outputs, the crisp output is then determined by taking the centroid, maximum, or mean of maximum.

The knowledge base comprises knowledge of the application domain and the modeling goals. From the knowledge base, the input and output fuzzy partitions and rule base are established. Aspects that are taken into account in construction of the membership functions and rule base are characteristics of human control behavior, development of process skills, individual differences between process operators, task factors affecting performance, and organization of the operator's behavior. The membership functions are selected to serve as meanings for the linguistic labels in the inference rules.

Normally, for a system designed exclusively on the basis of intuitive knowledge, fine tuning of the membership functions and rule base amounts to observing the results and then intuitively adjusting membership functions that were derived from intuition in the first place. For simple fuzzy models, such as those with two inputs, this trial and error method is adequate. However, with larger systems, the trial and error method is less feasible. Instead, neural networks may be utilized for efficient fine tuning.

11.10 REFERENCES

[1]. Zadeh, L. A. Fuzzy sets. *Information and Control, 8,* 338–353, 1965.

[2]. Zadeh, L. A. Outline of a new approach to the analysis of complex systems and decision processes. *IEEE Trans SMC, 3,* 28–44, 1973.

[3]. Mamdani, E. H. and Assilian, S. An experiment within linguistic synthesis with a fuzzy logic controller. *Int J Man-Machine Studies, 7,* 1–13, 1975.

[4]. Mamdani, E. H. Twenty years of fuzzy control: Experiences gained and lessons learnt. In *Proceedings IEEE International Conference on Fuzzy Systems.* San Francisco, CA, March, 1993. pp. 339–342.

[5]. Pedrycz, W. An identification algorithm in fuzzy relational systems. *Fuzzy Sets and Systems, 103,* 153–167, 1994.

[6]. Takagi, H. and Sugeno, M. Fuzzy identification of systems and its application to modeling and control. *IEEE Trans Sys Man Cyber, 15,* 116–132, 1985.

[7]. Takagi, H. Application of neural networks and fuzzy logic to consumer products. In *Proceedings International Conference on Industrial Electronics, Control, Instrumentation, and Automation. Power Electronics and Motion Control.* San Diego, CA, November, 1992. vol. 3, pp. 1629–1633.

[8]. Matsumoto, K. and Shikamori, T. Fuzzy control for fully automatic washer. *J Japan Soc for Fuzzy Theory and Systems, 2,* 492–497, 1990.

[9]. Schwartz, D. G., Klir, G. J., Lewis, H. W., and Ezawa, Y. Applications of fuzzy sets and approximate reasoning. *Proc IEEE, 82,* 482–498, 1994.

[10]. Keller, J. M. and Tahani, H. Backpropagation neural networks for fuzzy logic. *Info Sci, 62,* 205–221, 1992.

[11]. Babuska, R. *Fuzzy Modeling for Control.* Kluwer: Boston, 1998.

[12]. Kandel, A., Luo, Y., and Zhang, Y. Q. Stability analysis of fuzzy control systems. *Fuzzy Sets and Systems, 105,* 33–48, 1999.

[13]. Muller, R., and Nocker, G. Intelligent cruise control with fuzzy logic. In *Proceedings Intelligent Vehicles Symposium.* Detroit, MI, June, 1992. pp. 173–178.

[14]. Wang, D., Keller, J. M., Carson, C. A., McAdoo-Edwards, K. K., and Bailey, C. W. Use of fuzzy-logic-inspired features to improve bacterial recognition through classifier fusion. *IEEE Trans Sys Man Cyber, Part B, 28,* 583–591, 1998.

[15]. Baraldi, A. and Blonda, P. A survey of fuzzy clustering algorithms for pattern recognition—Part II. *IEEE Trans Sys Man Cyber, 29,* 786–801, 1999.

Further Reading

Fuzzy Logic:

Nguyen, H. T., and Walker, E. A. *A First Course in Fuzzy Logic.* CRC Press: Boca Raton, FL, 1997.

Kosko, B. *Fuzzy Thinking: the New Science of Fuzzy Logic.* Hyperion: New York, 1993.

Fuzzy Models:

Kandel, A. and Langholz, G., eds. *Fuzzy Control Systems.* CRC Press, Boca Raton, FL. 1994.

Fuzzy Pattern Recognition:

Bezdek, J. C., Keller, J., Krisnapuram, J. K., and Pal, M. *Fuzzy Models and Algorithms for Pattern Recognition and Image Processing.* Kluwer: Boston, 1999.

Artificial Intelligence:

Chen, C. H., ed. *Fuzzy Logic and Neural Network Handbook.* McGraw-Hill, New York: 1996.

Gupta, M. M., and Sinha, N. K. *Intelligent Control Systems: Theory and Applications.* IEEE Press: Piscataway, NJ, 1996.

11.11 RECOMMENDED EXERCISES

Fuzzy Logic: see Mendel, J. M. *Uncertain Rule-Based Fuzzy Logic Systems: Introduction and New Directions.* Prentice Hall PTR: Upper Saddle River, 2001. Exercises 1-2, 1-5, 1-12, and 1-17.

Fuzzy Control: see Mendel, J. M. *Uncertain Rule-Based Fuzzy Logic Systems: Introduction and New Directions.* Exercises 5-1, 5-6, 5-12, and 6-7.

12

CONTINUOUS NONINVASIVE BLOOD PRESSURE MONITORING: PROOF OF CONCEPT

In this chapter, we discuss the application of a fuzzy model to beat-to-beat control of a noninvasive sensor that continuously measures **blood pressure** (BP). During surgery, it is often desirable to measure blood pressure continuously to avoid the negative effects of anesthesia. However, with oscillometry, the noninvasive technology typically accepted in operating rooms, several beats are required for each updated blood pressure measurement. For a new noninvasive sensor, it is shown that the required pressure applied to this sensor over the radial artery to obtain accurate blood pressure readings may be continually updated using fuzzy control. A method for **continuous noninvasive blood pressure** (CNIBP) monitoring is considered one of the "holy grails" of medical instrumentation monitoring.

12.1 PHYSIOLOGY

As we discussed in Chapter 6, blood pressure originates in the left ventricle during the cardiac cycle. During systole, an **isovolumetric contraction period** and **ejection period** occur (Figure 6.2). At the onset of ventricular systole, the intraventricular pressure rises, causing immediate closure of the AV valves. At first, the arterial valves also remain closed, causing continual contraction around the incompressible contents. When the intraventricular pressure exceeds the diastolic pressure of ~ 80 mmHg, the arterial valves open and blood begins to be expelled. Initially, the intraventricular pressure continues to rise, until it reaches a maximum of ~ 130 mmHg. Toward the end of systole, the pressure begins to fall.

During diastole, an **isovolumetric relaxation period** and **filling period** occur. Initially, for ~ 50 ms, all the valves remain closed. The resulting relaxation is thus isovolumetric, as the intraventricular pressure falls rapidly to almost zero. When the pressure is lower than the atrial pressure, the AV valves open, causing the ventricle to fill in preparation for the next systole. During the filling period, the intraventricular pressure rises only slightly.

In principle, the right heart encounters these same four periods. However, because the vascular resistance is lower in the pulmonary circulation, the pressure the right heart must develop in systole is considerably lower.

12.1.1 Arterial Blood Pressure

The maximum pressure is called the **systolic pressure;** the minimum pressure is called the **diastolic pressure.** The integral of the pressure over one cardiac cycle is called the **mean pressure,** or **mean arterial pressure** (MAP). The difference between the systolic and diastolic pressures is known as the **pulse pressure.** As the pressure is transported from the left ventricle to the aorta, smaller arteries, and arterioles, it experiences damping from reflections in the arterial tree. The blood and vascular wall possess viscous properties that attenuate high frequency components in the pressure waveform. The waveform expands, obtaining a **dicrotic notch** (Figure 6.2). The pulse pressure increases, and the mean pressure decreases. The pressure is then transported to the capillaries and the venous system. Blood pressure measurements are usually acquired from the arterial system. The normal value of blood pressure is not 120/80 mmHg (systolic/diastolic pressure), but is dependent on the age and gender of an individual.

12.1.2 Vascular Resistance

As discussed in Chapter 2, we know from the mechanical analog of Ohm's law that blood pressure is obtained from the product of blood flow and **systemic vascular resistance** (SVR):

$$\text{pressure} = \text{flow} \cdot \text{resistance} \quad (12.1)$$

$$BP = CO \cdot SVR \quad (12.2)$$

where CO = cardiac output. At any site in the vasculature, the blood pressure is dependent on the local vascular resistance, which is in turn a function of vessel length (l), vessel radius (r), and the viscosity of the blood (η) (Figure 12.1). According to **Poiseuille's equation,**

$$\text{resistance} = \frac{8l\eta}{\pi r^4} \quad (12.3)$$

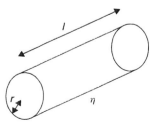

Figure 12.1 Blood vessel parameters in Poiseuille's equation.

Vascular resistance is controlled by the two divisions of the **autonomic nervous system**. The entire arterial tree is innervated by the **sympathetic division**. Additionally, the coronary, cerebral, and genital circulations are innervated by the **parasympathetic division**. In general, sympathetic nerve activation causes arterial **vasoconstriction** (decreased blood vessel radius) in all vascular beds. However, the arterioles in the skeletal muscle dilate at low concentrations of **epinephrine** and vasoconstrict with larger doses of epinephrine. Anesthetics such as **thiopental** reduce sympathetic nervous activity.

12.2 IN VIVO AND IN VITRO BLOOD PRESSURE MEASUREMENTS

Blood pressure may be measured invasively and noninvasively. The invasive gold standard of care is intraarterial measurement. Regardless of the method used, a blood pressure sensor should be positioned at heart level. Otherwise, the long columns of blood in the arterial and venous pressure systems contribute a hydrostatic pressure term to the total measured pressure. The hydrostatic pressure is the product of the blood density × the acceleration of gravity × the height above heart level.

12.2.1 Intraarterial Blood Pressure

In preparation for **intraarterial blood pressure** (IAP) measurement, a 20 gauge catheter is inserted into the artery of interest. Typically, blood pressure is monitored in the radial artery because it is highly accessible and easily visible. However, this site possesses a relatively high complication rate and high degree of disability if complication occurs. For example, the author has personally seen several surgeries delayed for over 30 minutes because an anesthesiology resident (anesthesiologist in training) punctured the radial artery at the wall directly opposite the insertion site. As a result, several minutes were required for the radial artery to "recover" before reinsertion could occur. Once the catheter is in place, it is connected by tubing to a pressure transducer. The tubing and catheter are filled with saline, in order to transmit the radial arterial pressure to the pressure transducer. The transducer is positioned at heart level to insure proper pressure readings, and zeroed to compensate for offset due to hydrostatic pressure differences. The tubing length is minimized to minimize damping of the signal (Figure 12.2). The transducer outputs continuous pressure readings that are displayed on a monitor.

Complications of IAP monitoring include **air embolism, thrombosis,** and infection. Air embolism is the introduction of air into the circulatory system. Venous air embolism may reduce or stop blood flow through the heart or may cause neurologic complications. Thrombosis involves the formation of clots at the end of a catheter that may be flushed into the circulation.

Intraarterial catheter measurement is one of the reference methods recommended by the Association for the Advancement of Medical Instrumentation (AAMI) for evaluation of blood pressure instrumentation accuracy [1].

12.2.2 Auscultatation

Auscultation is the second recommended AAMI blood pressure reference method [1]. It requires a **sphygmomanometer**—which consists of an inflatable bladder within a cuff, rubber bulb for cuff inflation, and **manometer** for pressure readings—as well as a **stetho-**

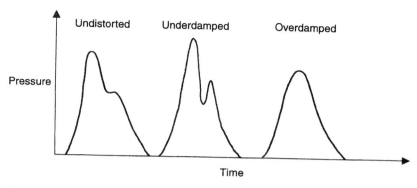

Figure 12.2 Typical undistorted, underdamped, and overdamped radial artery waveforms.

scope. Using this indirect method, a clinician inflates an external cuff surrounding the tissue over the brachial artery until the inflated pressure is above systole. It is assumed that the pressure applied to the arterial wall is equal to that of the external cuff. When a cuff of sufficient width and length is utilized, the cuff pressure is evenly transmitted to the underlying artery.

The clinician then slowly releases the pressure (at ~ 2–3 mmHg/sec), while listening for heart sounds below the level of the cuff. When the systolic peaks are higher than the cuff pressure, audible sounds called **Korotkoff sounds** are generated by the flow of blood and vibrations of the vessel under the cuff. The pressure reading at the first detection of the Korotkoff sounds indicates systolic pressure. As the cuff pressure is further decreased, the Korotkoff sounds eventually become muffled and disappear. The pressure reading during the transition from muffling to silence brackets diastolic pressure.

Several auscultatory measurements should be averaged to minimize the effects of normal respiratory and vasomotor waves. Because cuff size relative to arm circumference affects the accuracy of noninvasive measurements [2], the 1993 World Health Organization/International Society of Hypertension advised that a cuff for adults possess a bladder width of 13 to 15 cm and a bladder length of 30 to 35 cm. They further advised that larger cuffs be used for larger arms and smaller cuffs for children [3]. However, this method fails to provide accurate pressure measurements in infants and hypotensive patients. The failure of the auscultatory technique for hypotensive patients may be due to the low sensitivity of the human ear to the low frequency vibrations induced by low blood pressures [4].

12.2.3 Oscillometry

The most widely used automatic indirect blood pressure method is **oscillometry**. With this method, an external cuff is inflated above the systolic pressure over the brachial artery. The external pressure is then slowly decreased, while measuring the amplitude of pressure oscillations that are created by expansion of the arterial wall during each beat as blood is forced through the artery. At systolic pressure, the cuff pressure significantly increases in amplitude. At mean pressure, the cuff pressure reaches its maximum amplitude. Because there is no clear transition in amplitude during diastolic pressure, (Figure 12.3) proprietary algorithms are used to measure diastolic pressure. This method was developed

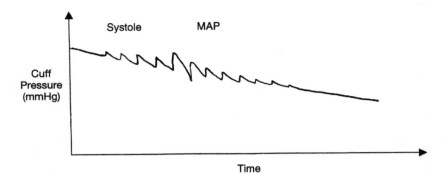

Figure 12.3 Cuff pressure oscillations during oscillometry.

by John Ramsey at Critikon in the 1970s [5]. Because it was the first mass-marketed automatic blood pressure monitor, the Critikon Dinamap monitor has traditionally been the most widely used monitor in hospitals and outpatient clinics.

As with auscultation, the cuff size relative to arm circumference affects measurement accuracy; therefore, the proper cuff size should always be used. Although the mechanism behind this phenomenon of pressure amplitudes is unknown, the MAP measurement is very robust. This property allows MAP to be measured reliably even in the case of hypotension with vasoconstriction and diminished pulse pressure. However, measurement of systolic pressure is less robust, and measurement of diastolic pressure may be questionable, depending on the proprietary algorithm utilized. Because most manufacturers do not disclose their proprietary algorithms and tend to modify algorithms without prior public notice [6–8], the guidelines of the British Hypertensive Society state that modifications to externally identical or indistinguishable versions of a model should be clearly indicated (i.e., change in device number), with full details of the modification provided and a new validation performed [9].

12.3 PROBLEM SIGNIFICANCE

If accurate, continuous NIBP would be monitored during surgery to avoid the negative effects of anesthesia. This monitoring would be especially useful in teaching hospitals, where anesthesiology residents are learning how to accurately titrate anesthesia agents. Additionally, it would be useful to track blood pressure in special populations, such as older subjects.

12.3.1 Clinical Significance

During preoperative evaluation of a physiologically compromised patient, an anesthesiologist determines the suitability of the patient to undergo surgery. Aware that about 10% of surgical patients experience a complication [10] and that the perioperative (during surgery) mortality rate for elective surgical procedures ranges from 0.0001 to 1.9% [11, 12, 13], an anesthesiologist may elect to provide more physiologic monitoring for a compromised patient. In patients receiving general anesthesia and undergoing elective surgery, hypertension is believed to not result in an increased risk, provided that two conditions hold. First, the diastolic blood pressure must be stable and not greater than 110 mmHg. Secondly, intraoperative and immediate postoperative blood pressures must be closely monitored and treated to prevent blood pressure changes of more than 50% above or 30% below the preoperative value for more than 10 minutes [14]. Due to the **vasodilative** (increased blood vessel radius) and vasoconstrictive effects of anesthestics, an anesthesiologist may elect to continuously monitor the radial arterial pressure of a hypertensive patient during surgery.

Additionally, it would be useful to monitor blood pressure continuously and noninvasively in the elderly. Steady state blood pressure is known to increase with advancing age [15], and blood pressure variability is known to increase with advancing age and higher blood pressure [16]. Thus, a continuous monitor would allow this variability to be tracked over the course of 24 hours, and the true blood pressure range determined.

12.3.2 Market Forces

In 1999, 41.3 million inpatient surgeries were performed in the United States [17]. If CNIBP monitoring were available, approximately 15% of these patients could have benefited from continuous blood pressure monitoring [18]. Assuming that the disposables for such an instrument cost $15 per application and that CNIBP monitoring replaced IBP monitoring, this could have resulted in $93 million annual U.S. income. The current income derived from blood pressure disposables is difficult to determine because many of the IAP catheters are also used for other purposes. Further, the current income derived from equipment for surgical patients receiving oscillometric monitoring is minimal because the blood pressure cuff is reused.

At the time this chapter went to press (2002), no commercial NIBP device had been clinically accepted in the United States.

12.4 PREVIOUS STUDIES

In this section, early commercial "continuous" blood pressure technologies are discussed. The accuracy of each device is evaluated against the AAMI standard of a mean difference $\leq 5 \pm 8$ mmHg between device measurements and intraarterial pressure measurements [1].

12.4.1 Finapres

In 1969, Jan Penaz of the University J. E. Purkyne v Brne in Czechoslovakia patented a CNIBP method that is commonly referred to as the **volume clamp method** or **method of Penaz** [19]. This method was subsequently developed by Karel Wesseling at the Academic Medical Center in Amsterdam and other groups [20, 21] in the 1980s. Wesselings' implementation is the basis of the Finapres (Finger artery pressure) monitor manufactured by TNO Biomedical Instruments and formerly marketed by Ohmeda [22, 23]. Using **photoplethysmography,** which is the basis of pulse oximetry (see Chapter 4), light from a **light emitting diode** (LED) is passed through a finger and detected. The detected light corresponds to absorbance of hemoglobin in the arterial blood; therefore, each period of a detected pulsation corresponds to one cardiac cycle. Simultaneously, a cuff is securely placed over the same finger, with the cuff pressure continuously adjusted to maintain a constant **transmural** (on both sides of the arterial wall) pressure across the **digital** (finger) arteries.

During self-calibration, the cuff pressure at which the pulsation amplitudes and therefore the finger blood volume are maximized is assumed to be an indication of the pressure during which the artery is in a relaxed state and the arterial diameter is maximal. After calibration, the monitor servo system attempts to control the pressure to this set point. At the set point, it is assumed that the transmural pressure equals zero and an approximately linear relationship exists between the intraarterial pressure and detected light. The systolic, mean, and diastolic finger pressures detected at this set point are assumed to equal the intraarterial radial systolic, mean, and diastolic pressures. The latest version of the Finapres is portable, and is called the Portapres (Figure 12.4).

In a recent review, Wesseling et al. reviewed 15 years of experience with the Finapres [24]. In 43 studies of 1031 subjects utilizing the Finapres, the weighted (weighted by total number of patients) mean differences in pressures were -1 ± 12 mmHg (sys), -2 ± 8 mmHg (MAP), and -2 ± 8 mmHg (dias). In these studies, the reference pressures used were brachial IAPs, radial IAPs, or auscultatory/oscillometric pressures. The poorest estimates were recorded for systolic pressure. These studies included both measures of steady state and transient pressure. However, if only the transient pressures are analyzed, mean differences as high in magnitude as -13 (sys) or -33 (dias) mmHg were recorded [24, 25]. In this particular study of transient pressure, phenylephrine was infused into the antecubital vein of the left or right arm for 16 minutes to cause graded vasoconstriction in eight hypertensive patients. Finapres pressures were compared to brachial IAPs [25].

Finapres inaccuracy may be traced to several causes. The transmission of the pressure pulse from the brachial to digital arteries adds an additional transfer function to the system, which degrades pressure measurement accuracy. Due to age-related hemodynamic changes, systolic pressure is lower in the brachial than in the digital arteries in younger subjects, whereas in older people the opposite is observed [26]. Moreover, the pressure gradient from the brachial to the digital arteries decreases mean and diastolic pressure measurements at this latter, more distal, site [27].

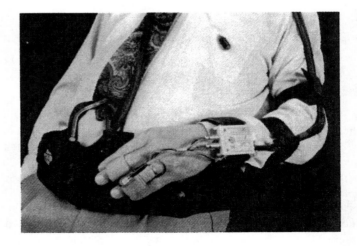

Figure 12.4 Portapres Model–2. Courtesy of TNO-TPD Biomedical Instruments, Affligem, Belgium.

12.4.2 Arterial Tonometry

Arterial tonometry inspired two commercial devices: the NCAT–500, manufactured by Colin and marketed by Nellcor; and the Vasotrac, manufactured by Medwave. Before discussing these monitors, we first discuss the history of arterial tonometry.

Arterial tonometry may be traced back to G. L. Pressman and P. M. Newgard at Stanford Research Institute, who postulated that the technique of **ocular tonometry** could be adapted to blood pressure [28]. With ocular tonometry, a sensor is applied to the cornea until its central area is flattened. This flattening indicates that the circumferential stresses in the corneal wall have been removed and that the internal and external pressures are equal. The final pressure with which this sensor is applied equals the intraocular pressure. Similarly, Pressman and Newgard believed that sufficient pressure could be applied to an artery such as the radial artery, which possesses sufficient bony support. With sufficient pressure, the transmural pressure would equal zero and the external pressure would equal the internal pressure under the following conditions:

1. The underlying vessel must be **applanated** (sufficient pressure applied) by deflecting it with a rigid, planar surface.
2. The sensor must lie in the tonometer surface and record only the force acting upon its sensing area.
3. The force sensor must sense an area equal to or less than the size of the flattened vessel wall area [28].

During his graduate work at the University of Pennsylvania [29] and later as a Professor at Rutgers University in New Jersey, Gary Drzewiecki conducted exhaustive studies of arterial tonometry. Drzewiecki demonstrated that in excised carotid arteries, at zero transmural pressure, the mean arterial circumference increased by 20% [30]. Based on this result, he postulated that at transmural pressure, the detected radial arterial pulse pressure during a sweep from over- to underapplanation (compression from above systolic pressure to below diastolic pressure) is maximal (Figure 12.5) [31].

12.4.3 NCAT

The first U.S. commercial arterial tonometry system—the NCAT N-500—was developed by Colin and marketed by Nellcor in the early 1990s. NCAT stood for noninvasive continuous arterial tonometry. Based on a sensor first investigated by Joseph Eckerle at Stanford Research Institute (SRI) International [33], a linear array of piezoelectric pressure sensors was positioned so that at least one sensor was positioned directly over the radial artery. Sufficient pressure was applied to the sensor by an inflatable bladder to flatten the artery, so that the arterial wall became parallel to the sensor surface. Theoretically, because the pressure sensor only measured those pressures perpendicular to the sensor surface, only the intraarterial pressure was measured without interference of the arterial wall stress. Practically, absolute intraarterial pressure was not measured due to differing skin, subcutaneous tissue, and arterial wall characteristics. To overcome this limitation, the measured systolic, mean, and diastolic pressures were calibrated regularly by oscillometry, using an external cuff.

Although initial results from the NCAT precursor—the Colin CBM 3000—were favorable [34–35], these results could not be duplicated in NCAT studies. Tonometry and

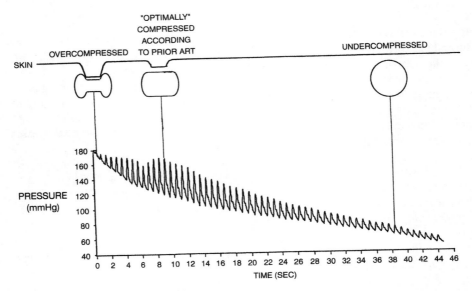

Figure 12.5 Illustration of traditional arterial tonometry principles, with transmural pressure detected at maximal pulse pressure. Based on [32].

intrarterial blood pressure comparisons were made in separate arms, with investigators observing before surgery that blood pressures between arms did not differ by more than 10 [36] or 5 mmHg [37]. In a study of 16 patients undergoing elective noncardiac surgery, 88,158 simultaneous MAP measurements using arterial tonometry and radial catheters were compared. The mean MAP difference was 1 ± 9 mmHg. A histogram of the percentage observation in each patient during which the mean difference was categorized into a specific range of 5 mmHg is given in Figure 12.6. Based on this histogram, the mean difference exceeded 5 mmHg for 53% of measurements and exceeded 10 mmHg for 25% of measurements [36].

In a study of nine patients requiring arterial cannulation for anesthetic management, 140 tonometric measurement periods, between calibrations, were analyzed. The duration of the measurement periods ranged from 0.5 to 9.6 minutes. From each measurement period, the systolic and diastolic tonometric and intraarterial measurements were taken from the waveforms with the highest systolic pressure value (highest sample group), lowest diastolic pressure value (lowest sample group), and last waveform. The mean differences were then calculated for each group (Table 12.1) [37].

In both NCAT studies, the results did not meet the AAMI standard of $\leq 5 \pm 8$ mmHg. Note that the results are probably optimally presented. A thorough analysis would have summarized the mean differences for beat-to-beat systolic, diastolic, and mean pressures. Further, the results are only presented for beats between calibration. A severe limitation of the NCAT was that, when set for automatic calibration rather than for a fixed calibration period, the monitor would recalibrate exactly during the time an anesthesiologist needed measurements most: as the blood pressure began to sharply change. This calibration period lasted from 0.3 to 7.6 minutes [37].

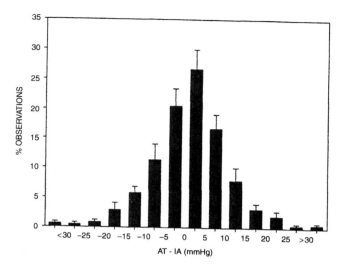

Figure 12.6 Beat-by-beat distribution of MAP differences, in terms of mean % observations in 16 patients. Each bin spans 5 mmHg. MAP difference = (arterial tonometry pressure - intraarterial pressure). AT = arterial tonometry pressure. IA = intraarterial pressure. The MAP difference exceeded 5 mmHg for 53% of the measurements and exceeded 10 mmHg for 25% of the measurements. Reprinted from [36] with kind permission from Lippincott, Williams and Wilkins, Philadelphia, PA.

12.4.4 Vasotrac

More recently, in the late 1990s, Medwave introduced the Vasotrac monitor. A nondisposable, hydraulically driven sensing device is placed over the radial artery, with enough pressure applied until "maximum energy transfer" is reached. As defined by Medwave, "maximum energy transfer" refers to the maximal pressure pulse signal with minimal sensor pressure over the radial pulse. Three beats during this period are used, with predetermined coefficients, to determine systolic, mean, and diastolic pressure values. Because 12–15 beats are required for a cycle of increasing to decreasing pressure to determine the maximal energy transfer period (Figure 12.7), this monitor is reported to output "continual," rather than continuous, NIBP.

In a multicenter trial of 80 surgical and intensive care unit patients, 17,468 simultaneous Vasotrac and radial intraarterial measurements were recorded. Before each study, it was determined that difference of ≤ 5 mmHg existed between oscillometric measurements recorded on each arm. The application of each sensor was supervised by either

Table 12.1 NCAT-500 Pressure differences for various sample groups. Based on [36]

Pressure difference	Highest sample group (mmHg)	Lowest sample group (mmHg)	Last sample group (mmHg)
Systolic	15 ± 21	4 ± 13	8 ± 17
Diastolic	−1 ± 12	−8 ± 12	−4 ± 13
Pulse	16 ± 17	12 ± 14	13 ± 15

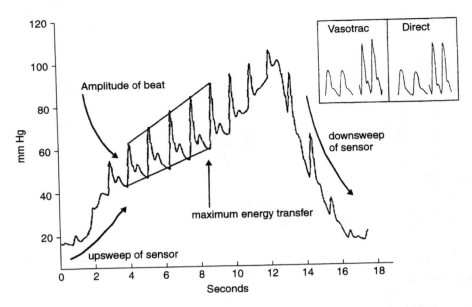

Figure 12.7 Vasotrac pressure pulse signal generated during compression and release over the radial artery. Reprinted from [38] with kind permission from Lippincott, Williams and Wilkins, Philadelphia, PA.

principal investigator Kumar Belani of the University of Minnesota or the Medwave Director of Clinical Research. The mean differences between Vasotrac and IAP pressures were: 0 ± 5 mmHg (sys), 0 ± 3 mmHg (MAP), 0 ± 4 mmHg (dias). Bland Altman plots are given in Figure 12.8 [38].

Although these results meet the AAMI standard, this monitor has not gained clinical acceptance. Displayed pressures are extremely position- and motion-sensitive. The investigators stated that "training and expertise is required to find the best point over the radial pulse for accurate readings." [38] Note that all sensors in this study, which was conducted in four countries and three continents, were carefully positioned under the supervision of only two investigators. Further, the investigators observed that "the system is extremely sensitive to motion artifact, and for accurate measurements and waveform display the arm has to be still." [38]

12.4.5 Early Work at VitalWave Corporation

When Gregory Voss and Alvis Somerville filed the first patent for their company, Vital-Wave Corporation, in 1996, they believed they had developed a practical solution to arterial tonometry. Their tonometry sensor assembly included a sensor base having an open cavity that contained the actual pressure sensor. It was claimed that a "coupling device urges the pressure sensor assembly to a position that partially compresses the blood vessel and provides a mean transmural vessel pressure of substantially zero". [39] To enable direct positioning of the sensor over the radial artery, **Doppler** ultrasound was used [40].

Unfortunately, in 12 measurements of eight subjects using geometric variations of this sensor assembly and various random position angles in two dimensions, mean tonometry

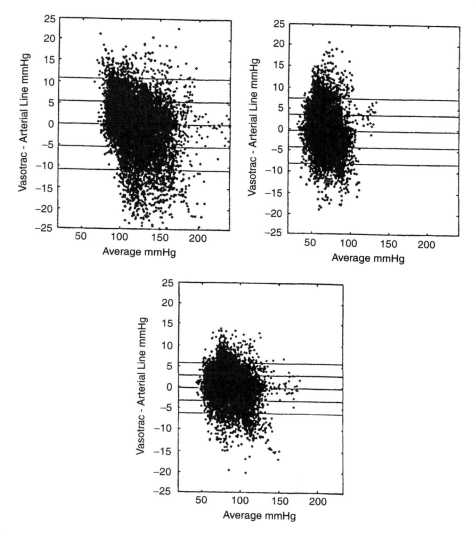

Figure 12.8 Bland–Altman plots for Vasotrac systolic, diastolic, and mean blood pressure. Reprinted from [38] with kind permission from Lippincott, Williams and Wilkins, Philadelphia, PA.

MAP measurements did not compare favorably to oscillometric MAP readings. An oscillometric measurement was taken immediately before and after each of the readings; the average of each set of oscillometric measurements was used as the pressure reference standard. The mean MAP differences were 15 ± 17 mmHg. [32]

12.5 VITALWAVE WORK BASED ON DIGITAL SIGNAL PROCESSING

As shown in the previous section, arterial tonometry has several limitations. First, as demonstrated by the Medwave and early VitalWave systems, tonometry is extremely position-sensitive. This position sensitivity is probably related to the limitation of the NCAT monitor, which was ascribed to differences in "skin, subcutaneous tissue, and arterial wall characteristics" [37]. The NCAT monitor attempted to overcome these limitation through external calibration with an oscillometric cuff. However, recalibration was required during the sharp blood pressure changes when an anesthesiologist most required the measurements, and this calibration was subject to error in the oscillometric estimates. Second, it is unclear during the course of an applanation sweep which blood pressure should be selected as the basis of a pressure estimate. Although Drzewiecki identified the beat with maximal pulse pressure as the beat possessing zero transmural pressure in excised carotid arteries, this has never been proven in vivo. Both the Medwave and early VitalWave system utilized the event marker of maximum pulse pressure. It is possible that skin and subcutaneous tissue overlying the radial artery obscure pulse transfer from the artery, causing maximal pulse transfer at a beat that does *not* possess zero transmural pressure. This would explain the large error in the early VitalWave system results.

12.5.1 An Orthogonal Signal

In order to overcome both limitations, the author decided to search for an event marker that was insensitive to sensor position and could correctly identify the onset of zero transmural pressure. Since pressure waveforms seemed to be affected by underlying tissue, she decided to search for a truly **orthogonal** event marker. An obvious candidate signal was the Doppler signal used for precise transverse positioning of the sensor over the radial artery.

The sensor transmits 8 MHz acoustic pulses, which are reflected as echoes via interactions with red blood cells present in the artery. The frequencies of these echoes are shifted due to the well known Doppler shift phenomenon. The absolute mean blood velocity, $|\overline{v}(k)|$, may be calculated as

$$|\overline{v}(k)| = \frac{f_d(k)c}{2 f_0 \cos \theta} \tag{12.4}$$

where f_0 is the transmission frequency, $f_d(k)$ is the reflected Doppler frequency, c is the speed of sound in water = 1500 m/sec, and θ is the transmission angle. When the sensor is positioned only over tissue, no Doppler signal is present. The Doppler signal is maximal when the sensor is directly positioned over the artery. This signal was used to determine event markers for MAP and diastolic pressure. The estimates of MAP and diastolic pressure were then used to determine a scale factor enabling systolic pressure estimation [32, 41].

12.5.2 MAP Estimation

The author calculated the pseudo-Wigner time-frequency distribution, $P_{PW}(k, f)$, (see Chapter 5) for the Doppler signal during decreasing applanation from over- to undercompression as

$$P_{PW}(k, f) = 2 \sum_{\tau=-L}^{+L} e^{-j4\pi f\tau/N} u^*(k-\tau)u(k+\tau) \tag{12.5}$$

12.5 VITALWAVE WORK BASED ON DIGITAL SIGNAL PROCESSING

where f = calculation frequency, $u(t)$ = absolute velocity, $u(t)$ and its complex conjugate are sample-limited to $\{-K/2, +K/2\}$, K is even, and $N = K + 1$. A rectangular window is used, so that $L = K/2 - |k| = 5$. Because this distribution over time for particular frequencies resulted in similar plots, a frequency of 0 Hz was selected, which simplified the calculation to

$$P_{PW}(k, 0) = 2 \sum_{\tau=-L}^{+L} u^*(k-\tau)u(k+\tau) \qquad (12.6)$$

To simplify calculations, Eq. (12.6) was divided by a factor of 2.

The pseudo-Wigner distribution at 0 Hz resulted in a plot with a characteristic peak (Figure 12.9). The characteristic peak was isolated by calculating the mean distribution

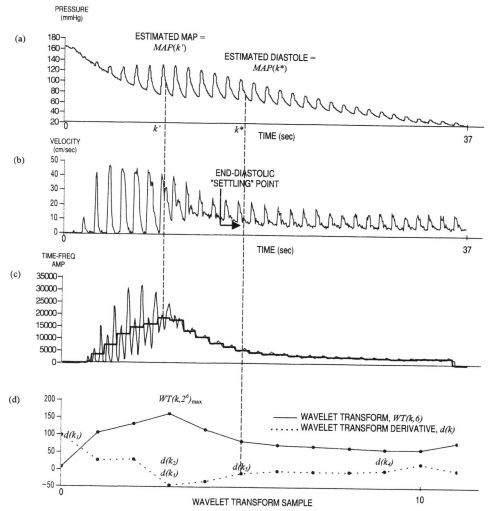

Figure 12.9 Illustration of maximal time-frequency and wavelet transform methods. These are typical plots during decreasing applanation of (a) tonometry sensor pressure, (b) blood velocity, (c) a Pseudo-wigner distribution of the blood velocity and mean distribution per beat, and (d) a wavelet transform and its derivative of the blood velocity. Based on [41].

during each blood pressure beat, and finding the maximum value. The maximal time-frequency value was identified as the MAP event marker and used to identify the beat possessing the true MAP. Using the same 12 data files that were used to assess the early VitalWave system, a mean difference of 3 ± 11 mmHg was calculated, which using the paired, two-sided t test was shown to be significantly different ($p \leq 0.02$) from the early VitalWave system result of 15 ± 17 mmHg [32].

This method was validated using an early prototype ultrasound circuit possessing comparatively low sensitivity. MAP was estimated in 156 decreasing applanation sweeps obtained from seven surgical patients. The sweeps were obtained during conditions possessing a variety of prevailing catheter MAP values, ranging from 48 to 132 mmHg. The reference IAP mean pressure was averaged from the first, middle, and last beats obtained during a decreasing applanation sweep. Using the early VitalWave system, the mean MAP difference was 11 ± 20 mmHg (Figure 12.10). In contrast, using the maximal time-frequency method, the mean MAP difference was 2 ± 15 mmHg (Figure 12.11). The squared correlation coefficients, r^2, for these estimates were 0.61 ($p = 0.034$) and 0.67 ($p = 0.030$), respectively. Although the squared correlation coefficients were both significant, only the time-frequency method resulted in low mean MAP differences. It is believed that the large standard deviation in the time-frequency measurement reflects the low signal-to-noise ratio of the velocity signal (Figure 12.9) [41]. For example, in modeling drug transport data, a required input to a compartmental model is an estimate of measurement noise, which is typically given in terms of standard deviation or fractional standard deviation [42]. The standard deviation is expected to decrease significantly when an ultrasound circuit with greater sensitivity is used [41].

The peak in the pseudo-Wigner distribution at 0 Hz may be related to a sudden increase in arterial diameter. In other arteries such as the brachial and femoral arteries, a similar peak trend in the **end-diastolic velocity** (baseline velocity) may be induced after complete arterial occlusion with a cuff for several minutes, followed by complete cuff re-

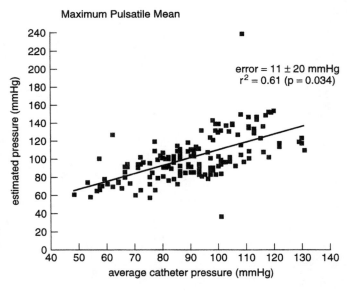

Figure 12.10 Early VitalWave results, with estimated MAP versus average catheter MAP plotted. The MAP event marker is maximum pulse pressure. Based on [41].

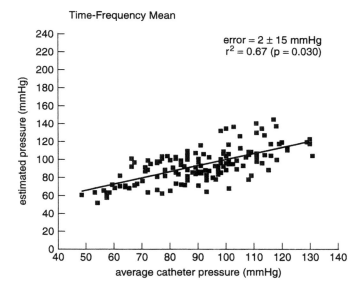

Figure 12.11 VitalWave digital signal processing results, with estimated MAP versus average catheter MAP plotted. The MAP event marker is the maximal time-frequency distribution. Based on [41].

lease. As shown in Figure 12.12, the transient increase in blood flow following a brief arterial occlusion is called **reactive hyperemia**. In the brachial artery, this transient increase in blood flow and end-diastolic velocity induces a transient increase of 19% [43]. Similarly, blood flow in the radial artery is completely occluded during an applanation sweep. As sensor compression decreases during a decreasing applanation sweep, reactive hyperemia and its signature peak trend in end-diastolic velocity may be induced. The accompanying transient increase in arterial diameter could occur transversely across the artery, but would be initially prevented **sagittally** (top to bottom) by the external pressure applied by the sensor. However, pressure within the artery could become sufficient so that the sagittal arterial diameter could also increase. This increase in sagittal arterial diameter could occur when the transmural pressure equaled zero (Figure 12.13).

The relative arterial diameter of any artery, $\nu(k)$, is proportional to the phase echo, which is in turn proportional to the integral of the Doppler frequency. The phase, $\phi(k)$, is a function of the time delay between reflections from the near and far arterial walls. Because the time delay depends only on the time difference between reflections from the two arterial walls, the measurement is insensitive to transmission angle. Only relative arterial diameter changes from an initial diameter value during overcomppression can be estimated.

The phase of the received echo is calculated from the summation of the discrete Doppler frequency (or integral of the continuous Doppler frequency) [44] as

$$\phi(k) = 2\pi \Sigma f_d(k) \tag{12.7}$$

The diameter is calculated as

$$\nu(k) = \frac{\phi(k)c}{4\pi f_0} \tag{12.8}$$

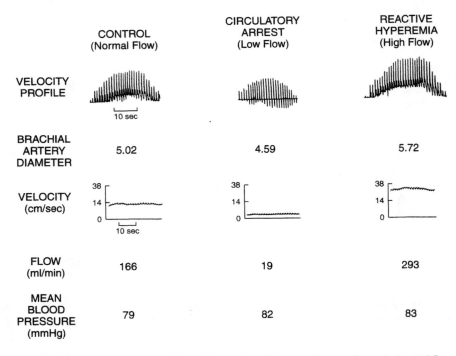

Figure 12.12 Velocity profile across the brachial artery, brachial artery diameter, flow velocity, and flow and mean blood pressure from one subject. Values are shown during control, distal circulatory arrest, and reactive hyperemia. Reprinted with kind permission from [43] by Lippincott, Williams and Wilkins, Baltimore, MD.

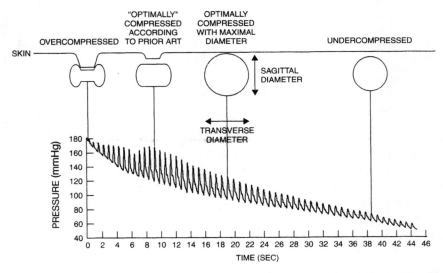

Figure 12.13 Illustration of proposed arterial tonometry mechanism, with transmural pressure detected at maximal arterial diameter. Based on [32].

Substitution of Eqs. (12.7) and (12.4) into (12.8) results in

$$u(k) = \cos\theta \Sigma |\overline{v}(k)| \qquad (12.9)$$

Comparing Eq. (12.9) to (12.6), and recognizing that "2" and "cos θ" are only scale factors, the pseudo-Wigner distribution at 0 Hz is equivalent to the proportional squared relative arterial diameter. Therefore, the peak distribution may occur at a sudden change in sagittal arterial diameter when true MAP is reached. Because the time-frequency distribution acts as a smoothing filter, the distribution is smooth, rather than discontinuous, at the peak [32]. The insensitivity of relative diameter change to sensor position could explain the insensitivity of the maximal time-frequency method to sensor position.

12.5.3 Diastole Estimation

In a manner similar to that described for MAP estimation, an event marker for diastolic pressure may also be determined from the Doppler signal. Specifically, the author investigated the time during which end-diastolic velocity first "settles" to its final value after transiently increasing. This setting point may mark the beat possessing the true diastolic pressure, and can be isolated using the wavelet transform. The wavelet transform is a **time-scale distribution,** which was first introduced in Chapter 5.

First, the approximation coefficients of a **Haar wavelet transform,** $WT_H(k, 2^j)$, of scale 2^6 are calculated as

$$WT(k, 2^6) = \frac{1}{\sqrt{2^6}} \sum_{i=0}^{N_s-1} u(i)\phi_h(k)\left(\frac{i-k}{2^6}\right) \qquad (12.10)$$

where $u(k)$ = the velocity signal, 2^j = the scale, N_s = the total number of samples of the velocity signal, and $\phi_h(k)$ is the Haar scaling function:

$$\phi_h(k) = \begin{cases} 1, & 0 \le k < 1 \\ \text{otherwise} \end{cases} \qquad (12.11)$$

Note that the total number of samples in the wavelet transform is $1/2^j$ the total number of samples in the input signal, N_s. The approximation coefficients of the wavelet transform effectively function as a lowpass filter (Figure 12.9). To simplify the calculation, the scale factor $1/\sqrt{2^6}$ is omitted. The maximum wavelet transform, $WT(k, 2^6)_{max}$, is then found.

Next, the derivative (difference of consecutive samples) of the wavelet transform, $d(k)$, is calculated. The following steps are then conducted (Figure 12.9). Starting at the first sample, the first instance where the derivative, $d(k)$, is greater than zero $[d(k_1) > 0]$ is determined. Starting from sample k_1, the first instance where $d(k_2) < 0$ is then found. Starting from sample k_2, the first local minimum at sample k_3 is found. Starting from sample k_3, the first instance where $d(k_4) > 0$ is found. Next, the smallest value of $d(k_5) > 0.15$ $WT(k, 2^6)_{max}$ is found within the range $k_3 < k < k_4$. The mean of the pressure waveform that includes sample $k^* = 2^6 k_5 - 2^{6-1}$, which is the diastolic pressure event marker, is then calculated. The estimated diastolic pressure is then determined as the value of MAP(k^*).

This wavelet method was validated using the same surgical data used to validate the time-frequency method. Diastolic pressure was estimated in 156 decreasing applanation sweeps obtained from seven surgical patients. The resulting mean diastolic pressure difference was 5 ± 14 mmHg (Figure 12.14). The squared correlation coefficient, r^2, for these estimates was 0.56 (p = 0.038). Again, it is believed that the large standard deviation in the wavelet measurement reflects the low signal-to-noise ratio of the velocity signal. Note that the standard deviation in the diastolic pressure difference is quite similar to that of the MAP difference (14 versus 15 mmHg, respectively) [41].

12.5.4 Scaling for Systolic Pressure Estimation

Given estimations of MAP and diastolic pressure, can the systolic pressure be inferred through scaling of the sensor-recorded waveform? In order to answer this question, the author obtained 10-second intervals of data from three different surgical patients. Using the method described in Section 12.6, the recorded sensor data tracked the mean IAP. The MAP measured for each of the subjects was 73, 126, and 83 mmHg, respectively. These means were subtracted from their respective data sets, and the data were fit to an **autoregressive exogeneous input (ARX) model** (see Chapter 7):

$$y(k) = -\sum_{n=1}^{N} a_n y(k-n) + \sum_{n=0}^{M} b_n u(k-n) \tag{12.12}$$

where a_n = feedback parameter, b_n = feedforward parameter, $u(k)$ = input IAP, $y(k)$ = output tonometry pressure, N = number of feedback parameters, and M = number of feedforward parameters. The model is termed "autoregressive" because it looks back on past values of itself (the feedback parameters). Various combinations of feedback and feedforward parameters were used to model the data (Table 12.2). The models were evaluated using the standard criteria of the **Akaike final prediction error criterion**, standard

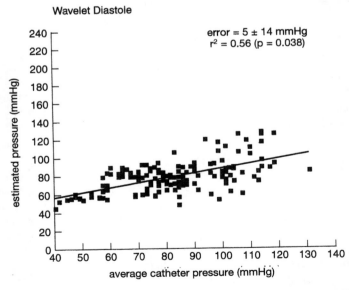

Figure 12.14 VitalWave digital signal processing results, with estimated diastolic versus average catheter diastolic pressures plotted. The diastolic pressure event marker is the blood velocity settling point. Based on [41].

Table 12.2 Combinations of model orders (N) and number of feedforward coefficients (M) used to assess mechanical impulse response. From [41]

N	M
4	3
10	9
2	1
1	1
0	1

deviations associated with identified parameters, and residuals between the estimated and observed waveforms (see Chapter 7).

For all three subjects, the optimum model was a zero-order model ($N = 0$) with a single feedforward coefficient ($M = 1$). The fit for the second subject is shown in Figure 12.15. The identified feedforward coefficients and their associated standard deviations are shown in Table 12.3. Therefore, the mechanical impulse response between the true invasive arterial pressure and the sensed tonometry pressure is merely a scaling factor. This scaling factor, F_1, may be estimated from the estimated MAP and diastolic pressures as

$$F_1 = \frac{1}{b_0} = \frac{MAP(k') - MAP(k^*)}{MAP(k') - Dias\_Pres(k')} \qquad (12.13)$$

where k' is the sample identified by the maximum time-frequency algorithm during which the measured MAP corresponds to the catheter MAP, and k^* is the sample identified by

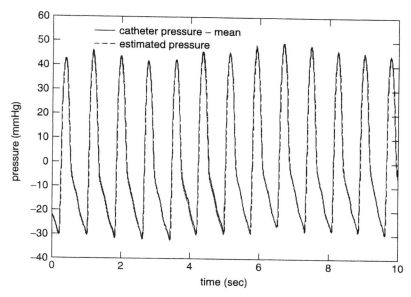

Figure 12.15 Estimation of catheter pressure using measured pressure as the input, and an ARX model with $N = 0$ and $M = 1$, in subject 2. Based on [41].

Table 12.3 Identified feedforward coefficients. From [41].

Subject	MAP (mmHg)	b_0	Standard deviation
1	77	0.75	0.003
2	121	0.86	0.001
3	81	0.80	0.003

the wavelet algorithm during which the measured MAP corresponds to the catheter diastolic pressure. This scaling factor is only calculated intermittently during an applanation sweep that acts as a calibration sweep. Thereafter, as the sensor tracks mean arterial pressure, the entire blood pressure waveform, $y_{servo}(k)$, may be scaled to

$$y_{scaled}(k) = F_1[y_{servo}(k) - MAP_{servo}] + MAP_{servo} \qquad (12.14)$$

where MAP_{servo} is the current sensor MAP during tracking.

Alternatively, when the catheter pressure is not changing significantly, the applanation may be fixed at a low, constant externally applied pressure. During this "steady state" condition, the measured MAP, MAP_{ss}, and measured diastolic pressure, $Dias\_Pres_{ss}$, will not change significantly. A second steady state scaling factor, F_2, can therefore be derived as

$$F_2 = \frac{MAP(k') - MAP(k^*)}{MAP_{ss} - Dias\_Pres_{ss}} \qquad (12.15)$$

The entire blood pressure waveform, $y_{ss}(k)$, may then be scaled to [41]

$$y_{scaled}(k) = F_2[y_{ss}(k) - MAP_{ss}] + MAP(k') \qquad (12.16)$$

12.6 CONTINUOUS BLOOD PRESSURE MEASUREMENT

In Section 12.5, we described the author's method for estimation of systolic pressure, diastolic pressure, and MAP during a decreasing applanation sweep. However, to be accepted by clinicians, a NIBP monitor must continuously estimate blood pressure. Continuous measurement was accomplished using fuzzy control.

12.6.1 Fuzzy Control

The fuzzy controller was based on the author's observation that the blood pressure beat possessing the true (IAP) MAP corresponds to the beat with the maximum end-diastolic blood velocity. The controller was implemented with two inputs: (1) the mean pseudo-Wigner distribution of the current beat, and (2) the difference between the current and last mean pseudo-Wigner distribution. The controller output was the number of applanation steps sent to the motor. This output ranged from −400 to +400 steps, in multiples of 50 steps. For the VitalWave motor, 38,400 steps equaled 1 inch. If the difference input was a positive value, the output signal directed the applanation motor to continue in the same direction for the calculated number of steps. If the difference input was a negative value, the output signal directed the applanation motor to change direction for a calculated number of steps. The input and output membership functions of the controller were typical functions of overlapping trapezoids. Unfortunately, these membership functions have not been publicly disclosed [41].

12.6.2 Data Acquisition and Processing

In an initial trial in one surgical patient, the controller was evaluated during an observed 50 mmHg drop in MAP over approximately 11 minutes. This severe pressure drop occurred in response to epidural administration of the anesthetic bupivicaine. In two anesthetized surgical patients, the fuzzy controller was tested during two continuous 20 minute intervals, with each test followed by a 5 minute resting interval. During each 20 minute measurement, one applanation pressure sweep was conducted, followed by continuous servo control. Tonometry pressures, IAPs, and blood velocity were simultaneously recorded using a National Instruments data acquisition card, DAQ Card AI-16E-4. The sampling frequency was 400 Hz; the data were digitized with 12 bit resolution.

The digitized data were decimated by a factor of 20 to obtain 20 Hz data. The pseudo-Wigner distribution at 0 Hz was then calculated using Eq. (12.6), with $L = 5$. The mean time-frequency signal was calculated for each beat. The mean pseudo-Wigner distribution and difference in pseudo-Wigner distribution were input to the fuzzy controller. The controller output determined the motor steps moved after each beat. On a beat-per-beat basis, the resulting tonometry MAPs and IAP MAPs were compared. The initial period of an applanation sweep was excluded from the analysis.

12.6.3 Data Analysis

During the initial trial, when the blood pressure dropped 50 mmHg in 11 minutes, the controller tracked 552 pressure beats, with a mean difference of 3 ± 4 mmHg (Figure 12.16). Figure 12.17 is a detail view of Figure 12.16, illustrating three 40 second windows of the catheter and servo data. Figure 12.18 is a 20 second interval snapshot of the data in Figure 12.16 that occurred at 6.5 minutes from the onset of data recording in Figure 12.16. As illustrated in Figure 12.18, a significant drop in the end-diastolic velocity was corrected within five beats.

During the four measurement periods, the catheter MAPs ranged from 69 to 106 mmHg. Over 3103 beats, the mean MAP difference was –3 ± 5 mmHg. Individual data set statistics are summarized in Table 12.4. In data from all three surgical patients, the AAMI standard of ≤ 5 ± 8 mmHg was met for MAP differences.

Unfortunately, these preliminary results are the only publicly disclosed fuzzy control data from the VitalWave system. However, even with only preliminary results, proof of concept of CNIBP has been demonstrated by the author [41]. It is possible to use arterial tonometry to continuously, noninvasively measure blood pressure, provided that an orthogonal signal such as blood pressure velocity is used as a blood pressure event marker. Because the blood velocity is orthogonal to the tonometry pressure, it is insensitive to positional sensor changes due to overlying tissue. Features in the end-diastolic velocity may be extracted using digital signal processing techniques to provide event

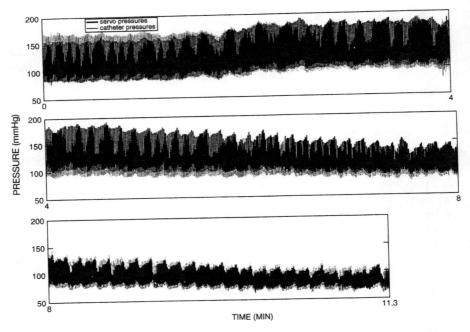

Figure 12.16 Continuous catheter and servo pressures in a surgical patient over 11.3 minutes. During this time, the catheter MAP dropped 50 mmHg. Based on [41].

markers for diastolic and mean pressure estimation and for use in continuously tracking MAP.

VitalWave Corporation continues to develop a blood pressure monitoring system. In 2000, VitalWave changed its name to Tensys Medical Systems.

12.7 SUMMARY

In this chapter, we have discussed the application of fuzzy control to continuous noninvasive blood pressure monitoring. During surgery, it is often desirable to measure blood pressure to avoid the negative effects of anesthesia. The required pressure applied to this sensor over the radial artery to obtain accurate blood pressure readings may be continually updated using fuzzy control.

Blood pressure originates in the left ventricle during the cardiac cycle. The maximum pressure is called the systolic pressure; the minimum pressure is called the diastolic pressure. The integral of the pressure over one cardiac cycle is called the mean pressure, or mean arterial pressure. The difference between the systolic and diastolic pressures is known as the pulse pressure. As the pressure is transported from the left ventricle to the aorta, smaller arteries, and arterioles, it experiences damping from reflections in the arterial tree. The blood and vascular wall possess viscous properties that attenuate high-frequency components in the pressure waveform. The waveform expands, obtaining a dicrotic notch. The pulse pressure increases, and the mean pressure decreases. The pressure is then transported to the capillaries and the venous system. Blood pressure measurements

12.7 SUMMARY **285**

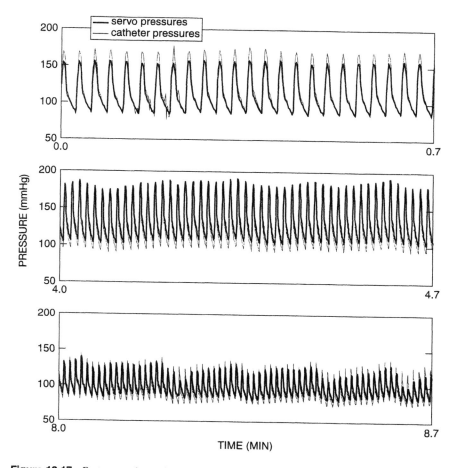

Figure 12.17 Forty second snapshots of the catheter and servo data in Figure 12.16. Based on [41].

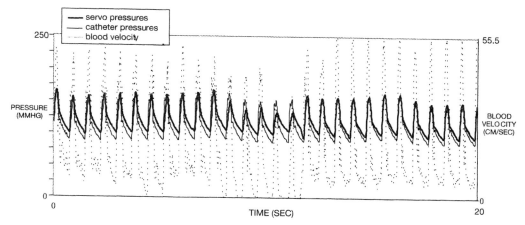

Figure 12.18 20 second snapshot of catheter and servo data at 6.5 minutes. During this time, a significant drop in the end-diastolic velocity was corrected within 5 beats. Based on [41].

Table 12.4 Preliminary results of servo control. From [41]

Patient	Data set	Pressure range (mmHg)	Number of beats	MAP difference (mmHg)
1	1	69–106	653	−1 ± 5
1	2	82–93	830	−6 ± 6
2	1	73–83	747	−1 ± 4
2	2	81–93	873	−4 ± 6
		Total:	3,103	−3 ± 5

are usually acquired from the arterial system. The normal value of blood pressure is not 120/80 mmHg (systolic/diastolic pressure), but is dependent on the age and gender of an individual.

Blood pressure may be measured invasively and noninvasively. The invasive gold standard of care is intraarterial measurement. In preparation for intraarterial blood pressure measurement, a 20 gauge catheter is inserted into the artery of interest. Typically, blood pressure is monitored in the radial artery because it is highly accessible and easily visible. Once the catheter is in place, it is connected by tubing to a pressure transducer. The tubing and catheter are filled with saline, in order to transmit the radial arterial pressure to the pressure transducer. Intraarterial catheter measurement is one of the reference methods recommended by the Association for the Advancement of Medical Instrumentation for evaluation of blood pressure instrumentation accuracy. The most widely used noninvasive measurements are auscultation, during which the systolic and diastolic pressures are determined through observation of Korotkoff sounds; and oscillometry, during which the systolic and mean pressures are determined through observation of the pulsatile pressures.

Because auscultation and oscillometry are intermittent methods, much research has been conducted to invent a continuous noninvasive method. Two of the three commercial U.S. systems (Nellcor NCAT–500 and MedWave Vasotrac) are based on the principle of arterial tonometry. Arterial tonometry may be traced back to G. L. Pressman and P. M. Newgard at Stanford Research Institute, who postulated that the technique ocular tonometry could be adapted to blood pressure. With ocular tonometry, a sensor is applied to the cornea until its central area is flattened. This flattening indicates that the circumferential stresses in the corneal wall have been removed and that the internal and external pressures are equal. The final pressure with which this sensor is applied equals the intraocular pressure. Similarly, Pressman and Newgard believed that sufficient pressure could be applied to an artery such as the radial artery, which possesses sufficient bony support. With sufficient pressure, the transmural pressure would equal zero and the external pressure would equal the internal pressure. Later, Drzewiecki demonstrated that in excised carotid arteries, at zero transmural pressure, the mean arterial circumference increased by 20%. Based on this result, he postulated that at transmural pressure, the detected radial arterial pulse pressure during a sweep from over- to underapplanation (compression from above systolic pressure to below diastolic pressure) is maximal.

Maximal pulse pressure did not result in accurate estimates of MAP during arterial tonometry. In both cases, the sensor was position sensitive, due to differences in skin, subcutaneous tissue, and arterial wall characteristics. Possibly, these differences cause maximal pulse transfer to occur at a beat that does not possess zero transmural pressure. In order to overcome both limitations, the author identified event markers for mean and diastolic pressure that were insensitive to sensor position and could correctly identify the onset of zero transmural pressure. In both cases, these orthogonal event markers were derived from the blood velocity during a pressure sweep, with features of the velocity detected using a time-frequency, psuedo-Wigner distribution and wavelet transform.

Because it was observed that the blood pressure beat possessing the true (IAP) MAP corresponds to the beat with the maximum end-diastolic blood velocity, a fuzzy controller was implemented with two inputs that identified the value of end-diastolic blood velocity: (1) the mean pseudo-Wigner distribution of the current beat, and (2) the difference between the current and last mean pseudo-Wigner distribution. The controller output was the number of applanation steps sent to the motor. This output ranged from –400 to +400 steps, in multiples of 50 steps. For the VitalWave motor, 38,400 steps equaled 1 inch. If the difference input was a positive value, the output signal directed the applanation motor to continue in the same direction for the calculated number of steps. If the difference input was a negative value, the output signal directed the applanation motor to change direction for a calculated number of steps. The input and output membership functions of the controller were typical functions of overlapping trapezoids.

During an initial trial in one surgical patient, when the blood pressure dropped 50 mmHg in 11 minutes, the controller tracked 552 pressure beats, with a mean difference from IAP of 3 ± 4 mmHg. During four measurement periods of 20 minutes in two other surgical patients, the catheter MAPs ranged from 69 to 106 mmHg. Over 3103 beats, the mean MAP difference was –3 ± 5 mmHg. In data from all three surgical patients, the AAMI standard of ≤ 5 ± 8 mmHg was met for MAP differences.

12.8 REFERENCES

[1]. Association for the Advancement of Medical Instrumentation. *ANSI/AAMI SP10-1992: Electronic or Automated Sphygmomanometers.* AAMI: Arlington, VA, 1992.

[2]. van Montfrans, G. A., van Der Hoeven, G. M. A., Karemaker, J. M., Wieling, W., and Dunning, A. J. Accuracy of auscultatory blood pressure measurement with a long cuff. *Br Med J, 295,* 354–355, 1987.

[3]. Guidelines Sub-Committee. 1993 guideines for the management of mild hypertension: Memorandum from a World Health Organization/International Society of Hypertension meeting. *J Hypertens, 11,* 905–918, 1993.

[4]. Geddes, L. A. *The Direct and Indirect Measurement of Blood Pressure.* Chicago: Year Book, 1970.

[5]. Ramsey, M., III. Noninvasive automatic determination of mean arterial pressure. *Med Biol Eng Comput, 17,* 11–18, 1979.

[6]. Hansen, K. W. and Orskov, H. A plea for consistent reliability in ambulatory blood pressure monitors: A reminder. *J Hypertens, 10,* 1313–1315, 1992.

[7]. O'Brien, E., Mee, F., Atkins, N., and O'Malley, K. Accuracy of the Takeda TM–2420/TM–2020 determined by the British Hypertension Society Protocol. *J Hypertens, 9,* 571–572, 1991.

[8]. O'Brien, E., O'Malley, K., Atkins, N., and Mee, F. A review of validation procedures for blood pressure measuring devices. In Waeber, B., O'Brien, E., O'Malley, K., and Brunner, H. R., eds. *Twenty-Four-Hour Blood Pressure Monitoring in Clinical Practice.* Raven Press: New York, 1994, pp. 1–32.

[9]. O'Brien, E., Petrie, J., Littler, W., de Swiet, M., Padfield, P. L., Altman, D. G., Bland, M., Coats, A., and Atkins, N. An outline of the revised British Hypertension Society protocol for the evaluation of blood pressure measuring devices. *J Hypertens, 11,* 677–679, 1993.

[10]. Pedersen, T., Eliasen, K., and Henriksen, E. A prospective study of risk factors and cardiopulmonary complications associated with anaesthesia and surgery: risk indicators of cardiopulmonary morbidity. *Acta Anaesthesiol Scand, 34,* 144–155, 1990.

[11]. Spence, A. A. The lessons of CEPOD (editorial). *Br J Anaesth, 60,* 753–754, 1988.

[12]. Farrow, S. C., Fowkes, F. G. R., Lunn, J. N., Robertson, I. B., and Samuel, P. Epidemiology in anaesthesia. II. Factors affecting mortality in the hospital. *Br J Anaesth, 54,* 811–817, 1982.

[13]. Fowkes, F. G. R., Lunn, J. N., Farrow, S. C., Robertson, I. B., and Samuel, P. Epidemiology in anaesthesia. III. Mortality risk in patients with co-existing disease. *Br J Anaesth, 54,* 819–825, 1982.

[14]. Goldman, L. and Caldera, D. L. Risks of general anesthesia and elective operation in the hypertensive patient. *Anesthesiology, 50,* 285–292, 1979.

[15]. Staeesen, J., Amery, A., and Fagard, R. Isolated systolic hypertension (editorial). *J Hypertens, 8,* 393–405, 1990.

[16]. Mancia, G., Di Rienzo, M., and Parati, G. Ambulatory blood pressure monitoring use in hypertension research and clinical practice. *Hypertension, 21,* 510–524, 1993.

[17]. U.S. Department of Health and Human Services website. www.cdc.gov/nchs/fastats/insurg.htm.

[18]. Dueck, R. (UCSD Professor of Anesethesiology.) Personal Communication. 2002.

[19]. Penaz, J. *Patentova Listina.* CISLO 133205, 1969.

[20]. Cejnar, M., Hunyor, S. N., Liggins, G. W., and Davis, R. J. Description of a new continuous non-invasive blood pressure monitoring instrument. *Proc Aust Physiol Pharmacol Soc, 19,* 90P, 1988.

[21]. Yamakoshi, K., Kamiya, A., Shimazu, H., Ito, H., and Togawa, T. Noninvasive automatic monitoring of instantaneous arterial blood pressure using the vascular unloading technique. *Med Biol Eng Comput, 21,* 557–565, 1983.

[22]. Molhoek, G. P., Wesseling, K. H., Settels, J. J., van Vollenhoven, E., Weeda, H. W. H., de Wit, B., and Arntzenius, A. C. Evaluation of the Penaz servo-plethysmo-manometer for the continuous, non-invasive measurement of finger blood pressure. *Basic Res Cardiol, 79,* 598–609, 1984.

[23]. Wesseling, K. H. Finapres, continuous noninvasive finger arterial pressure based on the method of Penaz. In Meyer-Sabellek, W., Anlauf, M., Gotzen, R., and Steinfeld, L., eds. *Blood Pressure Measurements.* Steinkopff Verlag: Darmstadt, Germany, 1990. pp. 161–172.

[24]. Imholz, B. P. M., Wieling, W., van Montfrans, G. A., and Wesseling, K. H. Fifteen years experience with finger arterial pressure monitoring: Assessment of the technology. *Cardiovasc Res, 38,* 605–616, 1998.

[25]. Imholz, B. P. M., Parati, G., Mancia, G., and Wesseling, K. H. Effects of graded vasoconstriction upon the measurement of finger arterial pressure. *J Hypert, 10,* 979–984, 1992.

[26]. O'Rourke, M. Arterial haemodynamics and ventricular-vascular interaction in hypertension. *Blood Press, 3,* 33–37, 1994.

[27]. Rowell, L. B., Brengelmann, G. L., Blackmon, J. R., Bruce, R. A. and Murray, J. A. Disparities between aortic and peripheral pulse pressures induced by upright exercise and vasomotor changes in man. *Circulation, 27,* 954–964, 1968.

[28]. Pressman, G. L. and Newgard, P. M. A transducer for the continuous external measurement of arterial blood pressure. *IEEE Trans Bio-med Elect, 10,* 73–81, 1963.

[29]. Drzewiecki, G. M., Melbin, J. and Noordergraaf, A. Arterial tonometry: Review and analysis. *J Biomechanics, 16,* 141–152, 1983.

[30]. Drzewiecki, G. M. and Moubarak, I. F. Transmural pressure-area relation for veins and arteries. In *Proceedings 14th Annual Northeast Bioengineering Conference.* Durham, NH, March 10–11, 1988.

[31]. Drzewiecki, G. M. Noninvasive assessment of arterial blood pressure and mechanics. In Bronzino, J. D., ed., *The Biomedical Engineering Handbook*. CRC Press: Boca Raton, FL, 1995. pp. 1196–1211.

[32]. Baura, G. D. *Method and Apparatus for the Noninvasive Determination of Arterial Blood Pressure*. U.S. Patent pending.

[33]. Eckerle, J. D. Tonometry, arterial. In Webster, J. G., ed., *Encyclopedia of Medical Devices and Instrumentation*. Wiley: NY, 1988. pp. 2770–2776.

[34]. Kemmotsu, O., Ueda, M., Otsuka, H., Yamamura, T., Winter, D. C., and Eckerle, J. S. Arterial tonometry for noninvasive, continuous blood pressure monitoring during anesthesia. *Anesthesiology, 75,* 333–340, 1991.

[35]. Kemmotsu, O., Ueda, M., Otsuka, H., Yamamura, T., Okamura, A., Ishikawa, T., Winter, D. C., and Eckerle, J. S. Blood pressure measurement by arterial tonometry in controlled hypotension. *Anesth Analg, 73,* 54–58, 1991.

[36]. Siegel, L. C., Brock-Utne, J. G., and Brodsky, J. B. Comparison of arterial tonometry with radial artery catheter measurements of blood pressure in anesthetized patients. *Anesthesiology, 81,* 578–584, 1994.

[37]. De Jong, J. R., Ros, H. H., and De Lange, J. J. Noninvasive continuous blood pressure measurement during anaesthesia: A clinical evaluation of a method commonly used in measuring devices. *Int J Clin Mon Comput, 12,* 1–10, 1995.

[38]. Belani, K., Ozaki, M., Hynson, J., Hartmann, T., Reyford, H., Martino, J., Poliac, M., and Miller, R. A new noninvasive method to measure blood pressure: Results of a multicenter trial. *Anesthesiology, 91,* 686–692, 1999.

[39]. Voss, G. I., and Somerville, A. J. *Apparatus and method for noninvasively monitoring a subject's arterial blood pressure*. U.S. Patent 5,848,970. Dec. 15, 1998.

[40]. Voss, G. I., Somerville, A. J, and Finburgh, S. E. *Apparatus and method for non-invasively monitoring a subject's arterial blood pressure*. U.S. Patent 6,176,831. Jan, 23, 2001.

[41]. Baura, G. D. *Method and Apparatus for the Noninvasive Determination of Arterial Blood Pressure*. U.S. Continuation-in-part. U.S. Patent pending.

[42]. Carson, E. R., Cobelli, C., and Finkelstein, L. *The Mathematical Modeling of Metabolic and Endocrine Systems: Model Formulation, Identificaiton, and Validation*. Wiley: New York, 1983. pp. 185–189.

[43]. Anderson, E. A. and Mark, A. L. Flow-mediated and reflex changes in large peripheral artery tone in humans. *Circulation, 79,* 93–100, 1989.

[44]. Hartley, C. J., Litowitz, H., Rabinovitz, R. S., Zhu, W., Chelly, J. E., Michael, L. H., and Bolli, R. An ultrasonic method for measuring tissue displacement: Technical details and validation for measuring myocardial thickening. *IEEE Trans Biomed, 38,* 735–747, 1991.

Further Reading

Blood Pressure Physiology: Patton, D. H., Fuchs, A. F., Hille, B., Scher, A. M., and Steiner, R., eds. *Textbook of Physiology*. 21st ed. Saunders: Philadelphia, 1989.

Blood Pressure Monitoring: Civetta, J. M., Taylor, R. W., and Kirby, R. R., eds. *Critical Care*. 3rd ed. Lippincott-Raven: Philadelphia, 1997.

12.9 INFUSION PUMP OCCLUSION ALARM EXERCISES

Engineer F at Startup Company has been asked to develop a cassette side occlusion (CSO) alarm for an infusion pump in which the drug is delivered from a plastic cassette. Having just heard a lecture on fuzzy logic, he is inspired to develop a fuzzy alarm system.

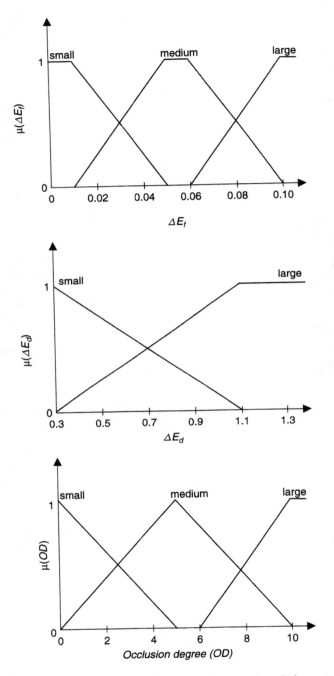

Figure 12.19 Input and output membership functions for occlusion exercise.

Table 12.5 Fuzzy logic exercise rule base

ΔE_f \ ΔE_d	Small	Large
Small	Small	Small
Medium	Medium	Small
Large	Large	Small

He wishes to base his alarm system on the change in energy required to fill a cassette, ΔE_f, and the change in energy required to deliver the dosage, ΔE_d. In general, when ΔE_f is large and ΔE_d is small, then a cassette side occlusion has occurred. We recreate his analysis below.

12.9.1 Useful Matlab Functions

The following function from the Fuzzy Logic Toolbox is useful for completing this exercise: fuzzy.

12.9.2 Exercises

1. Based on retrospective data, Engineer F determines the input and output membership functions (Figure 12.19). (a) Which of these membership functions represents a fuzzy partition? (b) What is the level of coverage for each group of membership functions?

2. Based on retrospective data, he also determines the rules (Table 12.5). (a) Determine the alarm output for $\Delta E_f = 0.04$ and $\Delta E_d = 0.9$, using Mamdani max–min inference. Repeat the calculation using the (b) product and (c) Lukasiewicz intersections. (d) How sensitive is the calculation to the type of intersection selected?

3. Repeat the calculation in Question 2a using (a) the maximum and (b) mean of maximum methods during defuzzification. (c) How sensitive is the calculation to the type of defuzzification method selected?

4. (a) Build this fuzzy system within Matlab, using the FIS, Membership Function, and Rule Editors. (b) Repeat Questions 2 and 3 using sigmoid membership functions.

5. Do you believe this system will accurately identify cassette side occlusions? Explain.

COMPARTMENTAL MODELS

In Part III of this textbook, we turn our attention to models that were specifically derived from mass balance considerations, in order to quantitatively study the kinetics of materials in physiologic systems. The application of these compartmental models differs substantially from the application of the models in Part II. First, it is assumed that the models in Part II may be applied to real-time systems, such as those that measure blood pressure, or any other signal that is readily converted to a current or voltage that may be sampled many times per second. This is certainly not the case for compartmental models. Many important physiologic signals such as proteins cannot be readily sampled, and are never sampled in real time. Instead, the signal's concentration in tissue, plasma, or another accessible fluid is sampled over intervals of several minutes. Once the signal's concentration over time is determined by a biochemical assay or imaging technique, the signal transport can be modeled with a compartmental model and used to understand physiologic function.

Second, in Part II it is assumed that real-time signals are sampled at equally spaced times (i.e., 200 Hz is equivalent to sampling every 5 msec). In contrast, because the number of samples in a compartmental modeling experiment is often limited (for example, only few blood samples may be obtained from a premature infant), these samples are acquired at appropriate, nonequispaced times during the experiment to obtain the best estimates of model parameters. Rather than sampling at a frequency above the Nyquist rate to avoid aliasing, in compartmental modeling experiments, samples may be obtained using an optimal sampling protocol to minimize the fractional standard deviations associated with parameter estimates.

In compartmental modeling nomenclature, the process of signal acquisition is referred to as experimental design. Although the real-time models in Part II are described using discrete time, the discussion in Part III utilizes continuous time because this is the traditional approach.

In Chapter 13, we discuss aspects of the linear compartmental model, including modeling identifiability, nonlinear least-squares estimation, sampling schedules, and model validation. A brief description is also given of the noncompartmental model. Additionally, the various assays by which signal concentrations are obtained are highlighted. In Chapter

14, the application of the linear compartmental model to drug delivery control systems is summarized.

In Chapter 15, we discuss two common nonlinear functions that are incorporated in nonlinear compartmental models: Michaelis–Menten dynamics and the bilinear relation. Michaelis–Menten dynamics describe drug transport that is facilitated by a drug receptor. The bilinear relation, a flux that is proportional to the product of two masses, has been used in the well-known minimal model of insulin sensitivity. In Chapter 16, the application of a nonlinear compartmental model to the development of an antiobesity drug is summarized.

Taken together, the filters of Part I, real-time models of Part II, and compartmental models of Part III comprise a multifaceted approach to biomedical signal processing.

13

THE LINEAR COMPARTMENTAL MODEL

In the first two parts of this textbook, we discussed methods for filtering and modeling physiologic signals that can be acquired in real time. These signals, such as the electrocardiogram and blood pressure, are readily converted to a current or voltage that may be sampled many times per second. However, many important signals—proteins, drugs, and other macromolecules transported throughout the body for particular tasks—cannot be readily sampled. Instead, the signal's concentration in tissue, **plasma** (the fluid portion of the blood), or other accessible fluids is sampled over intervals of several minutes. Once the signal's concentration over time is known, the signal transport can be modeled and used to understand physiologic function.

13.1 PROTEIN STRUCTURE

One of the primary signals of interest—protein—has many functions within the body. Some proteins carry out the transport and storage of small molecules; others make up a large part of the structural framework of cells. Some proteins are **enzymes**—the catalysts that promote the enormous variety of reactions that channel **metabolism** into essential pathways. As discussed in Chapter 4, metabolism is the totality of chemical reactions that occur in animal cells. In particular, proteins that are **hormones**, like insulin, act as chemical messengers at all levels of metabolic regulation.

A protein is composed of a series of **α-amino acids.** As shown in Figure 13.1, each α-amino acid contains a primary carbon ion, the α-carbon, which is attached to an **amide** group (NH_3^+), a **carboxyl** group (COO^-), a hydrogen ion, and a side chain, R. Twenty different amino acids, each with a different side chain, are coded for in the genes and incorporated into proteins. Amino acids can be covalently linked together by formation of an **peptide bond** between the amino and carboxyl groups, with water as a byproduct. The result of this bond is a **peptide.** An example of such a reaction is shown in Figure 13.2.

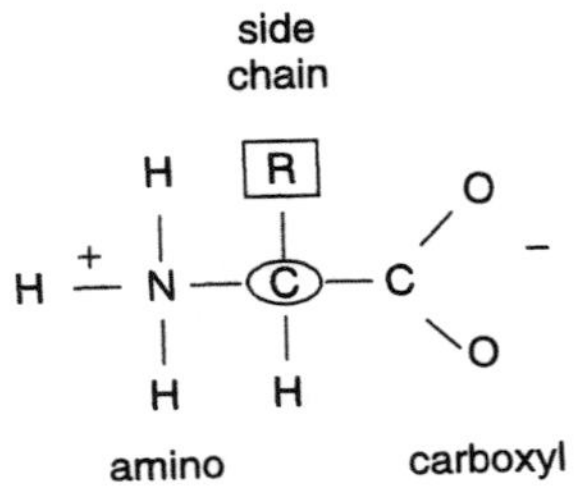

Figure 13.1 Generalized α-amino acid structure.

Note that this reaction still leaves an amide and carboxyl group available for future bonding. Once a long chain of amino acids is formed, it is called a **polypeptide.**

The **primary structure** of a protein is its amino acid sequence. Once the sequence chain is formed, it locally wraps into regions of helical structure, the **secondary structure** of the molecule. This local regular folding is due to **hydrogen bonding** between the amino and carboxyl groups. The helically coiled regions, in turn, fold into a specific compact structure for the entire polypeptide chain called the **tertiary structure** of the molecule (Figure 13.3). As we will see in Chapter 15, tertiary structure is an important component of protein function.

Many drugs are also proteins or amino acid derivatives. In Chapter 2, for example, we discussed **vasoactive** drugs such as **dopamine,** which enhance heart muscle contractility, thus increasing blood pressure. Dopamine is synthesized in the central nervous system from the amino acid tyrosine, and also plays a role in central nervous system transmis-

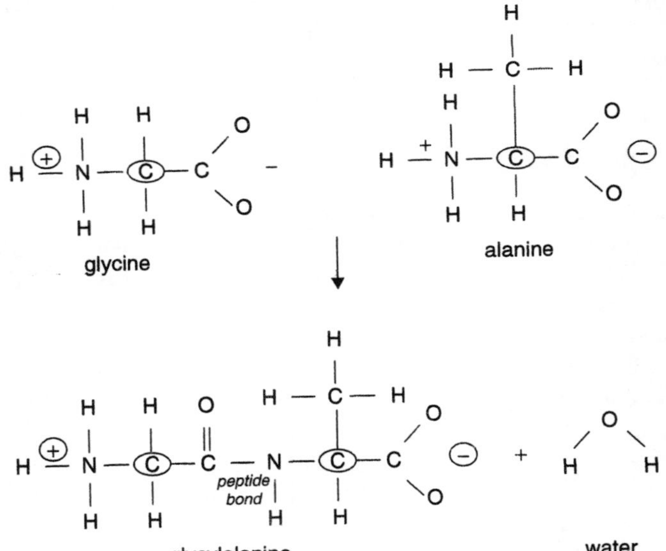

Figure 13.2 Peptide formation.

Figure 13.3 Tertiary structure of human bactericidal/permeability-increasing protein. Reprinted from [1] with permission of Elsevier Science.

sion. Dopamine levels are abnormally low in a particular region of the brain of patients with **Parkinsonism,** a severe neurological disorder.

13.2 EXPERIMENTAL DESIGN

In general, the concentration of a protein or drug in a sample is determined through a biochemical **assay** or **imaging** technique such as **positron emission tomography.** Some common assays are discussed below. Positron emission tomography, a technique for measuring the concentrations of positron-emitting radioisotopes within a three-dimensional object by the use of external measurements of the radiation, is beyond the range of our discussion.

13.2.1 High-Performance Liquid Chromatography

Proteins and other macromolecules may be isolated through **high-performance liquid chromatography** (HPLC). With this technique, a column is packed with a material that can selectively **adsorb** (accumulate on its surface) molecules on the basis of some difference in their chemical structure. Initially, the column is wetted with an appropriate buffer solution. The mixture of molecules to be separated is then placed on top of the column and washed through the column with buffer. In HPLC, pressures of 5000 to 10,000 psi are used to force the solutions rapidly through the material. With these pressures, separations that formerly required hours can be performed in minutes, with higher resolution. Below the column, fractions are collected, as this process of **elution** of the column continues. Because some kinds of molecules are adsorbed only weakly or not at all, they are eluted first. The most strongly adsorbed are eluted last. Typical graphs of concentration versus elution time for separation of five peptides are shown in Figure 13.4.

13.3.2 Radioimmunoassay

Because most hormones are present in extremely low concentrations in biological materials ($\sim 10^{-10}$ M), they cannot be measured by standard methods such as HPLC. Further, a hormone is usually surrounded by excessive amounts of chemically related substances, which would interfere with standard analytical techniques. Instead, hormone concentra-

298 THE LINEAR COMPARTMENTAL MODEL

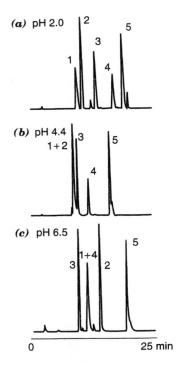

Figure 13.4 Elution of five common peptides at (a) pH 2.0, (b) 4.4, and (c) 6.5, with phosphate as the buffer. The x axis is time in min; the y axis is concentration. The peptides are: 1—bradykinin, 2—oxytocin, 3—angiotensin II, 4—neurotensin, 5—angiotensin I. All the peptides elute earlier at pH 4.4 than at pH 2.0. Courtesy of Vydac, Hesperia, CA.

tions are measured through **radioimmunoassay** (RIA), a technique developed by Yalow and Berson in the 1960s. Rosalind Yalow received the Nobel Prize in Physiology or Medicine in 1977 for this discovery.

Radioimmunoassay is based on the **immune response.** When a foreign substance, or **antigen,** such as a protein invades the tissues of a higher vertebrate, the organism defends itself by synthesizing and attaching an **antibody** to the invading substance. Specific cells then destroy the "marked" antigen. An antibody to any protein can be prepared by injection of the protein into an experimental animal.

The principle of RIA involves combining a hormone or other protein of unknown concentration (the antigen) with a fixed amount of radioactive antigen. A fixed amount of antibody is then added, which binds to both radio- and unlabeled antigen. The antigen and antibody molecules bind, and are then precipitated out of solution. The bound radioactive antigen is detected, and is inversely proportional to the unlabeled antigen present.

13.2.3 Radioactive and Stable-Isotope Tracers

Alternatively, a labeled molecule, a **tracer,** may be introduced into the body to study the behavior of its corresponding endogenous molecule, the **tracee.** The labeling may either use a radioactive isotope, which is not preferable in human studies, or a stable isotope such as [^{13}C] leucine. If the isotope is introduced from outside of the body, such as via the bloodstream, the process is called **exogenous labeling**. If a labeled precursor of the molecule of interest is used to label the tracer, such as infusion of a radiolabeled amino acid to label a protein, the process is called **endogenous labeling**. Over time, samples of fluid such as plasma or urine are collected, and stable samples are analyzed by **mass spectroscopy** to determine the tracer concentration. With this technique, the difference in masss-to-charge ratios of ionized atoms or molecules is exploited to separate them from each other. In theory, the analysis of radioactive and stable-isotope tracers should lead to the same kinetic results [2].

13.2.4 Optimal Sampling Schedule

In a real-time system, the requirement of sampling at a frequency of at least twice the bandwidth, the Nyquist rate, must be met to prevent aliasing. Similarly, in pharmacokinetic systems, an **optimal sampling schedule** should be followed. Optimal sampling is described in Section 13.5.

13.3 KINETIC MODELS

Once **kinetic** data have been obtained, their behavior can be modeled. Two main classes of kinetic models are the **noncompartmental model** and the **compartmental model**. A **compartment** refers to a well-mixed and kinetically homogeneous amount of material, and is not necessarily a physiologic space or well-delimited physical volume. Examples of compartments are the glucose in plasma and in the kidney. The compartment may be accessible or unaccessible for measurement.

13.3.1 Noncompartmental Model Definition

The noncompartmental model considers the input and output relationships of a system accessed through a single accessible compartment (usually the plasma). From Chapter 1, recall that the relationship between the input, $u(t)$, and output, $y(t)$, is the impulse response, $h(t)$, and that the output is the convolution of the input with the impulse response:

$$y(t) = u(t) * h(t) = \int_{-\infty}^{+\infty} u(\tau) h(t - \tau) \qquad (13.1)$$

Based on this equation, if any two of the three parameters is known, the third can be determined.

13.3.2 Commonly Measured Parameters of the Noncompartmental Model

It is common in the pharmacokinetic literature to utilize a noncompartmental model approach by determining model-independent parameters from data. The **clearance rate**, Cl, is the ratio of the intravenous dose to the **total area under the drug concentration in blood or plasma versus time curve**, AUC:

$$Cl = \frac{\text{intravenous dose}}{AUC} \qquad (13.2)$$

300 THE LINEAR COMPARTMENTAL MODEL

The **half-life of a molecule,** $t_{1/2}$, can be easily determined from a plot of the log of plasma concentration versus time. The half-life is the time required for the drug concentration at any point on the straight line to decrease by one-half. Since the eigenvalue is the slope of this line, the eigenvalue, λ, of an exponential i is related to the half-life as:

$$\lambda_i = \frac{0.693}{t_{1/2i}} \tag{13.3}$$

In a two-exponential system, the half-life associated with the larger eigenvalue is called the **distribution half-life,** $t_{1/2\alpha}$. The half-life associated with the smaller eigenvalue is called the **elimination half-life,** $t_{1/2\beta}$.

The **apparent volume of distribution,** V_D, is the proportionality constant relating drug concentration in blood or plasma to the amount of drug in the body. This parameter is confusing because no obvious relationship exists between the apparent and real volume of distribution of drug. Although in many cases this volume corresponds to the plasma, in some cases it is a volume that includes the plasma together with other distribution spaces. The apparent volume of distribution is calculated as the ratio of an intravenous dose to the initial drug concentration immediately after intravenous bolus injection but before any drug has been eliminated:

$$V_D = \frac{\text{dose}}{\text{initial concentration}} \tag{13.4}$$

For the remainder of this chapter, we will emphasize the widely used compartmental model.

13.3.3 Compartmental Model Definition

The compartmental model is governed by the law of conservation of mass, and is described using ordinary differential equations. In contrast to our discussion in Parts I and II, this model is a **multiple input, multiple output system** (Figure 13.5). For each com-

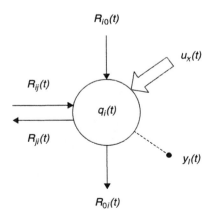

Figure 13.5 Compartment i of multiple input, multiple output system.

partment, the flux of molecules entering and exiting the compartment are denoted as $R_{ij}(t)$, the flux from compartment j to compartment i $(i \neq j)$, and are represented by arrows. **De novo** synthesis of molecules is represented as $R_{i0}(t)$; flux out of the system is represented as $R_{0i}(t)$. The flux is a function of mass, $q_i(t)$. All fluxes and masses are ≥ 0. An exogenous input is represented as an open arrow labeled $u_x(t)$. If the compartment is accessible to measurement, a desired output, $y_i(t)$, is represented as a dashed line with a bullet. Measurements are typically made in units of concentration, $c_i(t)$. The concentration is the mass within the volume of the compartment, $v_i(t)$, where

$$c_i(t) = \frac{q_i(t)}{v_i(t)} \tag{13.5}$$

Using the mass balance principle, the change in mass in each compartment for $t > 0$ can be calculated as the sum of fluxes and any exogenous inputs:

$$\dot{q}_i(t) = \sum_{j=0}^{N} R_{ij}(t) + \sum_{\substack{j=0 \\ j \neq i}}^{N} -R_{ji}(t) + u_x(t) \tag{13.6}$$

For a linear, time-invariant system, each flux is described as the product of a constant, k_{ij}, and the mass of the source compartment, resulting only from diffusion. This linear compartmental model is described as

$$\dot{q}_i(t) = \sum_{j=0}^{N} k_{ij} q_j(t) + \sum_{\substack{j=0 \\ j \neq i}}^{N} -k_{ji} q_i(t) + u_x(t) \tag{13.7}$$

In Eq. (13.7), we assume that $q_0(t) \equiv q_i(0)$, which represents the mass involved in de novo synthesis. If we then divide each mass, $q_i(t)$, by its corresponding volume, $v_i(t)$, the linear system is simplified to

$$\dot{c}_i(t) = \sum_{j=0}^{N} k_{ij} c_j(t) - \sum_{\substack{j=0 \\ j \neq i}}^{N} -k_{ji} c_i(t) + \frac{u_x(t)}{v_i(t)} \tag{13.8}$$

Each constant, k_{ij}, is referred to as a rate constant, and has the unit of inverse time.

In the simple case of a two-compartment system, where it is assumed that $k_{01} = 0$, the eigenvalues, λ_1 and $\lambda_2 (\lambda_1 < \lambda_2)$, can be used to calculate the rate constants as:

$$k_{12} = \frac{\lambda_1(\lambda_2 - \lambda_1)}{\lambda_1 + \lambda_2} + \lambda_1 \tag{13.9}$$

$$k_{01} = \frac{\lambda_1 \lambda_2}{f_{12}} \tag{13.10}$$

$$k_{21} = \frac{\lambda_1 \lambda_2}{f_{12}} \lambda_1 + \lambda_2 - k_{01} - k_{12} \tag{13.11}$$

The eigenvalues can also be used to calculate the two exponential apparent volume of distribution:

$$V_D = \frac{\text{dose}\, f_{12} f_{21} c_1(0)}{\lambda_1 \lambda_2} \tag{13.12}$$

The smaller eigenvalue is associated with a **central compartment** that represents highly perfused organs and tissues such as the liver and kidney, in rapid distribution equilibrium with the blood. Similarly, the larger eigenvalue is associated with a **peripheral compartment** that represents poorly perfused tissues and fluids of the central compartment and the poorly perfused or less readily accessible tissues.

In special cases, a **forcing function** may be associated with a compartment. When this occurs, the normal equations associated with the compartment of interest [Eqs. (13.7) and (13.8)] are replaced by a specified set of equations. In this way, the rest of the system is "forced" to utilize this function. For example, the sampled plasma concentration of a protein over time, which is affected by its disposition in the kidneys, may define the plasma compartment that is in contact with the cerebrospinal fluid (CSF) compartment in the brain, as proteins move across the blood–CSF barrier.

As an example, dopamine kinetics can be modeled with a typical two-compartment system, as shown in Figure 13.6. Here compartment 1 represents the central compartment and compartment 2 represents the peripheral compartment. If dopamine is injected, the exogenous input is an impulse function. If it is infused, $u_1(t)/v_1(t)$ is a step function. Rate constants k_{12} and k_{21} reflect transport between the two compartments. Rate constant k_{01} is associated with elimination from the central compartment.

13.3.4 Compartmental Model Properties

Because the masses and concentrations of each compartment are nonnegative for $t \geq 0$, the corresponding rate constants, k_{ij}, must be nonnegative. The eigenvalues of the system lie in the left half-plane, including the origin, and are therefore stable. In general, the eigenvalues are real if the compartmental structure satisfies the following conditions:

$$k_{i_1 i_2} k_{i_2 i_3} \cdots k_{i_j i_1} = k_{i_2 i_1} k_{i_3 i_2} \cdots k_{i_1 i_j} \tag{13.13}$$

for all $i_1, i_2, \ldots i_j$, distinct, $j = 1, 2, \ldots, N$.

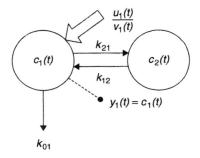

Figure 13.6 Dopamine compartmental model.

13.3.5 Example

In 1991, dopamine pharmacokinetics were measured in 17 infants and children recovering from shock or cardiac surgery at the Children's Hospital Los Angeles. The subjects ranged in age from 3 months to 13 years and weighed 5.0 to 73 kg. Dopamine hydrochloride was infused with an infusion rate within the range 0.5 to 3.0 ml/hr, which corresponded to 1 to 20 µg/kg-min. The infusion duration ranged from 1 to 6 days before the study date. The study was performed when the primary physician discontinued the infusion or standardized the dosage. Blood samples of 1 ml were collected at steady state during a step change in infusion, and at 1.0, 2.5, 5.0, 7.5, 10.0, 15.0, 20.0, 30.0, and 60.0 minutes after discontinuation of the infusion. Dopamine hydrochloride concentrations were determined by HPLC. Plasma concentration data were fit by a two-exponential model.

Based on this data, the mean clearance rate was 454 ± 900 ml/kg-min. The distribution half-life was 1.8 ± 1.0 minutes, whereas the elimination half-life was 26 ± 14 minutes. The apparent volume of distribution was 2952 ± 2332 ml/kg [2]. Using the mean half-lives, the eigenvalues can be calculated from Eq. (13.3), and used to calculated rate constants from Eqs. (13.9) to (13.11) as $k_{12} = 0.05$ min^{-1}, $k_{01} = 0.21$ min^{-1}, and $k_{21} = 0.16$ min^{-1}.

Substituting these values into Eq. (13.8) and assuming a constant infusion rate, the central compartment concentration step response can be determined (Figure 13.7). As shown in this figure, approximately 84 minutes are required to reach steady state, as approximated by 94% of the final value. Often, clinicians ignore the effect of the elimination half-life and estimate the time to steady state as 3.3 times the value of the distribution half-life, which for this study is only 5.9 minutes. A similar longer time to dopamine steady-state concentration was recently estimated in adult patients as 70 to 125 minutes, depending on the infusion rate. As the authors of the adult study noted, "ignoring this delay (true time to steady state versus assumed shorter time to steady state) could explain the lack of relation often observed between the infusion rate and the hemodynamic effects" [4].

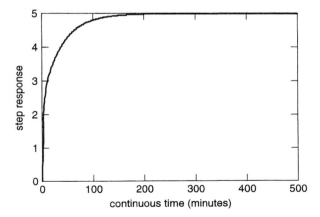

Figure 13.7 Dopamine step response in infants and children, based on mean parameters from [3].

13.4 MODEL IDENTIFIABILITY

Often, only limited information is available from studies of the intrinsic dynamics of pharmacokinetic systems under normal operating conditions. Therefore, it is important before conducting the experiment to determine if it is theoretically possible to obtain unique estimates of all the unknown model parameters. This problem is referred to as **theoretical** or **a priori identifiability**. Even if a proposed model is theoretically identifiable, the practical process of parameter estimation may not yield acceptable results. Problems may arise from different types of experimental error, leading to a loss of **practical** or **a posteriori identifiability**.

13.4.1 Theoretical Identifiability

A priori identifiability addresses the issue that a postulated model structure may be too complex for a particular set of ideal data. A model may be uniquely identifiable, **nonuniquely identifiable,** or **nonidentifiable**. If it is uniquely identifiable, the parameters may be uniquely determined. If it is nonuniquely identifiable, one or more of the parameters possesses more than one, but a finite, number of possible values. If it is nonidentifiable, one or more parameters possesses an infinite number of solutions.

One method for testing theoretical identifiability is the **transfer function matrix approach**. To utilize this approach, Eq. (13.8) is rearranged into **canonical form**, which requires an explicit definition of the outputs:

$$\dot{\mathbf{c}}(t) = \underline{\mathbf{A}}(\mathbf{k})\mathbf{c}(t) + \underline{\mathbf{B}}(\mathbf{k})\mathbf{u}(t) \tag{13.14}$$

$$\mathbf{y}(t) = \underline{\mathbf{C}}(\mathbf{k})\mathbf{c}(t) \tag{13.15}$$

where $\mathbf{c}(t)$ is the concentration vector with dimension $(n-1) \times 1$, $\dot{\mathbf{c}}(t)$ is the concentration derivative vector, $\mathbf{u}(t)$ is the exogenous input vector, $\mathbf{k}$ is the admissible parameter space of rate constants with dimension $p \times 1$, $\mathbf{y}(t)$ is the desired output vector, $\underline{\mathbf{A}}(\mathbf{k})$ is the feedback matrix containing rate constants, $\underline{\mathbf{B}}(\mathbf{k})$ is the feedforward matrix, and $\underline{\mathbf{C}}(\mathbf{k})$ is the relationship between the concentration and desired output vectors.

Continuing with our dopamine example, let the desired output equal the central compartment. The system equations in canonical form are then:

$$\begin{bmatrix} \dot{c}_1(t) \\ \dot{c}_2(t) \end{bmatrix} = \begin{bmatrix} -k_{01} - k_{21} & k_{12} \\ k_{21} & -k_{12} \end{bmatrix} \begin{bmatrix} c_1(t) \\ c_2(t) \end{bmatrix} + \begin{bmatrix} 1 \\ 0 \end{bmatrix} u(t) \tag{13.16}$$

$$y(t) = \begin{bmatrix} 1 & 0 \end{bmatrix} \begin{bmatrix} c_1(t) \\ c_2(t) \end{bmatrix} \tag{13.17}$$

Next the transfer function is calculated. Recall that the transfer function in the Laplace domain, of dimension $r \times m$, may be calculated as

$$\underline{\mathbf{H}}(s, \mathbf{k}) = \underline{\mathbf{C}}(\mathbf{k}) \left[s\underline{\mathbf{I}} - \underline{\mathbf{A}}(\mathbf{k}) \right]^{-1} \underline{\mathbf{B}}(\mathbf{k}) \tag{13.18}$$

where $\underline{\mathbf{I}}$ is the identity matrix. In the dopamine example, the transfer function is calculated as

$$\underline{\mathbf{H}}(s, \mathbf{k}) = \begin{bmatrix} 1 & 0 \end{bmatrix} \begin{bmatrix} s + k_{01} + k_{21} & -k_{12} \\ -k_{21} & s + k_{12} \end{bmatrix}^{-1} \begin{bmatrix} 1 \\ 0 \end{bmatrix} \quad (13.19)$$

$$\underline{\mathbf{H}}(s, \mathbf{k}) = \left[\frac{s + k_{12}}{s^2 + (k_{01} + k_{21} + k_{12})s + k_{12}k_{01}} \right] \quad (13.20)$$

Each element $\underline{\mathbf{H}}_{ij}(s, \mathbf{k})$ reflects an experiment performed on the system between input j and output i. Each element can be further subdivided into the coefficients of the numerator polynomial, $\beta_1^{ij}(\mathbf{k}) \ldots \beta_{n-1}^{ij}(\mathbf{k})$, and the coefficients of the denominator polynomial, $\alpha_1^{ij}(\mathbf{k}) \ldots \alpha_n^{ij}(\mathbf{k})$. From these coefficients, the $(2n-1)rm \times p$ **Jacobian matrix, $\underline{\mathbf{G}}(\mathbf{k})$,** is formed, such that

$$\underline{\mathbf{G}}(\mathbf{k}) = \begin{bmatrix} \frac{\partial \beta_1^{11}}{\partial k_1} & \cdots & \frac{\partial \beta_1^{11}}{\partial k_p} \\ \vdots & & \vdots \\ \frac{\partial \alpha_n^{11}}{\partial k_1} & \cdots & \frac{\partial \alpha_n^{rm}}{\partial k_p} \\ \vdots & & \vdots \\ \frac{\partial \beta_1^{rm}}{\partial k_1} & \cdots & \frac{\partial \beta_1^{rm}}{\partial k_p} \\ \vdots & & \vdots \\ \frac{\partial \alpha_n^{rm}}{\partial k_1} & \cdots & \frac{\partial \alpha_n^{rm}}{\partial k_p} \end{bmatrix} \quad (13.21)$$

The model is identifiable if and only if the rank of the Jacobian matrix equals p.

Assuming that $\mathbf{k} = [k_{21} \; k_{12} \; k_{01}]^T$ (the volume of the central compartment is considered as a parameter in one of the exercises at the end of the chapter), the Jacobian matrix of the transfer function in Eq. (13.20) is

$$\underline{\mathbf{G}}(\mathbf{k}) = \begin{bmatrix} 0 & 0 & 0 \\ 0 & 1 & 0 \\ 0 & 0 & 0 \\ 1 & 1 & 1 \\ 0 & k_{01} & k_{12} \end{bmatrix} \quad (13.22)$$

which has a rank of 3. In this example, $r = m = 1$ and $n = 3$. Therefore, the dopamine model is a priori identifiable. Also, it is clear from Eq. (13.20) that $\underline{\mathbf{H}}_{11}(s, \mathbf{k})$ gives k_{21} uniquely, $\underline{\mathbf{H}}_{21}(s, \mathbf{k})$ gives k_{01} uniquely, and $\underline{\mathbf{H}}_{ij}(s, \mathbf{k})$ gives k_{12} uniquely. Therefore, the model is uniquely identifiable.

The transfer function matrix approach was recently implemented as the software program GLOBI, an acronym for GLOBal Identification [5].

13.4.2 Practical Identifiability

Even if a system is theoretically identifiable, it may not be practically identifiable. Limitations such as imperfections in the test signal, measurement errors, disturbances, and errors in the model structure may lead to nonunique parameter estimates.

Normally, test signal errors are known explicitly or can be neglected if they are insignificant. The effects of measurement errors and disturbances are minimized through appropriate parameter estimation and optimum sampling schedules. Errors in the model structure are usually considered during **model validation.**

13.5 NONLINEAR LEAST SQUARES ESTIMATION

In contrast to the discussion in Chapter 7 of linear least squares estimation, compartmental models require **nonlinear least squares estimation** because they are nonlinear in the parameters, **k**. Only if the state parameters, **c**(t, **k**), can be observed can the problem be transformed into one of linear estimation.

13.5.1 Parameter Estimation

The parameter estimation problem can be formulated as

$$\dot{\mathbf{c}}(t, \mathbf{k}) = \mathbf{j}[\mathbf{c}(t, \mathbf{k}), \mathbf{u}(t), t; \mathbf{k}], \quad \mathbf{c}_0 = \mathbf{c}(t_0, \mathbf{k}) \tag{13.23}$$

$$\mathbf{y}(t, \mathbf{k}) = \mathbf{g}[\mathbf{c}(t, \mathbf{k}); \mathbf{k}] \tag{13.24}$$

$$z_l(t, \mathbf{k}) = y_l(t, \mathbf{k}) + e_l(t), \quad l = 1, 2, \ldots m \tag{13.25}$$

$$\mathbf{h}[\mathbf{c}(t, \mathbf{k}), \mathbf{u}(t), \mathbf{k}] \geq 0 \tag{13.26}$$

where $z_l(t, \mathbf{k})$ is the actual noisy measurement for each of the m measurable outputs [and is related to the noise-free measurable variable $y_l(t)$ by Eq. (13.25)], and $e_l(t)$ is the error associated with the measurement at sample t. Numerical values of the unknown parameter vector **k** are estimated from actual data by minimizing the difference between the model prediction and measured data. Note that the dynamics in Eqs. (13.23)–(13.26) are linear [see Eqs. (13.14) and (13.15)], but the model is nonlinear because **y**(t, **k**) is a nonlinear function of **k**.

In contrast to Chapter 7, we define the performance function as

$$\xi(\mathbf{k}) = \mathbf{e}(t)^T \underline{\mathbf{R}}^{-1}(t) \mathbf{e}(t) \tag{13.27}$$

where $\mathbf{e}(t)$ is the error vector and $\underline{\mathbf{R}}(t)$ is the output correlation matrix. Often, $\underline{\mathbf{R}}(t)$ is determined by the assumption that the measurement errors are white and stationary. With these assumptions, the errors are uncorrelated so that the off-diagonal elements of $\underline{\mathbf{R}}(t)$ are zero and the diagonal elements are equal such that

$$\underline{\mathbf{R}}(t) = \sigma^2 \mathbf{I} \tag{13.28}$$

where σ^2 is the measurement error variance. If, however, the measurement errors are known apart from a scale factor, then

$$\underline{\mathbf{R}}(t) = \underline{\mathbf{W}}(t) \sigma^2 \tag{13.29}$$

where $\underline{\mathbf{W}}(t)$ is a known weighting matrix and σ^2 is unknown. In both cases, it can be shown that these performance functions lead to linear and unbiased estimates that possess minimum variance.

For a performance function of the type shown in Eq. (13.27), the error is given by

$$\mathbf{e}(t) = \mathbf{z}(t, \mathbf{k}) - \mathbf{y}(t, \mathbf{k}) = \mathbf{z}(t, \mathbf{k}) - \mathbf{g}[\mathbf{c}(t, \mathbf{k}); \mathbf{k}] \tag{13.30}$$

Since the model is nonlinear in the parameters, the performance function is not of quadratic form as a function of the parameters. Therefore, an explicit analytical solution such as those shown in Chapter 7 is not possible; instead, an iterative approach is required.

13.5.2 Sensitivity Approach

One practical approach to solving this system of equations is the **Gauss–Newton iterative scheme,** which like Cauchy's method (recall Chapter 3) involves calculation of the gradient. In one possible approach, **sensitivity equations** based on partial derivatives are calculated, leading to exact numerical, but computational complex, calculation of derivatives. The sensitivity approach is implemented in the widely used computer program SAAMII [6].

For each iteration, the parameter vector, $\hat{\mathbf{k}}$, is computed and the covariance matrix, $\underline{\mathbf{V}}(\hat{\mathbf{k}})$, can be approximated as

$$\underline{\mathbf{V}}(\hat{\mathbf{k}}) \cong [\underline{\mathbf{S}}^T(t, \hat{\mathbf{k}}) \, \underline{\mathbf{R}}^{-1}(t) \, \underline{\mathbf{S}}(t, \hat{\mathbf{k}})]^{-1} \tag{13.31}$$

where

$$\underline{\mathbf{S}}(t, \hat{\mathbf{k}}) = \begin{bmatrix} \left.\dfrac{\partial \mathbf{y}(t, \hat{\mathbf{k}})}{\partial k_1}\right|_{k_0, t_1} & \cdots & \left.\dfrac{\partial \mathbf{y}(t, \hat{\mathbf{k}})}{\partial k_p}\right|_{k_0, t_1} \\ \vdots & & \vdots \\ \left.\dfrac{\partial \mathbf{y}(t, \hat{\mathbf{k}})}{\partial k_1}\right|_{k_0, t_n} & \cdots & \left.\dfrac{\partial \mathbf{y}(t, \hat{\mathbf{k}})}{\partial k_p}\right|_{k_0, t_n} \end{bmatrix} \tag{13.32}$$

The covariance matrix is utilized in model validation. Each diagonal element, $v_{ii}(t, \hat{\mathbf{k}})$, provides an estimate of the variance associated with the parameter estimate, $\hat{k}_i$. Therefore, the precision with which the parameter, $\hat{k}_i$, can be estimated may be expressed in terms of its standard deviation as

$$\text{precision}\,(\hat{k}_i) = \sqrt{v_{ii}(t, \hat{\mathbf{k}})} \tag{13.33}$$

Parameter precision may also be expressed as the **coefficient of variation** (CV), which is also known as the **fractional standard deviation** (FSD):

$$CV(\hat{k}_i) = \frac{\sqrt{v_{ii}(t, \hat{\mathbf{k}})}}{\hat{k}_i} \times 100 \tag{13.34}$$

13.5.3 Initial Parameter Estimates

In contrast to linear parameter estimation, it is necessary that initial parameter estimates k_0 of the unknown parameter vector, k, be provided. Since the performance function is no longer quadratic, the initial choice could lead to convergence to a local, rather than the desired global, minimum. Local minima may be avoided by iterating several times with different initial estimates and checking for convergence to the same minimum.

For example, the following data points in the adult dopamine study [4] were extrapolated from the highest magnitude plot in the graph shown in Figure 13.8: 70 ng/ml at 0 min, 55 ng/ml at 1 min, 44 ng/ml at 3 min, 39 ng/ml at 5 min, 29 ng/ml at 10 min, 21 ng/ml at 20 min, 12 ng/ml at 30 min, 11 ng/ml at 40 min, 9 ng/ml at 50 min, 8 ng/ml at 60 min. A measurement error coefficient of variation of 10% [4] was used as an input to the dopamine model, implemented in SAAMII. Rate constant values were calculated using three different sets of initial values for k_0: {0.21, 0.05, 0.16}, {0.5, 0.5, 0.5}, and {0.01, 0.01, 0.01}. A convergence criterion of less than 0.0001 difference in successive iteration values was assumed. In all three cases, the final values converged to {0.069, 0.71, 0.25} min^{-1}. Regardless of its initial value, the associated CV for each rate constant was $\leq$ 10%.

13.6 SAMPLING SCHEDULES

In traditional data acquisition, it is required that a sampling frequency of at least twice the analog bandwidth of the signal of interest be utilized. When this requirement, the Nyquist rate, is ignored, the resulting signal is aliased. In a similar fashion, the sampling schedule of data to be estimated by a compartmental model must be considered in order to avoid parameter estimates with large associated coefficients of variation. An **optimal sampling schedule** (OSS) refers to the sampling schedule with which the maximum precision of model parameter estimates is obtained, under the practical constraints of sample acquisition. For example, there is a physical limit to the amount of blood samples that can be obtained from a neonate (premature infant).

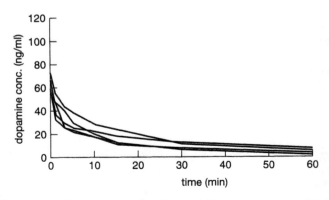

Figure 13.8 Individual dopamine plasma concentrations (ng/ml) versus time curves after discontinuation of intravenous infusions of 3 µg/kg-min in five patients. Reprinted from [5] with permission from Lippincott Williams & Wilkins.

13.6.1 Optimal Sampling Schedule

To calculate an optimal sampling schedule, we assume that the compartmental model described by Eqs. (13.23) to (13.26) describes a system being sampled several times. Replicate samples at the same time point are considered independent. Further, we assume that the parameters are uniquely identifiable and that some a priori estimates are available. The covariance matrix of unbiased parameter estimates, $\underline{V}(k)$, possesses the inverse of the **Fisher information matrix**, $\underline{J}(k, \sigma^2, SS)$, as a lower bound [7]. As shown in its nomenclature, this matrix is a function of the parameter values, error variance, and **set of sampling schedules**, SS, where $SS = \{SS_1, SS_2, \ldots, SS_m\}$ and $SS_l = (t_1, \ldots, t_n)$. Each element of $\underline{J}(k, \sigma^2, SS)$ is calculated by definition as

$$\underline{J}_{ij}(k, \sigma^2, SS) = \sum_{l=1}^{m} \sum_{o=1}^{n} \frac{1}{\sigma_l^2(t_o)} \frac{\partial y_l(t_o)}{\partial k_i} \frac{\partial y_l(t_o)}{\partial k_j}, \quad i,j = 1 \ldots p \quad (13.35)$$

This calculation is asymptotically correct for arbitrary distribution of the error under weighted least squares estimation. The design of a fixed optimal sampling schedule is a minimization problem for some scalar function of the inverse of $\underline{J}(k, \sigma^2, SS)$.

While several criteria have been proposed, we will consider the **D-optimal design approach** because of favorable theoretical properties of independence of the choice of units for the parameters or any **nondegenerate reparameterization** (transformation to forms such as the canonical form). A D-optimal design uses the **determinant**, *det*, of the inverse of $\underline{J}(k, \sigma^2, SS)$, and minimizes the joint asymptotic confidence region for the parameter estimates, thus minimizing the associated parameter estimate coefficients of variation.

Using the search algorithm proposed by Cobelli et al. [8], the determinant of the Fisher information matrix, the $det[\underline{J}(k, \sigma^2, SS)]$ is maximized. Note that maximization of $det[\underline{J}(k, \sigma^2, SS)]$ equals minimization of $1/\{det[\underline{J}(k, \sigma^2, SS)]^{-1}\}$. This search algorithm employs **Fibonacci** techniques [9]. Geometrically, for a single output search, a sequence of one dimensional searches in directions parallel to the coordinate axes of the feasible search space are conducted, with the sample points constrained to lie on a closed observation interval, the duration of the experiment. Each univariate search is performed using the Fibonacci scheme, by which a third search interval is the sum of the previous two search intervals.

For multiple outputs, the search is also univariate, in that one sample time for the first output, $y_1(t)$, is allowed to vary over a user-defined grid of sample points along the direction of increasing $det[\underline{J}(k, \sigma^2, SS)]$. Although the first output varies, all remaining samples are fixed at their current values until $det[\underline{J}(k, \sigma^2, SS)]$ is maximized. This optimization is repeated for each time in the first and subsequent outputs until the algorithm converges. Convergence is reached when the difference between subsequent iterations of $det[\underline{J}(k, \sigma^2, SS)]$ is below a fixed threshold.

For each output, the variance structure of the measurement error must be input to the algorithm. Convergence to D-optimality is not guaranteed because of the existence of local maxima. Therefore, the optimality procedure should be repeated with different initial sampling schedules to explore different regions of the design space. Model outputs and output sensitivities $[\partial y_l(t_o)/\partial k_i]$ are evaluated by integration of the system differential equations and the associated first-order linear sensitivity system (partial derivatives with respect to the parameters).

To test this algorithm, Cobelli et al. optimized the sampling schedule design in a two-

310 THE LINEAR COMPARTMENTAL MODEL

compartment model of insulin kinetics (Figure 13.9). In this model, a rectangular insulin infusion pattern as test input was investigated. The equations of this model are

$$\dot{q}_1(t) = -k_{21}q_1(t) + k_{12}q_2(t) + R + u(t) \quad (13.36)$$

$$q_1(0) = q_{10} \quad (13.37)$$

$$\dot{q}_2(t) = k_{21}q_1(t) - (k_{12} + k_{02})q_2(t) \quad (13.38)$$

$$q_2(0) = q_{20} \quad (13.39)$$

$$y(t) = \frac{1}{V_1}q_1(t) \quad (13.40)$$

where $q_1(t)$ and $q_2(t)$ are the insulin mass in plasma and in a peripheral compartment, respectively (μU); $u(t)$ is the test input (mU/min); k_{ij} are rate constants (1/min); R is a steady state insulin infusion of 26.7 mU/min; q_{10} and q_{20} are the steady state conditions of 79,590 and 146,700 μU, respectively; $y(t)$ is the insulin concentration in plasma (μU/ml); and V_1 is the volume of compartment 1 (ml). The unknown model parameters are three rate constants and the compartment volume. Mean values of a normal human population were assumed to be $k_{21} = 0.529$, $k_{12} = 0.0105$, $k_{02} = 0.182$, and $V_1 = 3312$.

Measurement error variance was experimentally determined and ranged from 34 to 4% CV for insulin concentrations varying from 5 to 1000 μU/ml. The chosen study observation interval was 1–100 min; the chosen sample resolution was 0.5 min. Four different unit step infusions, all for the same dose of 4500 mU, were used as test inputs: 9000 mU/min for $0 \le t \le 0.5$ min, 225 mU/ml for $0 \le t \le 20$ min, 75 mU/ml for $0 \le t \le 60$ min, 45 mU/ml for $0 \le t \le 100$ min (Figure 13.10).

Using D-optimal design, optimal sampling schedules were determined for four samples. These sample times, values of $\log\{det[\mathbf{J}(\mathbf{k}, \sigma^2, \mathbf{SS})]\}$, and the associated parameter estimate coefficients of variation are shown in Figure 13.10 and Table 13.1. Note that the largest $\log\{det[\mathbf{J}(\mathbf{k}, \sigma^2, \mathbf{SS})]\}$ for a given input also possesses the lowest parameter estimate coefficients of variation; this is the optimal sampling schedule.

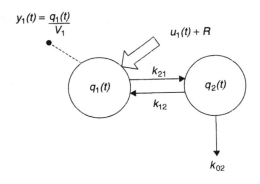

Figure 13.9 Insulin compartmental model. Based on [8].

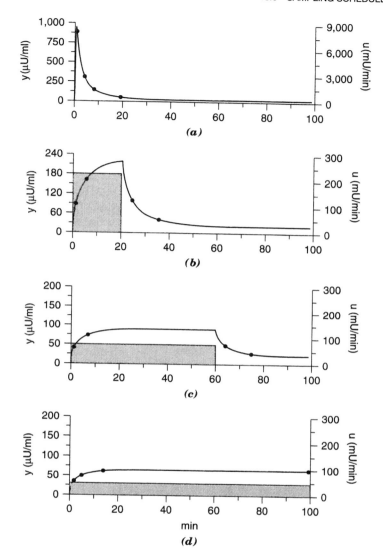

Figure 13.10 Optimal sampling schedules for insulin kinetics. The four samples of output $y(t)$ are denoted by •, and (a), (b), (c), and (d) refer to the four equidose inputs $u(t)$ (shaded area) lasting, respectively, 0–0.5, 0–20, 0–60, and 0–100 min. From [8].

13.6.2 Minimal Versus Practical Sampling Schedules

For a wide class of models with one output, the number of optimal sampling times is often equal to the number of a priori model parameters. This is the minimum number of sample times required to fit a given model to data obtained with an arbitrary sampling schedule [10, 11]. Box was the first to observe this minimal sampling schedule [12]. Similarly, the OSS example above gave as a result four data samples for identification of four parameters.

Table 13.1 Optimal sampling schedules for the insulin kinetics model. From [8]

Input duration (min)	Sampling schedule (min)				$\log_{10}[det(\mathbf{J})]$	Coefficient of variation (%)			
						k_{21}	k_{12}	k_{02}	V_1
0–0.5	1.0	3.0	8.0	19.5	5.5233	15	34	16	9
0–20	1.0	5.5	24.0	35.5	3.9666	32	80	30	12
0–60	1.0	6.5	64.0	74.5	2.2891	49	168	72	15
0–100	1.0	4.5	13.0	100.0	0.1685	139	668	264	26

However, for practical applications, to minimize the effects of unforeseen noise sources present during an experiment, we recommend taking as many samples as is practically possible. For example, in studies of insulin transport across the blood–brain barrier, it is known that canine cerebrospinal fluid can be sampled in 1.8 ml quantities once every 20 minutes without disturbing the cerebrospinal fluid production rate [13]. As a rule of thumb, because compartmental models are composed of sums of individual exponentials, samples should be acquired during the greatest changes (steepest slope) in the output function [14]. Such sampling leads to minimization of the number of possible exponentials that may fit the steeply changing output curve. Further, samples should also be acquired as steady state is reached. Again, such sampling leads to minimization of the number of exponentials that fit the tail of the output curve. Note that in the insulin example above, the sampling schedule with the lowest coefficients of variation contains samples acquired during the steepest curve changes and approximate steady state.

13.7 MODEL VALIDATION

Assuming that a model has been determined to a priori identifiable, how is it validated? Model validation involves analyzing the results of parameter identification and model plausibility, in order to select an "optimum" model.

13.7.1 Results of Parameter Identification

Once parameters have been identified for a chosen model from experimental data, the results may be analyzed to determine the model's utility using three validity measures. As described above, the covariance matrix contains estimates of the variance associated with each identified parameter. These estimates are used to determine the associated coefficient of variation for each parameter estimate. When the CVs of estimated parameter values are unreasonably large (i.e., greater than 100%), the model is typically considered invalid. Large CVs may arise from limitations in the experimental data such as a small number of measurements or large measurement errors. Large CVs may also arise from utilization of a model that is too complex for the available experimental data.

A second validity measure is the **goodness of fit.** As discussed in Chapter 7 for linear autoregressive moving average models, the best of several candidate models may be determined using **Akaike's Information Criterion,** AIC[15]. Because a more complicated model with more parameters may better fit experimental data, the number of model parameters is weighed against the number of data points and **residual sum of squares,** where residual refers to the difference between the observed and estimated data:

$$AIC = N \ln \sum_{l=0}^{N} \frac{1}{\sigma^2(t_l)} [z(t_l) - y(t_l, \hat{\mathbf{k}})]^2 + 2p \qquad (13.41)$$

where N is the number of data points, and p is the number of parameters. The best model is that which yields the lowest value of AIC.

The third validity measure is **residual statistics.** The residuals are an estimate of the noise in the system. If this noise is assumed to be white, Gaussian, and of zero mean, then the residuals should display these properties. If the residuals do not meet the assumptions, a systematic error in model identification may be present.

13.7.2 Model Plausibility

After the model fit has been analyzed, the model plausibility should be examined. First, the estimated parameters must lie within a reasonable physiologic range, or the model is invalid. Second, other values that can be estimated from the estimated parameters, such as the clearance rate, must also lie within a reasonable physiologic range. Most importantly, each compartment should represent a plausible anatomic space, compatible with current physiologic knowledge. The number of reasonable compartments required for adequate representation of a system is often an issue of contention between physiologists, who are more comfortable with a single compartment, and engineering scientists, who may observe a need for several compartments to fit features in the experimental data.

13.8 SUMMARY

In this chapter, we have reviewed the basic concepts that comprise the linear compartmental modeling process. First, the signal of interest, often a protein, must be sampled over time. An imaging technique such as positron emission tomography may be used for this acquisition. Alternatively, assays such as high-performance liquid chromatography, radioimmunoassay, or a tracer system may be used to acquire these sample curves.

The compartmental model used should be analyzed to determine if it is theoretically possible to obtain unique estimates of all the unknown model parameters. One method for testing theoretical identifiability is the transfer function matrix approach, which essentially looks at all input/output combinations. If this model is not a priori identifiable, then accommodations in the model should be made before proceeding in the modeling process.

Nonlinear, rather than linear, least squares estimation is required for parameter calculation since the model is nonlinear in the parameters. Because an explicit analytical solution is not possible, an iterative approach such as the sensitivity approach, based on partial derivatives, is utilized. Because the performance function is not quadratic, the initial choice of parameter estimates could lead to convergence to local, rather than the desired global, minimum. Local minima may be avoided by iterating several times with different initial estimates and checking for convergence to the same minimum.

The sampling schedule of data to be estimated must be considered, in order to avoid parameter estimates with large associated coefficients of variation. Theoretically, an optimal sampling schedule may be calculated by minimizing some scalar function of the inverse of the Fisher information matrix. One such minimization approach is the D-optimal design approach, based on the determinant of the inverse of the Fisher information matrix. Practically speaking, coefficients of variation may be minimized by acquiring samples during the steeply changing output curve and as steady state is reached. Such sampling leads to minimization of the number of exponentials that fit the acquired data.

The identified parameters should be analyzed for model validity. For a valid model, the associated coefficients of variation should be reasonable in size (less than 100%), the model should follow Akaike's information criterion, and the residuals should resemble random noise. The estimated parameters, derived parameters, and number of compartments must also be plausible for the physiologic system under study.

13.9 REFERENCES

[1]. Beamer, L. J., Carroll, S. F., and Eisenberg, D. The three-dimensional structure of human bactericidal/permeability-increasing protein. *Biochem Pharm, 57,* 225–229, 1999.

[2]. Ikewaki, K., Rader, D. J., Schaefer, J. R., Fairwell, T., Zech, L. A., and Brewer, H. B., Jr. Evaluation of apo A-I kinetics in humans using simultaneous endogenous stable isotope and exogenous radiotracer methods. *J Lipid Res, 34,* 2207–2215, 1993.

[3]. Eldadah, M. K., Schwartz, P. H., Harrison, R., and Newth., C. J. L. Pharmacokinetics of dopamine in infants and children. *Crit Care Med, 19,* 1008–1011, 1991.

[4]. Le Corre, P., Malledant, Y., Tanguy, M., and Le Verge, R. Steady-state pharmacokinetics of dopamine in adult patients. *Crit Care Med, 21,* 1652–1657, 1993.

[5]. Audoly, S., D'Angio, L., Saccomani, M. P., and Cobelli, C. Global identifiability of linear compartmental models—A computer algebra algorithm. *IEEE Trans Biomed, 45,* 36–47, 1998.

[6]. SAAM II User Manual. SAAM Institute: Seattle, WA. 1998.

[7]. Ljung, L. *System Identification: Theory for the User,* 2nd ed. Prentice Hall PTR: Upper Saddle River, NJ, pp. 217–218, 1999.

[8]. Cobelli, C., Ruggeri, A., Distefano, J. J., III, and Landaw, L. Optimal design of multioutput sampling schedules—Software and applications to endocrine-metabolic and pharmacokinetic models. *IEEE Trans Biomed, 32,* 249–256, 1985.

[9]. Distefano, J. J., III, Algorithms, software and sequential optimal sampling schedule designs for pharmacokinetic and physiological experiments. *Math Comput Simul, 24,* 531–534, 1982.

[10]. Distefano, J. J., III, Optimized blood sampling protocols and sequential design of kinetic experiments. *Am J Physiol, 240,* R259–R265, 1981.

[11]. D'Argenio, D. Z. Optimal sampling times for pharmacokinetic experiments. *J Pharm Biopharm, 9,* 739–754, 1981.

[12]. Box, M. J. Improved parameter estimation. *Technometrics, 12,* 219–228, 1970.

[13]. Baura, G. D., Foster, D. M., Porte, D., Jr., Kahn, S. E., Bergman, R. N., Cobelli, C., and Schwartz, M. W. Saturable transport of insulin from plasma into the central nervous system of dogs in vivo: A mechanism for regulated insulin delivery to the brain. *J Clin Invest, 92,* 1824–1830, 1993.

[14]. Cobelli, C and Ruggeri, A. Optimal design of sampling schedules for studying glucose kinetics with tracers. *Am J Physiol, 257,* E444–E450, 1989.

[15]. Akaike, H. A new look at the statistical model identification. *IEEE Trans Auto Cont, AC-19,* 716–723, 1974.

Further Reading

Pharmacokinetic Models:

Gibaldi, M. and Perrier, D. *Pharmacokinetics,* 2nd ed., vol. 15. Marcel Dekker: New York, 1982.

Rowland, M. and Tozer, T. N. *Clinical Pharmacokinetics: Concepts and Applications,* 3rd ed. Lippincott Williams & Wilkins: Philadelphia, 1995.

Compartmental Models:

Carson, E. R., Cobelli, C., and Finkelstein, L. *The mathematical modeling of metabolic and endocrine systems: Model formulation, identification, and validation.* John Wiley: New York, 1983.

Cobelli, C., Foster, D., and Toffolo, G. *Tracer Kinetics in Biomedical Research: From Data to Model.* Plenum: New York, 2001.

Jacquez, J. A. *Compartmental Models in Biology and Medicine,* 2nd ed. University of Michigan Press: Ann Arbor, 1985.

Biochemical Assays: Matthews, C. K. and van Holde, K. E. *Biochemistry.* Benjamin/Cummings: Redwood City, CA, 1990. Tools of Biochemistry sections.

13.10 RECOMMENDED EXERCISES

Compartmental Models. See: Shargel, L. and Yu, A. B. C. *Applied Biopharmaceutics and Pharmacokinetics,* 3rd ed. Appleton & Lange: Norwalk, Conn., 1993. Questions 3.4; 3.6; 5.2-5.6; 5.7c,d,e.

Model Identifiability. See: Cobelli, C. and Caumo, A. Using what is accessible to measure that which is not: necessity of model of system. *Metabolism, 47,* 1009–1035, 1998. Determine if the two models in Figure 12 are a priori identifiable using the transfer function matrix approach. Note that each model includes the parameter V_1, the volume of compartment 1.

14

PHARMACOLOGIC STRESS TESTING USING CLOSED-LOOP DRUG DELIVERY

In this chapter, we discuss the application of pharmacokinetics to **closed-loop drug delivery** (CLDD). During closed-loop drug delivery, a drug is administered from a drug delivery device such as an infusion pump, while the drug infusion rate is continuously adjusted by a closed-loop algorithm or controller that is based on control theory techniques.

Closed-loop drug delivery is only applicable to certain drugs. Typically, new drugs are designed to treat nearly all patients without toxicity. In human drug testing, a therapeutic dosing window that is typically effective in 99% and toxic in 1% of a population is determined. A wide therapeutic window eliminates the need for closed-loop drug delivery [1]. For this reason, drug delivery is mainly applied to drugs without this window, such as insulin for diabetics and classic cancer chemotherapy agents. Drug delivery may also be used for short half-life drugs such as **sodium nitroprusside** that may be cyanide toxic and possess a therapeutic effect (i.e., change in blood pressure) that can be monitored in real time.

14.1 PHARMACOKINETICS AND PHARMACODYNAMICS

In Chapter 13, we discussed kinetic aspects of **pharmacokinetics**—the transport of drugs to the site of action. After systemic absorption, the drug is transported throughout the body by the general circulation. The majority of the drug dose reaches unintended target tissues, in which the drug is passively stored, produces an adverse effect, or is eliminated. A fraction of the dose reaches the target site, or **effect compartment**, and establishes equilibrium. The time course of drug delivery in the effect compartment determines whether the onset of the pharmacologic response is immediate or delayed. This drug delivery is affected by the rate of blood flow, diffusion, and partition properties of the drug and molecules that interact with the drug, the **receptors**. The duration of drug activity is affected by the elimination half-life. For example, to improve antibiotic therapy with **penicillin** and **cephalosporin** antibiotics, clinicians may intentionally prolong the drug's action by giving a second drug, **probenecid,** which competitively inhibits renal excretion of the antibiotic. The relationship between drug concentrations at the site of action and subsequent pharmacologic response is called **pharmacodynamics.**

14.1.1 Drug Receptors

The drug receptor is the binding or recognition component of a macromolecular complex at which a drug interacts to produce its therapeutic or toxic action. Drug interaction with the receptor results in a specific effect that provides a rational basis for drug therapy. Receptor molecules include proteins that are **enzymes** (catalysts that assist in regulating cell chemistry), hormone or **neurotransmitter** (influencing cell-to-cell communication) receptors, and **ion channels** (ion-selective pores in the cell membrane). Other receptors are nucleic acids or miscellaneous molecules such as stomach acid (antacids bind to excess hydrogen ions), **cholesterol** (cholestyramine resin binds and reduces cholesterol concentration), and the **lipid bilayer** of the cell membrane.

The interaction between a drug and its specific receptor typically has a high potency, due to a good structural fit between the molecules. This fit enables noncovalent interaction between multiple portions of the drug and receptor molecules. Nanomolar to micromolar (10^{-9} to 10^{-6} M) drug concentrations are sufficient to observe the effects. Nonspecific drugs such as bulk laxatives often have much lower potencies (10^{-3} M). Specific structural requirements are necessary for drug activity; minimal changes in structure drastically alter the drug's potency. The drug–receptor interaction often shows **stereoselectivity,** with one **optical isomer** (mirror image of a molecule) generally 10–100 times more powerful than the other.

As an example, the **adrenergic receptors** in smooth muscle and the heart bind with multiple drugs, resulting in contraction or relaxation of the muscle. These contraction receptors are classified into the subtypes α_1, α_2, β_1, and β_2, depending on the drugs that bind, tissue location, and mechanism of action. For example, the drug **arbutamine** binds strongly to the β_1 receptor and weakly to the α_1 receptor (Figure 14.1).

14.1.2 Drug-Receptor Theory

The nature of the drug–receptor interaction is generally considered to be reversible, resulting from noncovalent interactions such as **ionic bonds, hydrogen bonds, Van der Waals interactions** (transient dipoles induced in neighboring nonpolar groups), or **hydrophobic interactions** (association of nonpolar molecules in aqueous solution). According to the **law of mass action,** a signal drug molecule, D, interacts with a receptor with a single binding site, R, to produce a pharmacologic response. This theory, **occupation theory,** is represented by the concentration equation:

$$[D] + [R] \leftrightarrow [DR] \rightarrow \text{Response} \tag{14.1}$$

where [ ] represents concentration.

For this theory, it is assumed that a drug molecule combines with a receptor molecule as a bimolecular association, and that the resulting drug–receptor complex, DR, dissoci-

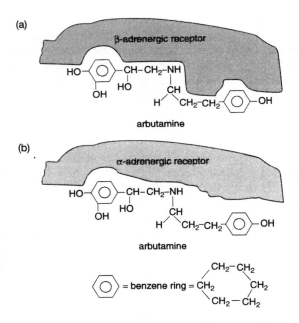

Figure 14.1 Diagramatic binding of arbutamine to (a) β-, but not (b) α-, adrenergic receptor.

ates as a unimolecular entity. It is also assumed that the binding of the drug with the receptor is fully reversible and that a single type of binding site with one site per molecule exists. Further, it is assumed that as more receptors are occupied by drug molecules, a greater response is obtained until a maximal response is reached.

As shown in Figure 14.2(a), the dose–response curve is approximately linear for very small doses, nonlinear for larger doses, and saturates at the maximal response. The drug concentration that elicits a half-maximal response is referred to as EC_{50}. For two drugs acting on the same receptor, the drug with the lower potency possesses the higher EC_{50}. EC_{50} can be more readily visualized in the sigmoidal log dose–response curve [Figure 14.2(b)].

The rate of association of formation of the drug–receptor complex is described by

$$\text{rate of association} = k_1[D][R] \qquad (14.2)$$

where k_1 is the association rate constant expressed in inverse units of the product of concentration and time. Similarly, the rate of dissociation of the drug–receptor complex is described by

$$\text{rate of dissociation} = k_2[DR] \qquad (14.3)$$

where k_2 is the dissociation rate constant expressed in units of inverse time. At equilibrium, or steady state, the rate of association equals the rate of dissociation. The equilibrium constant, K_D, can then be determined from the mass action equation in Eq. (14.1) as

$$K_D = \frac{k_2}{k_1} = \frac{[D][R]}{[DR]} \qquad (14.4)$$

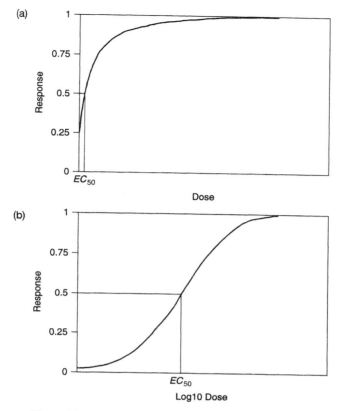

Figure 14.2 (a) Dose–response and (b) log dose–response curves.

Using occupation theory, three types of pharmacologic responses may be described at the receptor site. First, an **agonist** is a drug molecule that interacts with the receptor and elicits a maximal pharmacologic response. Second, a **partial agonist** is a drug that elicits a partial (lower than maximal) response. Third, an **antagonist** is a drug that elicits no response, but reversibly or irreversibly inhibits the receptor interaction of a second agent. For the β_1-adrenergic receptor, **isoproterenol** and arbutamine are agonists and **metoprolol** is an antagonist. For the α_1-adrenergic receptor, arbutamine is a partial agonist. The common term "beta blocker" refers to a drug that is an antagonist of the β-adrenergic receptor.

Because occupation theory is not consistent with all kinetic observations, alternative theories have been proposed. Newer theories based on observations of in vitro studies are difficult to extrapolate to in vivo conditions. The more accepted **rate theory** states that the pharmacologic response is not dependent on the drug–receptor complex concentration but depends on the rate of association of the drug and receptor.

14.1.3 Pharmacodynamic Feedback Signals

Pharmacologic responses are not easily continuously quantified. Therefore, the vital signs of heart rate, blood pressure, temperature, respiration, arterial oxygen saturation, and car-

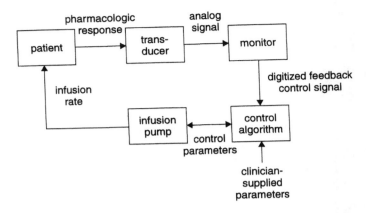

Figure 14.3 Typical drug delivery control system.

diac output are typically used as feedback signals for closed-loop drug delivery. As shown in Figure 14.3, a typical drug delivery system consists of the patient, a transducer, a monitor, a control algorithm, and an infusion pump. The patient receives a drug dosage from the infusion pump and provides a pharmacologic response. The response is sensed by the transducer, which converts the response to a proportional analog electrical current. The electrical current is digitized within the monitor, and provides a feedback signal for the control algorithm. The control algorithm uses the feedback signal and other clinician-supplied parameters such as cumulative drug dose to determine the next drug dosage infusion rate. A control algorithm supervisor may ensure that the closed-loop system remains stable and does not become a nonminimum phase or positive feedback system.

14.2 CONTROL THEORY

In control theory, the output, $y(t)$, of a plant, $h(t)$, is compared to an input reference signal, $u(t)$. Using some type of control algorithm within a controller, $g_c(t)$, the error between the two signals is minimized (Figure 14.4). The adaptive filter theory described in Chapter 3 can be applied to an adaptive controller. Similarly, the fuzzy logic model described in Chapter 11 can be applied to a fuzzy logic controller. Both control strategies, as well as that of the **proportional–integral–derivative (PID) controller**, are typically used in CLDD systems.

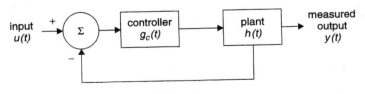

Figure 14.4 Closed-loop system.

14.2.1 Proportional–Integral–Derivative Control

The PID controller is a method utilized in classical control theory. The classical period of control theory began during World War II at the Massachusetts Institute of Technology. Classical control is based on analog methods and utilizes the Laplace transform function. The following discussion is based on the textbook *Modern Control Systems,* by Dorf and Bishop [2].

As stated by its name, a PID controller possesses a transfer function, $G_c(s)$, containing a proportional gain, K_p, an integral term, K_i/s, and a derivative term, $K_d s$:

$$G_c(s) = K_p + \frac{K_i}{s} + K_d s \tag{14.5}$$

This transfer function can alternatively be described as

$$G_c(s) = \frac{K_d(s + f_1)(s + f_2)}{s} \tag{14.6}$$

where f_1 and f_2 are the roots of the second order function, $s^2 + (K_p/K_d)s + (K_i/K_d)$. Therefore, this controller introduces one pole at the origin and two zeros that can located anywhere in the left-hand plane of the Laplace domain. The PID controller is often used in applications because of its robust performance in a wide range of operating conditions and functional simplicity. A closed-loop system consisting of a PID controller and a plant, $H(s)$, will be stable if all the poles of the system transfer function have negative real parts.

14.2.2 Robust Control Systems

The PID controller can be used to design a **robust control system.** A robust control system exhibits desired performance in the presence of significant plant uncertainty [2]. A general robust control system incorporates an unpredicted disturbance input, $V(s)$, sensor noise, $N(s)$, and a plant, $H(s)$, with potentially unmodeled dynamics or parameter changes (Figure 14.5). A prefilter, $G_p(s)$, is also included. Assuming that noise and disturbances are not present, the closed-loop transfer function between input and output is calculated by determining expressions for the error, $E(s)$, and the output, $Y(s)$, as a function of error:

$$E(s) = U(s)G_p(s) - Y(s) \tag{14.7}$$

$$Y(s) = E(s)G_c(s)H(s) \tag{14.8}$$

Substituting Eq. (14.7) into (14.8) leads to the closed-loop transfer function

$$\frac{Y(s)}{U(s)} = \frac{G_p(s)G_c(s)H(s)}{1 + G_c(s)H(s)} \tag{14.9}$$

Often, a second-order plant is analyzed. A second-order plant may be characterized by its **damping ratio,** ς, and **natural frequency,** ω_n. The damping ratio controls the overshoot of the system response, with a smaller coefficient resulting in a larger overshoot. The step re-

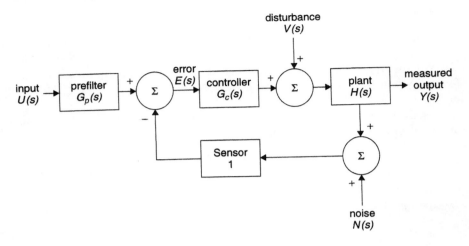

Figure 14.5 Robust control system.

sponse of a second-order system is shown in Figure 14.6. The **settling time**, T_s, may be defined as the time by which a system settles to within 2% of its final value (Figure 14.6). This is equivalent to four time constants of the dominant roots of a characteristic equation:

$$T_s = \frac{4}{\varsigma \omega_n} \qquad (14.10)$$

Selecting the three PID coefficents, K_p, K_i, and K_d, for the robust controller entails searching a three-dimensional space. As with previous algorithms, the system is minimized with respect to a performance function, $\xi(t)$. A common performance function is the **integral of time multiplied by absolute error (ITAE) function**,

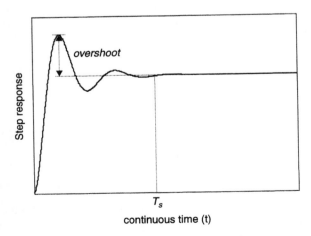

Figure 14.6 Step response of second-order system.

$$\xi_{ITAE}(t) = \int_0^T t|e(t)|dt \qquad (14.11)$$

Other methods for selecting the PID coefficients, which are beyond the range of our discussion, include the root locus method and frequency response analysis using Bode plots.

For a second-order plant and PID controller (one pole), the resulting characteristic equation is third order. Given a step input, the optimum coefficients of the characteristic equation based on the ITAE performance function are [2]:

$$s^3 + 1.75\omega_n s^2 + 2.15\omega_n^2 + \omega_n^3 \qquad (14.12)$$

For this performance function, it is assumed that the transfer function contains no zeros, and that the numerator is ω_n^3. Using the ITAE performance function produces very little overshoot in the transient response to a step or ramp.

The steps for designing a robust control system using a PID controller are:

1. Specify the system settling time, and select ω_n.
2. Use the ITAE performance index to calculate the PID coefficients.
3. Select a prefilter, $G_p(s)$, to eliminate zeros in the closed-loop system transfer function.

14.2.3 Robust PID Control Example

Let us assume a plant with transfer function

$$H(s) = \frac{1}{s^2 + 2s + 1} \qquad (14.13)$$

and a known damping ratio of 0.8. Without prefiltering, the closed-loop transfer function can be determined from Eqs. (14.5) and (14.9) as

$$\frac{Y(s)}{U(s)} = \frac{\left[K_p + \frac{K_i}{s} + K_d s\right]\frac{1}{s^2 + 2s + 1}}{1 + \left[K_p + \frac{K_i}{s} + K_d s\right]\frac{1}{s^2 + 2s + 1}} \qquad (14.14)$$

$$\frac{Y(s)}{U(s)} = \frac{K_d s^2 + K_p s + K_i}{s^3 + s^2[2 + K_d] + s[K_p + 1] + K_i} \qquad (14.15)$$

If we require a settling time of 0.5, we can calculate the natural frequency from Eq. (14.10) as $\omega_n = 10$. Substituting this value into ITAE performance function in Eq. (14.12) leads to the characteristic equation:

$$s^3 + 17.5s^2 + 215s + 1000 \qquad (14.16)$$

By equating this characteristic equation with the denominator of Eq. (14.15), we obtain the PID coefficients $K_p = 214$, $K_i = 1000$, and $K_d = 15.5$.

Substituting these coefficients into the closed-loop transfer function without prefiltering results in

$$\frac{Y(s)}{U(s)} = \frac{15.5s^2 + 214s + 1000}{s^3 + 17.5s^2 + 215s + 1000} \qquad (14.17)$$

$$\frac{Y(s)}{U(s)} = \frac{15.5(s + 6.9 + j4.1)(s + 6.9 - j4.1)}{s^3 + 17.5s^2 + 215s + 1000} \qquad (14.18)$$

As shown in Figure 14.7, the system requires 0.62 seconds to settle and possesses an overshoot of 34%. To achieve the desired ITAE response, the closed-loop numerator must equal ω_n^3. The zeros in Eq. (14.18) are cancelled with the prefilter

$$G_p(s) = \frac{64.5}{(s + 6.9 + j4.1)(s + 6.9 - j4.1)} \qquad (14.19)$$

which gives the desired response of

$$\frac{Y(s)}{U(s)} = \frac{G_c(s)H(s)G_p(s)}{1 + Y(s)} = \frac{1000}{s^3 + 17.5s^2 + 215s + 1000} \qquad (14.20)$$

As shown Figure 14.7, the system requires 0.32 seconds to settle, and possesses an overshoot of 2%.

14.2.4 Control Supervisor

At a software level above the control algorithm, a control supervisor is implemented to oversee controller conditions and provide intelligence to avoid overaggressive or sluggish control caused by nonphysiologic disturbances. These disturbances include an arterial line flush (bolus of saline), a blood draw, or change of drug concentration. The supervisor

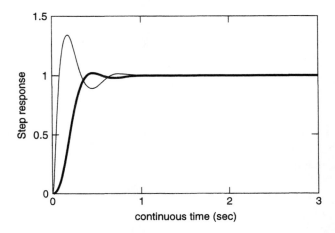

Figure 14.7 Step responses of robust control example with (thick line) and without (thin line) prefiltering.

may also switch between different control laws if available, inject learning signals to identify plant dynamics, and provide rules for controller output (i.e., maximum infusion rate) as defined by the manufacturer's drug-specific labeling.

14.3 PROBLEM SIGNIFICANCE

Closed-loop drug delivery has been applied to anesthesiology, cardiology, and endocrinology problems. In this chapter, we concentrate on the first combination drug/closed-loop drug delivery system to obtain **Food and Drug Administration** (FDA) approval: the GenESA system developed by Gensia Automedics. This system provided **pharmacologic stress testing.**

14.3.1 Clinical Significance

Up to 25% of the annual death rate in the western world is attributed to **coronary artery disease** (CAD) and its attendant complications. In 1993, the American Heart Association estimated the economic cost of treating cardiovascular disease in the United States as $117 billion [4]. Early detection of CAD, before complications occur, is an important aim in cardiology. In a study of 2365 clinical healthy men who were initially examined and given an exercise stress test and were then questioned 5 years later, the presence of two or more test-defined symptoms during the stress test was a predictor of primary coronary heart disease [5].

Exercise stress testing, using a bicycle or treadmill, is a well-established and safe method [6] of provoking **myocardial ischemia,** or decreased blood flow to the heart tissue. During exercise stress testing, heart rate and systolic blood pressure increase, causing the myocardial oxygen demand to increase and coronary blood flow to decrease. Under these conditions, ischemia is induced in areas where blood flow is insufficient. Ischemia is detected by appearance of severe chest pain or a change in the electrocardiogram. During the stress test, the functional performance of the heart is evaluated by electrocardiography, **echocardiography** (ultrasound for detection of wall-motion abnormalities), or **radionuclide perfusion imaging** (use of radioisotope for detection of blood flow defects).

Even though the majority of patients are able to complete an exercise stress test, a sizable minority (estimated as up to 30% [7]), are not able to achieve adequate stress (sufficient heart rate increase) because of anxiety, poor motivation, peripheral vascular disease, and muscular or neurological disorders [8]. Moreover, imaging techniques cannot be performed optimally with conventional exercise testing. Although stress testing using **atrial pacing** (pacing through a catheter inserted into the esophagus, which is in the proximity of the left atrium) has been proven a successful alternative to exercise, this invasive technique is inherently riskier and less tolerable. Another alternative to exercise is the use of a pharmacologic agent.

14.3.2 Pharmacologic Stress Agents

Isoproterenol, **dobutamine,** and **dipyridamole** have all been tested as pharmacologic stress agents. Isoproterenol is a **catecholamine** developed in the 1940s that is a potent nonselective β-adrenergic agonist with very low affinity for α-adrenergic receptors. Although it possesses high **sensitivity** (rate of identifying patients at risk) and **specificity** (rate of identifying patients not at risk) as a cardiac stress agent, isoproterenol may also induce cardiac arrhythmias and cardiac collapse [9, 10]. Further, in normal volunteers, it causes chest pain and palpitations [11].

Dobutamine is also a synthesized cathecholamine; it interacts with both the β- and α-adrenergic receptors. Dobutamine is more effective in increasing cardiac contractility than in increasing heart rate. Frequently, **atropine** must be coadministered with dobutamine to achieve the target heart rate response [12]. The **vasodilator** (blood vessel dilator) dipyridamole is used with radionuclide perfusion imaging to assess flow defects. Dipyridamole lacks sensitivity as a cardiac stress agent and may cause severe **hypotension** (low blood pressure) [13].

The catecholamine arbutamine was developed as an alternative stress agent to induce a more balanced **chronotropic** (affecting speed or rate) and **inotropic** (affecting muscular contractions) stimulation relative to dobutamine, in a dose-dependent manner without the hypotension observed with isoproterenol. It was intended that this drug rapidly resolve its response when its infusion is terminated [14].

To determine the dose–response curve of arbutamine with respect to isoproterenol (the most potent β_1 agonist) in β_1-adrenergic receptors, the left atria of rats (n not published) were removed, suspended in solution, and stimulated electrically with square wave pulses of 5 msec duration at a voltage of approximately double threshold at a rate of 3 Hz. The force of contraction was measured for each stimulus, at various concentrations of arbutamine or isoproterenol. The resulting log dose–response curves are shown in Figure 14.8.

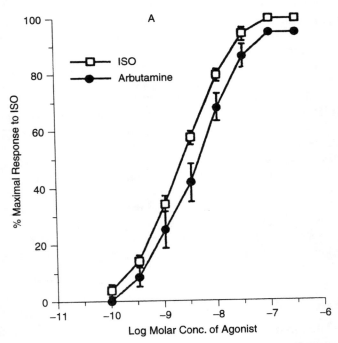

Figure 14.8 Dose–response curves of arbutamine and isoproterenol (ISO), indicating their relative potencies. Responses are expressed as a fraction of the maximal response to ISO ($5 \leq n \leq 14$). Reprinted from [15] with kind permission from Kluwer Academic Publishers.

Note that the y axis is the percentage maximal response of isoproterenol. The EC_{50} values for arbutamine and isoproterenol were 3.5×10^{-9} M and 2.8×10^{-9} M, respectively, with arbutamine about 0.8 times less potent. In a second set of experiments, beating paired atria of rats (n not published) were prepared as above, with a stimulus applied to the left atrium. Heart rate recordings from the spontaneously beating right atrium were calculated directly from the tension signal at constant chart speed. The resulting EC_{50} values for arbutamine and isoproterenol were 1.0×10^{-9} M and 1.5×10^{-9} M, respectively, with arbutamine about 1.5 times more potent than isoproterenol in its chronotropic action [15].

14.3.3 Commercial Closed-Loop Drug Delivery Systems

To increase the safety and efficacy of arbutamine as a pharmacologic stress agent, the Gen-ESA closed-loop drug delivery system (CLDD) was developed. CLDD research began in the 1970s when Bernard Widrow and Christy Schade used adaptive control (see Chapter 3) to regulate mean arterial pressure in canines while infusing catecholamines [16]; it reached its zenith in the 1980s. During this time, the presentations from a special conference held at the University of Washington in October, 1985 were collected as a Special Issue on Adaptive Control and Drug Delivery in the August, 1987 issue of *IEEE Transactions on Biomedical Engineering* [17]. Moreover, in a report prepared by a special IEEE EMBS task force that was submitted to the National Research Council in March, 1985, it was predicted that "today, we are close to being able to provide artificial closed-loop control systems to replace failed physiologic functions of the body" [18]. Many of the CLDD systems were developed to regulate blood pressure, since a transducer was available for feedback.

Surprisingly, only five systems were ever commercialized. While the Gen-ESA system was diagnostic in nature, the other four systems were therapeutic. In the United Kingdom, Hook Lane Nyetimber developed an oxytocin device to initiate labor. The feedback signal was intrauterine pressure, measured every 15 minutes. The control algorithm was a PID controller. This device was able to initiate labor with 20% of the total dose of oxytocine when compared with manual intervention [19].

Two systems, developed by Life Sciences Instruments Division of Miles Labs in the United States and Esaote Biomedica in Italy, infused insulin to regulate glucose in type-I diabetics. The feedback signal was blood glucose. In both systems, the control algorithm was based on the Toronto prototype [20], in which the future glucose concentration was estimated, and then used as an input to two nonlinear functions which determined the rate of dextrose infusion and rate of insulin infusion [21].

The only CLDD system commercialized to regulate mean arterial blood pressure was developed by IVAC Corporation in the United States. This system titrated sodium nitroprusside. The feedback signal was blood pressure; the control algorithm was based on an algorithm patented by the Cleveland Clinic. This empirical algorithm decreased the infusion rate if the patient's blood pressure decreased, increased the duration for which infusion rate increases were blocked if the patient's blood pressure stabilized, and increased the infusion rate if the patient's blood pressure increased [22]. In a multicenter trial during which postoperative hypertensive patients were managed by either manual nitroprusside titration ($n = 532$) or the IVAC system ($n = 557$), the automated group showed a 67% reduction in the number of hypertensive episodes per patient and a 25% reduction in the number of hypotensive episodes per patient [23].

14.3.4 Market Forces

Exercise stress tests are performed routinely in the United States. In 1994 alone, 875,780 treadmill exercise tests were charged to Medicare. Further, during this year, 213,404 echocardiography and 889,319 perfusion imaging procedures were charged to Medicare [24]. In 1998, Medicare reimbursed $98.89 for an exercise stress test; the typical office charge for the test was $175 to $250 [25]. If we assume that 30% of CAD patients are unable to participate in exercise stress testing [7] and that the average test charge is $213, then the potential exists to generate $80 million annually in pharmacologic stress tests from Medicare patients alone. The expected revenue would increase from patients not receiving Medicare reimbursement. In 1997, it was predicted that the pharmacologic stress test market had the potential to achieve annual sales of "$100 million in three to five years" [26].

With this large potential market in mind, Gensia Automedics developed their GenESA system for pharmacologic stress testing. Requests for device approval in Europe and the United States were submitted in January, 1994. Unlike the other devices highlighted in the application chapters of this textbook, which were submitted to the FDA for **510(k) approval**, the GenESA system was submitted for **premarket approval** (PMA). 510(k) approval is much easier to obtain because the submitted device is claimed to be similar in function to a previously FDA-approved device. In contrast, premarket approval is requested for a new type of device, in this case, a combination drug/product closed-loop system combining a new drug with a new device. Although this system received European approval in April, 1995, it was initially denied approval by the FDA during the same month [27]. After the PMA application was amended with further data, the GenESA system received PMA approval on September, 1997 [26]. Upon receiving FDA approval [28], the product began to be sold in the United States.

The GenESA product and Gensia Automedics were discontinued in June, 1998. Gensia Automedics was wholly owned by an investment group. The product was discontinued at the discretion of the investors because "the speed of product sales increase did not meet business expectations." However, it was believed that "the market opportunity continued to be present" [29].

14.4 CLOSED-LOOP DRUG INFUSION IN PHARMACOLOGICAL STRESS TESTS

The GenESA CLDD system was developed to deliver the pharmacologic stress agent arbutamine to a patient, while maintaining a heart rate response within a desired trajectory window specified by the operator. The control algorithm was developed in parallel with preclinical and clinical trials, and tuned with each new data set. Preclinical testing was conducted in canines. Phase 1 of clinical testing involved testing 18 healthy volunteers for drug tolerability and pharmacodynamic and pharmacokinetic effects. Phase 2 involved testing the diagnostic effects on 83 patients, and establishing a dosing regimen. Phase 3 involved proving the safety and efficacy of arbutamine in a statistically significant manner in more than 1000 patients. The preclinical and clinical trials were conducted between 1990 and 1994 [7].

14.4.1 Pharmacodynamic Model

The basic pharmacodynamic response to arbutamine was characterized in eight normal volunteers. Each subject received a constant infusion (0.18 μg/kg-min) for 32 minutes. A typical response of the change in heart rate is shown in Figure 14.9. A dose–response curve was constructed from the steady-state data (Figure 14.10). It was observed that at low infusion rates (< 0.05 μg/kg-min), no heart rate increase greater than 10 beats per minute (bpm) occurred.

Based on this information, a basic mathematical model was developed to describe the heart rate response to arbutamine infusion:

$$\alpha_0 HRa(t) + \alpha_1 HRa(t-1) = \beta \cdot inf(t-d) \qquad (14.21)$$

where $HRa(t)$ represents the measured heart rate increase above baseline, filtered to eliminate noise in the signal; inf equals the infusion rate, d is the onset delay, α_0 is assumed to equal 1, and α_1 and β are constants. The model order of one was determined using **Akaike's information criterion** [30]. For this first-order model, time constant, τ, and Gain were calculated as

$$\tau = \frac{\ln 0.5}{\ln \alpha_1} \qquad (14.22)$$

$$\text{Gain} = \frac{\beta}{1 + \alpha_1} \qquad (14.23)$$

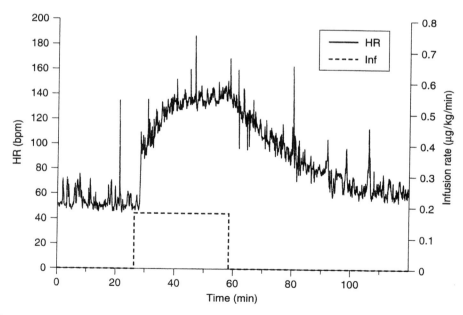

Figure 14.9 Typical heart rate response to single-step arbutamine infusion (solid line = heart rate, dashed line = infusion rate). From [7].

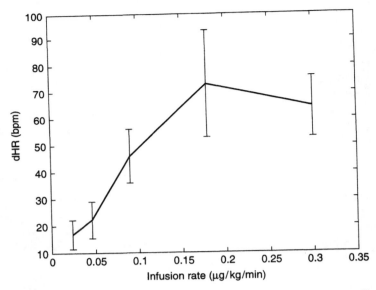

Figure 14.10 Dose–response curve for heart rate response to arbutamine infusion rate. From [7].

Based on analysis of clinical data, it was determined that a significant difference in the parameters of normal subjects and cardiac patients existed. Therefore, two sets of model parameters (α_1, β, and d) were maintained, as were two sets of gain (bpm-kg-min/μg), onset time (min), and onset delay (sec). Compensation for saturation in the dose–response curve was implemented as a linear decrease from the original gain at heart rate, $HRs1$, to half the original gain at heart rate, $HRs2$ (Figure 14.11). These two sets of parameters are summarized in Table 14.1. These parameters were estimated from data obtained in Phase 1 and 2 clinical studies, using an unspecified nonlinear gradient method. From placebo tests without drug infusion and with a constant heart rate, the noise was determined to possess a normal distribution and "small magnitude" of 2.6 bpm [7]. With drug infusion, the noise increased to 4.7 bpm.

Although the model in Eq. (14.21) may fit the data, it should be noted that it is not physiologically sound. Recalling the theory in Chapter 13, a delay such as the delay of several minutes between the initial increase in infusion and the initial increase in heart

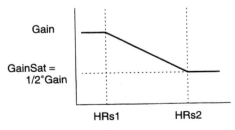

Figure 14.11 Representation of nonlinear gain in pharmacodynamic model. $HRs1$ and $HRs2$ represent heart rates at which saturation occur. From [7].

14.4 CLOSED-LOOP DRUG INFUSION IN PHARMACOLOGICAL STRESS TESTS

Table 14.1 Controller gains for different modes. From [7]

Gain	Level	Ramp	Target	Hold
K_i	4.0	4.0	2.0	2.0
K_p	0.4	0.4	1.0	2.0

rate should be modeled as a compartment, rather than as a delay itself. The data in Figure 14.9 is probably better fit by two compartments and exponentials, rather than one. A more physiologically sound model is

$$\alpha_0 HRa(t) + \alpha_1 HRa(t-1) + \alpha_2 HRa(t-2) = \beta \cdot \mathit{inf}(k) \qquad (14.24)$$

which is based on two exponentials.

14.4.2 Data Acquisition and Processing

Before arbutamine was infused to a patient, an operator attached electrocardiography (ECG) leads to monitor heart rate and a blood pressure cuff to monitor blood pressure to the patient. An intravenous catheter was inserted for infusion of arbutamine in a prefilled glass syringe, using a syringe pump. The operator also entered patient age and weight into the device (Figure 14.12). During this first predrug phase, the heart rate was monitored to establish a baseline value.

ECG was sampled with a frequency of 0.2 Hz. Heart rate, $HR(t)$, was determined from detection of the R wave peaks in the ECG signal. Because of the high noise level in the signal, the measured heart rate, $HRa(t)$, was filtered nonlinearly before use in the control algorithm:

$$HRa(t+1) = HRa(t) + \mathrm{Gain}(t)[HRa(t+1) - HRa(t)] \qquad (14.25)$$

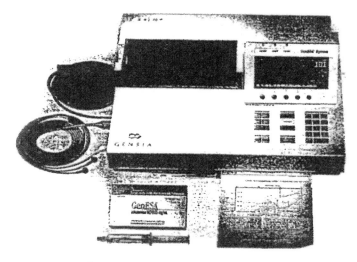

Figure 14.12 GenESA system. From [28].

$$\text{Gain}(t+1) = \frac{\text{Gain}(t)}{1+\text{Gain}(t)} + 0.04 \qquad (14.26)$$

where $\text{Gain}(0) = 1$, $HRa(0) = HR(0)$, and $[HRa(t+1) - HRa(t)]$ was clipped to remain between -8 and $+6$ bpm. Clipping this difference eliminated transient heart rate outliers from erroneously affecting the infusion rate. Blood pressure was sampled every 2 minutes.

The second phase of the GenESA test, the drug phase, was initiated by the operator by pushing the start key. The control algorithm calculated the infusion rate based on three parameters selected by the operator: Slope, the desired rate of heart rate increase; Target_D, the maximum desired heart rate increase; and (if selected during drug delivery) suspension of heart rate increase for a period of constant heart rate. The selected slope range was 4 to 12 bpm/min. The calculated desired infusion rate in units of µg/kg-min was combined with the patient's weight to obtain an infusion rate in typical units of ml/hr. The desired infusion rate was limited to a maximum value of 0.8 µg/kg-min and step size of 0.002 µg/kg-min. Total dose was limited to 10 µg/kg, the toxicology delivery limit.

The third phase of the GenESA test, the post-drug phase, began when the operator stopped drug delivery, an alarm condition occurred, or Target_D was reached. In the post-drug phase, no control actions were necessary as the short half-life of arbutamine resulted in an exponential heart rate decrease.

14.4.3 Control Algorithm

The objective of the control algorithm was to track the Slope and achieve the Target_D as closely as possible. The control supervisor switched between six control modes, based on measured heart rate, operator inputs, and alert/alarm conditions. These six control modes were: initialization control, level control, ramp control, hold control, target control, and restart.

To begin control in the initialization control mode, the onset delay and threshold level of action of arbutamine were overcome with a constant infusion of 0.1 µg/kg-min delivered over 1 minute. In the level control mode, the infusion was titrated to a constant algorithm target of 20 bpm above baseline. The initial infusion rate was 0.1 µg/kg-min, the infusion rate during initialization. This mode was maintained for a maximum of 3 minutes or until a heart rate increase of 20 bpm was observed. The infusion rate, $u(t)$, was updated every 15 seconds, based on a modified **proportional–integral** (PI) scheme:

$$u(t) = \sum_{i=t-1}^{t-12} \lambda_i u(i) - (K_i - K_p)[HRa(t) - \text{Target}_A(t)] + K_p[HRa(t-1) - \text{Target}_A(t-1)] \qquad (14.27)$$

where Target_A is an internal algorithm target; the weighting coefficients, $\lambda(t)$, were "selected to provide a stable but responsive infusion rate" [7]; $K_i = 4.0$; and $K_p = 0.4$.

In the ramp control mode, the infusion was titrated every 15 seconds using the same PI formula described in Eq. (14.27) and same controller gains. Target_A was reinitialized to the average heart rate value when the ramp mode began. If heart rate saturation occurred, the PI controller was overridden and the infusion was increased. If the slope increased above the desired slope set by the operator, the PI controller was overridden and the infusion was scaled back by a factor proportional to the excess slope.

At any time, the operator could place the system in a hold mode that caused an infu-

14.4 CLOSED-LOOP DRUG INFUSION IN PHARMACOLOGICAL STRESS TESTS

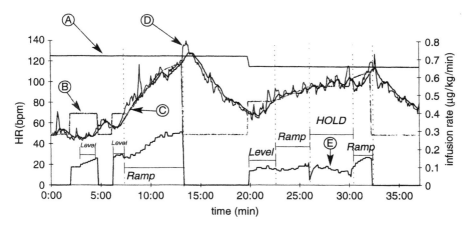

Figure 14.13 GenESA test protocol canine example. The test includes a hold period, two separate $Target_D$, and a restart event. The selected slope for the first ramp period was 12 bpm/min ($Target_D$ = 125 bpm) and 4 bpm/min for the second ramp period ($Target_D$ = 115 bpm). The hold period is annotated. The heart rate target was reached at $t = 14$ and $t = 32$ minutes. A = TrgtD; B = TrgtA; C = HRa; D = HR; E = Inf. From [7].

sion rate that maintained heart rate at a constant level for at most 5 minutes. The executed infusion rate was based on the PI formula. In target mode, the control algorithm attempted to reach $Target_D$ without overshoot. "Because of significant input–output delay" [7], empirical rules caused the infusion rate to decrease or stop as $Target_A$ was reached. Finally, in the restart mode, drug delivery was restarted by the operator if an alarm was remedied or the test was to be continued. An example of control mode execution in a canine is shown in Figure 14.13 [7, 31].

14.4.4 Implementation and Data Analysis

The control algorithm described above evolved over Phases 1 through 3 of clinical testing. Revision 4 of the algorithm (before the alert and alarm criteria were refined) was tested in 70 European patients in a randomized, crossover study with desired slopes of 6 or 10 bpm/min. Only patients with known CAD were enrolled in the study; the hold control mode was not tested. The control algorithm stopped the infusion because the target was reached in target control mode in only 35% (24/70) of all tests. However, it was expected that in the majority of cases for this patient population, the operator would stop drug delivery before the target was achieved because of signs of ischemia. As shown in Table 14.2, arbutamine was delivered with a small undershoot of 2 bpm, independent of slope.

Table 14.2 Evaluation of control algorithm in human trial. From [7]

Parameter	Slope = 6 bpm/min		Slope = 10 bpm/min	
Achieved heart rate slope (bpm/min)	5.5 ± 1.7*	n = 64	8.1 ± 2.2	n = 59
Heart rate overshoot (bpm)	−2.6 ± 2.6	n − 20	−2.2 ± 2.6	n = 21
Total dose (μg/kg)	3.4 ± 2.2	n = 61	3.3 ± 2.2	n = 61

*$p < 0.001$

The deviation between desired and achieved slope increased with higher slope, due to heart rate saturation and limits in physiologic response [7].

In a multicenter study of 210 patients with symptoms and angiographic evidence of CAD, the patients underwent both GenESA and exercise stress tests in randomly assigned order, with 20 hours to 14 days between tests. In the GenESA tests, the desired slope was 8 bpm/min and $Target_D$ was $(200 - age) \times 0.85$ bpm/min. GenESA testing was discontinued if heart rate saturation occurred. In this study, the mean increases in heart rate and systolic blood pressure evoked by arbutamine and exercise were 51 and 53 bpm ($p > 0.05$) and 36 and 44 Hg ($p < 0.0001$), respectively. The sensitivity for detecting ischemia by either **angina** (severe chest pain) or a change in the ECG **ST segment** (≥ 0.1 mV horizontal or downsloping depression between the S and T waves) was 84% using arbutamine versus 75% using exercise ($p \leq 0.014$) [32].

In both the early and multicenter studies, it is unclear if $Target_D$ was ever reached in another control mode besides target control mode. A more physiologically sound pharmacodynamic model could have reduced the variability in heart rate prediction. One such model would be based on two compartments, with the second compartment accounting for the system delay, and would include a nonlinear saturation term. Nonlinear compartmental models are discussed in Chapter 15. Further, if the control algorithm had been based on a nonlinear model such as fuzzy logic, rather than on the linear PI model with modifications of gain to account for heart rate saturation, less empirical rules for overriding the control algorithm could have been required. Because the heart rate varies on a per beat basis, it must be filtered sufficiently before it can be used as a feedback signal.

14.5 SUMMARY

In this chapter, we have discussed pharmacodynamics, the relationship between drug concentrations at the site of action and subsequent pharmacologic response. The molecules that interact with the drug are called receptors, and include proteins that are enzymes, hormone or neurotransmitter receptors, and ion channels. The nature of the drug–receptor interaction is generally considered to be reversible, resulting from noncovalent interactions such as ionic bonds, hydrogen bonds, van der Waals interactions, or hydrophobic interactions. According to the law of mass action, a signal drug molecule interacts with a receptor with a single binding site to produce a pharmacologic response. For this occupation theory, it is assumed that a drug molecule combines with a receptor molecule as a bimolecular association, and that the resulting drug–receptor complex dissociates as a unimolecular entity. It is also assumed that the binding of the drug with the receptor is fully reversible and that a single type of binding site with one site per molecule exists. Further, it is assumed that as more receptors are occupied by drug molecules, a greater response is obtained until a maximal response is reached. A dose–response curve, a representation of the drug response to increasing drug concentration, is approximately linear for very small doses, nonlinear for larger doses, and saturates at the maximal response.

In a closed-loop drug delivery system, the goal is to infuse a drug dosage that accounts for patient pharmacokinetics and pharmacodynamics and produces a desired response. A typical drug delivery system consists of the patient, a transducer, a monitor, a control algorithm, and an infusion pump. The patient receives a drug dosage from the infusion pump and provides a pharmacologic response. Typical responses are the vital signs of heart rate, blood pressure, temperature, respiration, arterial oxygen saturation, and cardiac output. The response is sensed by the transducer, which converts the response to a proportional analog electrical current. The electrical current is digitized within the monitor and provides a feedback signal for the control algorithm. The control algorithm uses the feedback signal and other clinician-supplied parameters such as cumulative drug dose to determine the next drug dosage infusion rate. Typical control algorithms are based on the PID controller, adaptive control, or fuzzy control. A control algorithm supervisor may ensure that the closed-loop system remains stable and does not become a nonminimum phase or positive feedback system.

To date, only five closed-loop systems have been commercialized. Four are therapeutic in nature, and have been used to infuse oxytocin to initiate labor, insulin to regulate glucose in type-I diabetics (two systems), and sodium nitroprusside to regulate blood pressure. The fifth device, the GenESA system, is diagnostic in nature, and was used to infuse arbutamine during pharmacologic stress testing to increase heart rate.

Both exercise and pharmacologic stress tests are used to provoke myocardial ischemia and identify coronary artery disease patients. Even though the majority of patients are able to complete an exercise stress test, a sizable minority are not able to achieve adequate stress and are candidates for pharmacologic stress testing. The market for pharmacologic stress tests has been estimated as $100 million annually in the United States. The GenESA system used a PI control algorithm and many empirical rules to regulate the increase of heart rate during arbutamine administration.

Unfortunately, the GenESA system was the first combination drug/product closed-loop system submitted to the Food and Drug Administration for premarket approval. Although a device that is similar in function to a previously FDA-approved device may require only 3 months to receive approval, the GenESA system was approved in 44 months. This unexpected approval delay resulted in a delay in product release in the United States. Although investors in Gensia Automedics, the inventors of the GenESA system, were willing to wait for FDA approval, they expected a large increase in market sales after product launch. Because the speed of product sales increase did not meet business expectations, the GenESA product and Gensia Automedics were dissolved within a year after product release [29].

14.6 REFERENCES

[1]. Woodruff, E. A. Clinical care of patients with closed-loop drug delivery systems. In *The Biomedical Engineering Handbook*, edited by J. D. Bronzino. CRC Press: Salem, MA, 1995. pp. 2447–2458.

[2]. Dorf, R. C., and Bishop, R. H. *Modern Control Systems,* 8th ed. Addison Wesley: Menlo Park, CA, 1998.

[3]. Graham, D., and Lathrop, R. C. The synthesis of optimum response: Criteria and standard forms, Part 2. *Trans AIEE, 72,* 273–288, 1953.

[4]. American Heart Association. *AHA Cardiovascular Statistics.* National Center: Dallas, TX, 1993.

[5]. Bruce, R. A., DeRouen, T. A., and Hossack, K. F. Value of maximal exercise tests in risk assessment of primary coronary heart disease events in healthy men: Five years' experience of the Seattle Heart Watch Study. *Am J Cardio, 46,* 371–378, 1980.

[6]. Gibbons, L., Blair, S. N., Kohl, H. W., and Cooper, K. The safety of maximal exercise testing. *Circulation, 80,* 846–852, 1989.

[7]. Valcke, C. P., and Chizeck, H. J. Closed-loop drug infusion for control of heart-rate trajectory in pharmacological stress tests. *IEEE Trans Biomed, 44,* 185–195, 1997.

[8]. Borer, J. S., Brensike, J. F., Redwood, D. R., Itscoitz, S. B., Passamani, E. R., Stone, N. J., Richardson, J. M., Levy, R. I., and Epstein, S. E. Limitations of the electrocardiographic response to exercise in predicting coronary-artery disease. *N Engl J Med, 293,* 367–371, 1975.

[9]. Wexler, H., Kuaity, J., and Simonson, E. Electrocardiographic effects of isoprenaline in normal subjects and patients with coronary atherosclerosis. *Brit Heart J, 33,* 759–764, 1971.

[10]. Lockett, M. F. Dangerous effects of isoprenaline in myocardial failure. *Lancet, 2,* 104–106, 1965.

[11]. Combs, D. T. and Martin, C. M. Evaluation of isoproterenol as a method of stress testing. *Am Heart J, 6,* 711–715, 1974.

[12]. Baptista, J., Arnese, M., Roelandt, J. R. T. C., Fioretti, P., Keane, D., Escaned, J., Boersma, E., Di Mario, C., and Serruys, P. W. Quantitative coronary angiography in the estimation of the functional significance of coronary stenosis: Correlations with dobutamine-atropine stress test. *J Am Coll Cardiol, 23,* 1434–1439, 1994.

[13]. Marangelli, V., Iliceto, S., Piccinni, G., De Martino, G., Sorgente, L., and Rizzon, P. Detection of coronary artery disease by digital stress echocardiography: Comparison of exercise, transesophageal atrial pacing and dipyridamole echocardiography. *JACC, 24,* 117–124, 1994.

[14]. Young, M., Pan, W., Wiesner, J., Bullough, D., Browne, G., Balow, G., Potter, S., Metzner, K., and Mullane, K. Characterization of arbutamine: A novel catecholamine stress agent for diagnosis of coronary artery disease. *Drug Dev Res, 32,* 19–28, 1994.

[15]. Abou-Mohamed, G., Nagarajan, R., Ibrahim, T. M., and Caldwell, R. W. Characterization of the adrenergic activity of arbutamine, a novel agent for pharmacological stress testing. *Cardiovasc Drugs Ther, 10,* 39–47, 1996.

[16]. Schade, C. M. An automatic therapeutic control system for regulating blood pressure. In *Proceedings of San Diego Biomedical Symposium,* 1973. p.47.

[17]. Chizeck, H. J., Jelliffe, R. W., and Cheung, P. W. Guest editorial. *IEEE Trans Biomed, 34,* 565–566, 1987.

[18]. Potvin, A. R., Crosier, W. G., Fromm, E., Lin, J. C., Neuman, M. R., Pilkington, T. C., Robinson, C. J., Schneider, L. W., Strohbehn, J. W., Szolovits, P., and Tompkins, W. J. Report of an IEEE task force—An IEEE opinion on research needs for biomedical engineering systems. *IEEE Trans Biomed, 33,* 48–59, 1986.

[19]. Westenskow, D. R. Automating patient care with closed-loop control. *M.D. Computing, 3,* 14–20, 1986.

[20]. Brunetti, P., Benedetti, M. M., Calabrese, G., and Reboldi, G. P. Closed-loop delivery systems for insulin therapy. *Int J Art Organ, 14,* 216–226, 1991.

[21]. Albisser, A. M., Leibel, B. S., Ewart, T. G., Davidovac, Z., Botz, C. K., Zingg, W., Schipper, H., and Gander, R. Clinical control of diabetes by the artificial pancreas. *Diabetes, 23,* 397–404, 1974.

[22]. Petre, J. H., and Cosgrove, D. M. Infusion pump controller. *U.S. Patent 4,392,849.* July 12, 1983.

[23]. Chitwood, W. R., Cosgrove, D. M., III, Lust, R. M., and the Titrator Multicenter Study Group. Multicenter trial of automated nitroprusside infusion for postoperative hypertension. *Ann Thorac Surg, 54,* 517–522, 1992.

[24]. Gibbons, R. J., Balady, G. J., Beasley, J. W., Bricker, J. T., Duvernoy, W. F. C., Froelicher, V. F., Mark, D. B., Marwick, T. H., McCallister, B. D., Thompson, P. D., Winters, W. L. Jr, and Yanowitz, F. G.. ACC/AHA guidelines for exercise testing: A report of the American College of Cardiology/American Heart Association Task Force on Practice Guildelines (Committee on Exercise Testing). *J Am Coll Cardiol, 30,* 260–315, 1997.

[25]. Darrow, M. D. Ordering and understanding the exercise stress test. *Am Fam Phys, 59,* 401–413, 1999.
[26]. FDA allows use of device in diagnosing heart disease. *Wall Street Journal, 401,* B4, September 15, 1997.
[27]. FDA denies approval. *Wall Street Journal, 372,* B7, April 11, 1995.
[28]. Gensia, Inc., GenESA System. *FDA Application Number P940001.* September 12, 1997.
[29]. Bochenko, W. (Former Gensia Automedics Senior Director, Drug Delivery Systems.) Personal Communication. August 23, 1999.
[30]. Akaike, H. A new look at the statistical model identification. *IEEE Trans Auto Cont, AC-19,* 716–723, 1974.
[31]. Valcke, C. P., Bochenko, W. J., and Hillman, R. S. Method and apparatus for closed loop drug delivery. *U.S. Patent 5,733,259.* March 31, 1998.
[32]. Dennis, C. A., Pool, P. E., Perrins, E. J., Mohiuddin, S. M., Sklar, J., Kostuk, W. J., Muller, D. W. M., and Starling, M. R. Stress testing with closed-loop arbutamine as an alternative to exercise. *J Am Coll Cardiol, 26,* 1151–1158, 1995.
[33]. Bergman, R. N., Ider, Y. Z., Bowden, C. R., and Cobelli, C. Quantitative estimation of insulin sensitivity. *Am J Physiol, 236,* E667–E677, 1979.
[34]. SAAM Institute, Seattle, WA. Data from Cold_Minimal_Model.stu example file, 1999.
[35]. SAAM II User Manual. SAAM Institute: Seattle, WA. 1998.

Further Reading

Pharmacodynamics: Tallarida, R. J. and Jacob, L. S. *The Dose-Response Relation in Pharmacology.* Springer-Verlag: New York, 1979.

14.7 PERIPHERAL INSULIN KINETICS EXERCISES

Engineer G at Startup Company has been asked to design a control algorithm to regulate insulin infusion in a new implantable insulin pump. Before he can design such an algorithm, he needs to understand glucose and insulin kinetics. He is aware that a minimal model has been validated as representative of glucose and insulin kinetics [33], and collaborates with an endocrinologist to obtain intravenous glucose tolerance test data that can be used with this model. This model is described in great detail in Section 15.2.

The equations of the minimal model for insulin sensitivity are

$$\dot{g}(t) = -[S_G + x(t)]g(t) + S_G g_{SS} \qquad (14.28)$$

$$\dot{x}(t) = -p_2\{x(t) - S_I[i(t) - i_{SS}]\} \qquad (14.29)$$

$$g(0) = G_0 \qquad (14.30)$$

$$x(0) = 0 \qquad (14.31)$$

where $g(t)$ is the plasma glucose concentration, $x(t)$ is a state variable based on a remote compartment of insulin, $i(t)$ is the plasma insulin concentration, g_{ss} is the steady state glucose concentration, and i_{ss} is the steady state insulin concentration. Here, S_G is the glucose effectiveness, S_I is insulin sensitivity, p_2 is an insulin action parameter, and G_0 is the glucose concentration extrapolated at time 0. In this model, the glucose profile is fitted by using insulin as a forcing function. We recreate his analysis below.

14.7.1 Clinical Data

The glucose and insulin data in Figure 14.14 were sampled by Engineer G in response to a glucose injection of 305 mg/dl at $t = 0$ min. The measurement error in the glucose was 2%; the measurement error in the insulin radioimmunoassay was assumed to $\leqslant 1\%$.

These data were generously provided by the SAAM Institute in Seattle, Washington [34]. This exercise will be conducted using a demonstration version of the program SAAMII [35]. The demonstration version is identical to the original version, except that file saving and printing functions have been disabled.

14.7.2 SAAMII Instructions

To download a demonstration version of SAAMII, go to the internet site *http://www.saam.com/download/download.htm*. Read the Test Drive License Agreement and specify your acceptance. Enter your name, address, and e-mail address. Read the down-

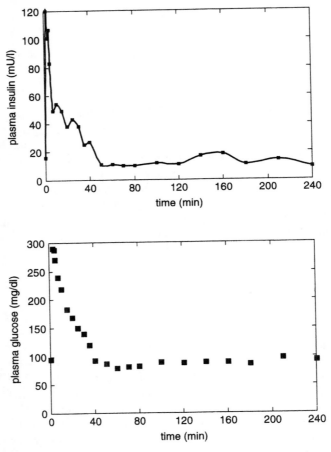

Figure 14.14 Plasma insulin and glucose data for exercise. From [34].

load directions, and click the *Download* tab. Run *s2demov12.exe* to uncompress program files. These files are saved in the *Saam* directory.

Please move the *Cold_Minimal_Model.stu* file from *ftp://ftp.ieee.org/uploads/press/baura* to your *saamIIv1.2demo* directory. From this directory, start the *Compwin.exe* program. Using the menu, *File/Open csf1.stu*. The compartmental model will appear. Click the *Experiment* tab in the Toolbox. The plasma insulin forcing function is represented by compartment 3; the state variable related to a remote compartment of insulin is represented by compartment 2. Plasma glucose is represented by compartment 1 and is sampled. To view the data values, use the menu to *Show/Data*. When finished, close the *Data* window to enable other functions. To view the plasma glucose plot, use the menu to *Show/Plot*. The first time you use the plot function, you will have to select "s1:g."

14.7.3 Exercises

1. Observe the initial values of the parameters by selecting Show/Parameters from the menu. (a) Solve for Eqs.(14.28) and (14.31) by selecting Compute/Solve from the menu. (b) How well do these initial values fit the data (check the plot)?
2. Iterate the fit by selecting *Compute/Fit* from the menu. (a) Has the fit improved? (b) Are the residuals random?
3. Evaluate the statistics by selecting *Show/Statistics* from the menu. (a) What is the coefficient of variation associated with each identified parameter? (b) What is the Akaike information criterion (choose the Objective box)?
4. In this model, plasma glucose kinetics are simplified to represent a single compartment that is fit by one exponential. However, at least one other fast exponential is present in the glucose data. The model accommodates this simplification by unweighting the first 6 data samples that are taken. Unweighting of data is represented by "(-)" in the data file. Uncover the effect of the fast exponential by reweighting the data [remove "(-)"]. Within the *Parameter* window, use the Edit and Save tabs to reinitialize the parameters to their previous initial values. Fit and iterate the data. (a) Are the residuals random? (b) Have the coefficients of variation and Akaike information criterion changed?
5. Is it sufficient to assess model validity by evaluating the magnitude of the coefficients of variation?

15

THE NONLINEAR COMPARTMENTAL MODEL

In Chapter 13, we discussed how drug transport may be described using a linear model. Such a model assumes that pharmacokinetic drug parameters do not change for different doses of a drug. However, at higher drug doses, deviations from linear pharmacokinetics may be observed. This nonlinear behavior is termed **dose-dependent kinetics**.

The basic equation describing drug transport, as determined by the mass balance principle, is

$$\dot{q}_i(t) = \sum_{j=0}^{N} R_{ij}(t) + \sum_{\substack{j=0 \\ j \neq i}}^{N} -R_{ij}(t) + u_x(t) \tag{15.1}$$

As stated in Chapter 13, $q_i(t)$ is the mass in compartment i, $R_{ij}(t)$ is the flux from compartment j to compartment i ($i \neq j$), and $u_x(t)$ is an exogenous input. Whereas the flux in the linear model was previously assumed to be dependent on a rate constant times mass, $k_{ij}q_i(t)$, indicating transport primarily resulting from diffusion, it may be replaced by various functions that are incorporated in nonlinear compartmental models.

Two common nonlinear functions are **Michaelis–Menten dynamics** and the **bilinear relation** (Figure 15.1). The flux resulting from Michaelis–Menten dynamics saturates at high doses. The flux resulting from the bilinear relation is based on two masses. Although other nonlinear functions have been described in the literature (such as Langmuir dynamics, the proportional plus derivative function, and the sigmoidal relation) [1], these functions are beyond the range of our discussion because they have not been widely applied.

15.1 MICHAELIS–MENTEN DYNAMICS

Michaelis–Menten dynamics is based on an analysis of the kinetics between an **enzyme**, E, its **substrate**, S, and a resulting product, P. A protein enzyme is a **catalyst** (see Section 13.1) that increases the rate, or velocity, of a reaction without itself being modified in the

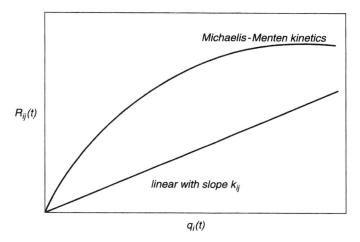

Figure 15.1 Flux functions. A linear flux is proportional to $q_i(t)$. A nonlinear flux may possess Michaelis–Menten kinetics. The bilinear flux, which is proportional to the product of $q_i(t)$ and $q_a(t)$, is not pictured because the function shape is variable.

overall process. The substance that is acted on by the enzyme is called the substrate. During this process, an enzyme (1) binds to a subtrate or subtrates, (2) lowers the energy of the transition state, and (3) directly promotes the catalytic event that produces a product or products.

15.1.1 Definition

The concentration equation describing the interaction among an enzyme, single substrate, and single product is

$$[E] + [S] \underset{k_4}{\overset{k_3}{\longleftrightarrow}} [ES] \xrightarrow{k_{cat}} [E] + [P] \tag{15.2}$$

where $[ES]$ is the enzyme–substrate complex, k_3 is the association rate constant expressed in inverse units of the product of concentration and time, k_4 is the dissociation rate constant expressed in units of inverse time, and k_{cat} is the catalyst rate constant expressed in units of inverse time. In this relationship, it is assumed (termed the **Briggs–Haldane analysis**) that as more ES is formed, ES dissociates faster. Therefore, $[ES]$ first builds up and then quickly reaches a steady state at which it remains constant.

Recalling Section 14.1.2, the rate of change of $[ES]$ with time is then

$$\frac{d[ES]}{dt} = k_3[E][S] - k_4[ES] - k_{cat}[ES] \tag{15.3}$$

The first term of the right side represents the rate of $[ES]$ formation. The second and third terms represent breakdown of $[ES]$ by two routes. At steady state, when $d[ES]/dt = 0$, we have

$$[ES] = \left(\frac{k_3}{k_4 + k_{cat}}\right)[E][S] \tag{15.4}$$

$$[ES] = \frac{1}{K_M}[E][S] \tag{15.5}$$

where K_M is the **Michaelis constant.**

Now, the concentration of enzyme, $[E]$, is the concentration of free enzyme. However, in experiments, the **total concentration of enzyme,** $[E_T]$, is the quantity usually measured. Since

$$[E] = [E_T] - [ES] \tag{15.6}$$

this relationship may be substituted into Eq. (15.5) as

$$[ES] = \frac{[E_T][ES]}{K_M} - \frac{[ES][S]}{K_M} \tag{15.7}$$

Rearranging yields

$$[ES]\left(1 + \frac{[S]}{K_M}\right) = \frac{[E_T][S]}{K_M} \tag{15.8}$$

$$[ES] = \frac{[E_T][S]}{[S] + K_M} \tag{15.9}$$

By definition, the reaction rate, or **velocity,** V, of the product is

$$V = k_{cat}[ES] \tag{15.10}$$

Substituting Eq. (15.9) into Eq. (15.10) yields

$$V = \frac{k_{cat}[E_T][S]}{[S] + K_M} \tag{15.11}$$

A plot of this function (Figure 15.2) resembles the dose–response curve of Chapter 14. At high substrate concentrations, $V_{max} = k_{cat}[E_T]$ is approached because the enzyme molecules become saturated. During **saturation,** every enzyme molecule is occupied by substrate and is carrying out the catalytic step. Since the limit of V is V_{max}, Eq. (15.11) can be rewritten as

$$V = \frac{V_{max}[S]}{[S] + K_M} \tag{15.12}$$

In terms of a function for our nonlinear compartmental model, we may describe the flux as velocity (both have units of mass per unit time), and use the same equation form as Eq. (15.12):

15.1 MICHAELIS–MENTEN DYNAMICS

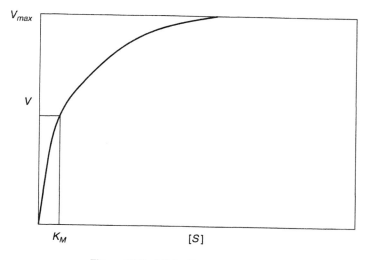

Figure 15.2 Michaelis–Menten curve.

$$R_{ij}(t) = \frac{V_{\max} q_i(t)}{q_i(t) + K_M} \tag{15.13}$$

It is reasonable to assume this alternate nonlinear transport description since drug transport may be facilitated by a receptor, and this facilitated transport mechanism may become saturated.

15.1.2 Parameter Estimation

Unfortunately, when one of the fluxes in Eq. (15.1) is replaced by the Michaelis–Menten dynamics of Eq. (15.13), estimation of the system parameters becomes ill-conditioned. For a first-order system, the solution of the equation is bounded by two first-order solutions of linear equations, accounting for the extreme values of $V_{\max} q_i(t)/[q_i(t) + K_M]$. However, parameter values in an *infinite region* in the parameter space have solutions that also lie within those two bounds. Therefore, a minor change of the data of the solution or the initial estimates of the parameters may cause a drastic change in their final estimate [2].

An alternate method of estimating $V_{\max}$ and K_M involves using the linear compartmental model. For each individual drug dose, $u_x(t)$, the nonlinear flux, $R_{ij}(t)$, may be approximated as $k_{ij} q_i(t)$. Once k_{ij} has been identified for the range of drug doses in the experiments, it may be plotted as a function of the mean mass in compartment i over time, $\bar{q}_i(t)$. If Michaelis–Menten kinetics are present in the system, then k_{ij} will initially be identified as a high value for a low dose/mean mass and eventually decrease to a very low estimated value for a high dose/mean mass (Figure 15.3). Further, if $R_{ij}(t) = k_{ij} \bar{q}_i(t)$ is plotted as a function of $\bar{q}_i(t)$ and Michaelis–Menten kinetics are present, the resulting plot will resemble Figure 15.3. These data may then be fit to Eq. (15.13) to directly estimate values of $V_{\max}$ and K_M with reasonable associated coefficients of variation.

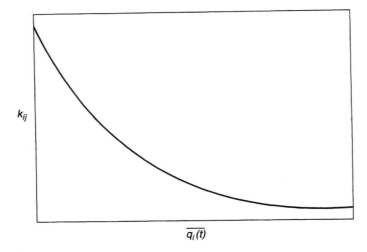

Figure 15.3 Rate constant variation for increasing mean compartmental mass, as evidence of saturation.

15.1.3 Lineweaver–Burk Plot

It should be mentioned that Eq. (15.12) is often transformed by taking reciprocals on both sides:

$$\frac{1}{V} = \frac{K_M}{V_{max}} \frac{1}{[S]} + \frac{1}{V_{max}} \qquad (15.14)$$

This transformation is performed to obtain a linear plot called the **Lineweaver–Burk plot** (Figure 15.4). Although the resulting plot may be linear, the measured error associated with the data has also been transformed by a reciprocal. Because the transformed error is much larger than in its original state, it is not recommended that data be transformed and fit to the model in Eq. (15.14). Further, with this transformation, K_M must be extrapolated at the x-intercept (Figure 15.4), which also increases the error in the estimate of K_M.

15.1.4 Saturation Example

As an example of Michaelis–Menten dynamics, let us analyze insulin kinetics with respect to its effect on glucose. In animals, an important regulatory function is maintenance of blood glucose levels within very narrow limits. Maintenance control within this range is accomplished through the hormones insulin and **glucagon.** When glucose levels rise shortly after a meal, insulin is secreted by the pancreas, which promotes uptake of glucose into cells and its utilization by tissues. Similarly, when glucose levels fall several hours after a meal, glucagon is secreted by the pancreas, which promotes both glucose release from intracellular stores and **gluconeogenesis** (glucose synthesis). Through the actions of these hormones, a normal level of glucose is maintained.

In the 1980s, Richard Bergman at the University of Southern California postulated that insulin transport was a site of regulation in the insulin-glucose feedback system. According to his single gateway hypothesis, insulin transport from plasma to **lymph,** which is derived

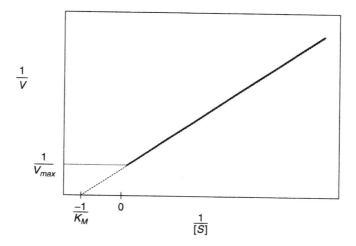

Figure 15.4 Lineweaver–Burk plot.

from the fluid between cells known as **interstitial fluid** (ISF), was facilitated by the **peripheral** (as opposed to the **central** nervous system, which is associated with the brain) insulin receptor. If this hypothesis were true, it could be the site associated with pathological states. In other words, defects in the rate of transport could explain the pathogenesis of **glucose intolerance** (inability to process blood sugar), diabetes, and other diseases in which **insulin resistance** (inefficient processing of insulin) is manifest [3, 4, 5].

To examine this hypothesis, Bergman's group modeled plasma and lymph data according to the model:

$$\dot{c}_2(t) = k_{21}c_1(t) - (k_{02} + k_{12})c_2(t) \tag{15.15}$$

where the concentration is the ratio of mass to volume, $c_1(t)$ is the plasma insulin concentration forcing function, $c_2(t)$ is the ISF insulin concentration as measured in lymph, k_{21} is the rate constant associated with insulin transport to ISF from plasma, k_{12} is the rate constant associated with insulin transport to plasma from ISF, and k_{02} is insulin clearance from ISF (Figure 15.5). It can be shown using the transfer function method described in Chapter 13 that this model is a priori identifiable for the parameters k_{21} and fraction of total ISF insulin clearance, $(k_{12} + k_{02})$.

According to this model, plasma insulin is transported to interstitial fluid with rate constant k_{21}, and interstitial fluid insulin is transported to plasma with rate constant k_{12}. If this transport occurs by diffusion alone, both k_{21} and k_{12} are nonzero constants. However, if transport from plasma to ISF is facilitated by the peripheral insulin receptor, then $k_{12} = 0$ and k_{21} decreases with increasing insulin dosage. Insulin degradation after binding to its receptor in ISF is represented by k_{02}.

To test this hypothesis, plasma and lymph insulin data were obtained from 16 male mongrel canines under anesthesia. Plasma insulin concentrations were determined from a variation of the radioimmunoassay method, using arterial samples (3 ml) obtained from the carotid artery. Plasma glucose was assayed using an automated analyzer. Lymph insulin concentrations were also determined from a variation of the radioimmunoassay

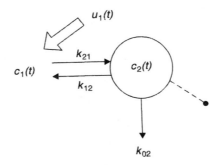

Figure 15.5 First-order compartmental model of plasma insulin transport to lymph. $c_1(t)$ is a forcing function of the plasma insulin concentration. $c_2(t)$ represents the ISF insulin concentration. $u_1(t)$ is the insulin dosage infused into plasma. k_{21}, k_{12}, and k_{01} are rate constants.

method, using hindlimb muscle lymphatic fluid (300–600 μl) sampled by a catheter inserted into a lymph vessel. The experimental protocol consisted of four phases: a basal (initialization) phase (−90 to −60 min), an insulin replacement phase (−60 to 0 min), an insulin activation phase (0–180 min), and an insulin deactivation phase (180–300 min).

During the insulin replacement phase, **somatostatin** was infused to suppress endogenous insulin release, and the endogenous insulin secretion was replaced with a systemic insulin infusion (0.2 mU/min-kg). Somatostatin and insulin replacement infusions were continued during the other two phases. During the insulin activation phase, a primed low (1.0 mU/min-kg, 5 mU/kg bolus, $n = 8$) or high (18 mU/min-kg, 325 mU/kg bolus, $n = 8$) dose infusion was administered. During the insulin deactivation phase, the primed insulin infusion was terminated. Glucose was infused during the last two phases to maintain arterial glucose at the basal replacement concentration.

Plasma insulin samples were collected at $t = −90, −80, −70, −30, −20, −10, −2, 1, 2, 3, 4, 5, 8, 10, 12, 14, 16, 20, 25, 30, 35, 40, 50, 60, 70, 80, 100, 120, 140, 160, 178, 181, 182, 183, 184, 185, 188, 190, 195, 200, 205, 210, 220, 230, 240, 250, 260, 280,$ and 300 min. ISF insulin samples were collected at $t = −90, −80, −70, −30, −20, −10, −2$, every 3 min until 30 min, $35, 40, 45, 50, 55, 60, 70, 80, 90, 100, 120, 140, 160, 178.5, 183, 186, 189, 192, 195, 198, 201, 204, 207, 210, 215, 220, 225, 230, 240, 250, 260, 280,$ and 300 min. Rate constants were identified using the software program MLAB, which was developed by Civilized Software in Bethesda, MD.

Based on the samples obtained, plasma glucose levels were not significantly different between the low- and high-insulin experiments during the basal, insulin activation, and insulin deactivation phases. However, plasma glucose was slightly lower during the insulin replacement phase in the high-dose group (119 ± 5 to 104 ± 4 mg/dl, $p = 0.01$). Basal and replacement plasma insulin concentrations were also not significantly different between the two groups. During the activation and deactivation phases, lymph insulin required a longer time interval to reach steady state compared to plasma insulin, in both low- and high-dose experiments (Figure 15.6).

Comparing low- and high-dose experiments, the rate constant associated with plasma insulin transport to lymph, k_{21}, actually increased 41% from $1.37 \pm 0.18 \times 10^{-2}$ to $1.93 \pm 0.24 \times 10^{-2}$ min^{-1} ($p = 0.09$). Concurrently, the rate constants associated with total ISF insulin clearance, $(k_{12} + k_{02})$, did not change significantly, as identified as $2.53 \pm 0.0026 \times$

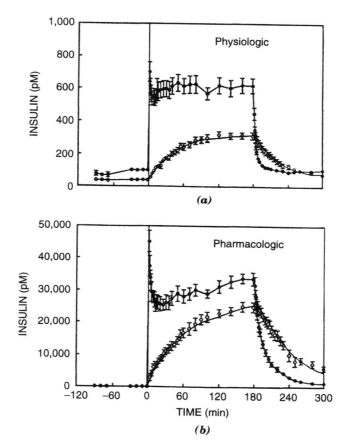

Figure 15.6 Arterial (●) and hindlimb lymph (○) insulin concentrations during (a) low (physiologic) and (b) high (pharmacologic) insulin infusions. Solid line through hindlimb lymph data represents the average fit obtained with Eq. (15.15). Note that 1 pM = 0.15 mU/liter. From [6]. With permission from the American Society for Clinical Investigation, Thoroughfare, NJ; permission conveyed through Copyright Clearance Center, Inc.

10^{-2} and $2.34 \pm 0.0028 \times 10^{-2}$ min^{-1}. Because k_{21} did not decrease between low and high doses and the total ISF clearance did not change significantly, plasma insulin transport to lymph is not saturable and is not facilitated by the peripheral insulin receptor [6].

15.2 BILINEAR RELATION

Alternatively, nonlinear flux may be described by the bilinear relation. Using this function, the flux is proportional to the product of two masses, $q_a(t)$ and $q_i(t)$, such that

$$R_{ij}(t) = b_{ij} q_a(t) q_i(t) \tag{15.16}$$

where b_{ij} is a constant. This function has been used to describe the dependence of glucose transport on insulin.

15.2.1 Example

In 1979, Richard Bergman (then at Northwestern University, now at USC) and Claudio Cobelli of the University of Padova, Italy published the first of many papers describing what has come to be known as the **minimal model of insulin sensitivity** [7]. The purpose of this model was to quantify the influence of decreased peripheral insulin sensitivity on the impairment of the patient's ability to tolerate a standard glucose load. The model is based on insulin and glucose plasma samples obtained after an intravenous glucose injection (also known as an **intravenous glucose tolerance test,** IVGTT). It attempts to characterize the abrupt, multiphase insulin secretory response to a rapid intravenous injection of glucose, which is associated with a rapidly decreasing glucose concentration.

The blood glucose regulating system may be simplified by excluding glucagon effects and separating the system into two subsystems (Figure 15.7). In the first subsystem, the plasma glucose concentration, $g(t)$, is considered the input to the beta cells in the pancreas

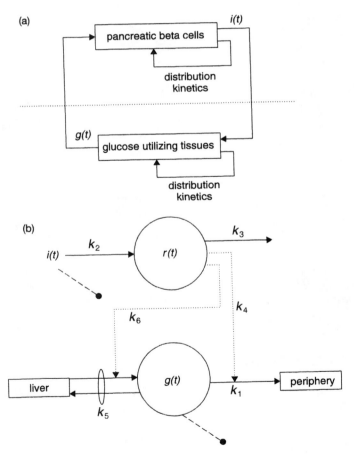

Figure 15.7 Blood glucose regulating system. (a) two simplified subsystems. $g(t)$ = glucose concentration, $i(t)$ = insulin concentration. (b) Compartmental model of glucose and insulin in a remote compartment, $r(t)$, representing glucose subsystem. k_i are rate constants.

which secrete, distribute, and metabolize insulin. In the second subsystem, the plasma insulin concentration, $i(t)$, is considered the input to the tissues that metabolize glucose. The minimal model of insulin sensitivity is based on the second subsystem, thus eliminating the difficulty of modeling beta cell function.

For this model, it is assumed that insulin as a forcing function enters a remote compartment, $r(t)$, with rate constant k_2, and is cleared from this compartment with rate constant k_3. It is also assumed that glucose uptake into tissues is directly dependent on the insulin in this remote compartment, $k_4 r(t) g(t)$. This bilinear relation is consistent with the existence of a remote compartment intimately involved in the action of insulin, increasing the mobility of glucose across the cell membrane.

To further simplify the model, insulin-independent tissues are separated from insulin-dependent, glucose-utilizing tissues, and are grouped as the "liver" and periphery, respectively. The rate of change of glucose is the difference between the net hepatic glucose balance, $b(t)$, and the disappearance of glucose into the peripheral tissues only, $u_p(t)$. In previous work, Bergman et al. showed that the hepatic glucose balance varies according to

$$b(t) = B_0 - [k_5 + k_6 r(t)] g(t) \quad (15.17)$$

where B_0 is the net balance expected when plasma glucose concentration is extrapolated to 0, k_5 is the net transport of glucose from liver to the glucose compartment, and k_6 is the rate constant representing the direct dependence of glucose production in liver on insulin in the remote compartment [8]. Glucose utilization is assumed to depend on insulin in the remote compartment, as detailed above, and on the glucose concentration:

$$u_p(t) = [k_1 + k_4 r(t)] g(t) \quad (15.18)$$

where k_1 is the rate constant associated with transport of glucose to the periphery.

The rate of change in glucose is then

$$\dot{g}(t) = b(t) - u_p(t) \quad (15.19)$$

$$\dot{g}(t) = B_0 + [-(k_1 + k_5) - (k_4 + k_6) r(t)] g(t) \quad (15.20)$$

By modeling only IVGTT data, with $g(0) \neq 0$, the extremes of hypoglycemia and hyperglycemia are avoided and renal glucose loss may be ignored.

From Figure 15.7, the rate of change of insulin in the remote compartment is

$$\dot{r}(t) = k_2 i(t) - k_3 r(t) \quad (15.21)$$

Multiplying both sides of Eq. (15.21) by the sum $(k_4 + k_6)$ results in

$$(k_4 + k_6) \dot{r}(t) = (k_4 + k_6) k_2 i(t) - (k_4 + k_6) k_3 r(t) \quad (15.22)$$

If we define a new state variable, $x(t)$, such that

$$x(t) = (k_4 + k_6) r(t) \quad (15.23)$$

Eq. (15.22) may be rewritten as

$$\dot{x}(t) = p_2 x(t) + p_3 i(t) \tag{15.24}$$

where $p_2 = -k_3$ and $p_3 = k_2(k_4 + k_6)$. Similarly, Eq. (15.20) may be rewritten as

$$\dot{g}(t) = [p_1 - x(t)]g(t) + p_4 \tag{15.25}$$

where $p_1 = -(k_1 + k_5)$ and $k_4 = B_0$. We also assume that p_5 represents the initial glucose concentration, $g(0)$. Eqs. (15.24) and (15.25) represent the minimal model. Using the transfer function method of Chapter 13, it can be shown that the minimal model is a priori identifiable for the parameters

$$\mathbf{p} = [p_1 p_2 p_3 p_4 p_5]^T \tag{15.26}$$

The minimal model is used to assess insulin sensitivity. Glucose effectiveness, $E(t)$, is defined as the quantitative enhancement of glucose disappearance due to an increase in the plasma glucose concentration:

$$E(t) \equiv -\frac{\partial \dot{g}(t)}{\partial g(t)} \tag{15.27}$$

where $\dot{g}(t)$ is the sample rate of change of the plasma glucose concentration. Combining Eqs. (15.25) and (15.27), glucose effectiveness may be calculated from

$$E(t) = x(t) - p_1 \tag{15.28}$$

Insulin sensitivity, S_I, is then defined, in terms of steady state (SS) conditions, as the quantitative influence of insulin to increase the enhancement of glucose of its own disappearance:

$$S_I \equiv \frac{\partial E_{SS}}{\partial i_{SS}} \tag{15.29}$$

Based on Eq. (15.24), at steady state, $\dot{x}(t) = 0$,

$$x_{SS} = \frac{-p_3}{p_2} i_{SS} \tag{15.30}$$

Combining Eqs. (15.28) and (15.30) results in

$$E_{SS} = x_{SS} - p_1 \tag{15.31}$$

$$E_{SS} = \frac{-p_3}{p_2} i_{SS} - p_1 \tag{15.32}$$

From Eq. (15.29), insulin sensitivity is then

$$S_I = \frac{-p_3}{p_2} \tag{15.33}$$

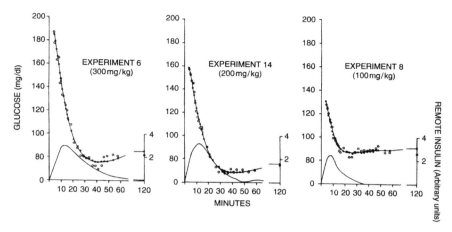

Figure 15.8 Compartmental model fits of plasma glucose concentration data. Closed circles represent fit of model with best parameter values for each experiment; open circles represent measured glucose data beginning at 4 min. Bottom curves are time course of insulin in a compartment remote from plasma that represents time course of insulin action. Glucose concentration at 120 min was assumed to be same as preinjection basal glucose. From [7], with permission from the American Physiological Society, Bethesda, MD.

The minimal model was validated in canines. As shown in Figure 15.8, this model enables plasma glucose data to be fit with plasma insulin as the input. The results are associated with reasonable coefficients of variation and random residual errors [7]. With a modification to the protocol, the minimal model was later adapted to humans [8]. In 1993, it was shown by Daniel Porte's group at the University of Washington that fasting insulin levels are related to insulin sensitivity, S_I, in a hyperbolic relationship. In 93 human subjects, greater degrees of insulin resistance (lower S_I) were associated with increased fasting insulin levels, whereas increasing insulin sensitivities (higher S_I) were associated with lower fasting insulin levels (Figure 5.9). Because of the hyperbolic relationship, when insulin sensitivity is high, large changes in insulin sensitivities would be expected to be associated with relatively small changes in fasting insulin. Similarly, when insulin sensitivity is low, small changes in insulin sensitivity would be expected to be associated with relatively large changes in fasting insulin [9].

15.3 SUMMARY

In this chapter, we have reviewed how compartmental models may incorporate nonlinear fluxes. A nonlinear compartmental model would typically incorporate a nonlinear function such as Michaelis–Menten dynamics or the bilinear relation. The flux resulting from Michaelis–Menten dynamics saturates at high doses. The flux resulting from the bilinear relation is based on two masses.

Michaelis–Menten dynamics is based on an analysis of the kinetics between an enzyme, its substrate, and a resulting product. The nonlinear flux, $V_{\max}q_i(t)/[q_i(t) + K_M]$, uses the same form as this enzyme kinetics equation. It is reasonable to assume this alternate nonlinear transport description since drug transport may be facilitated by a receptor, and this facilitated transport mechanism may become saturated.

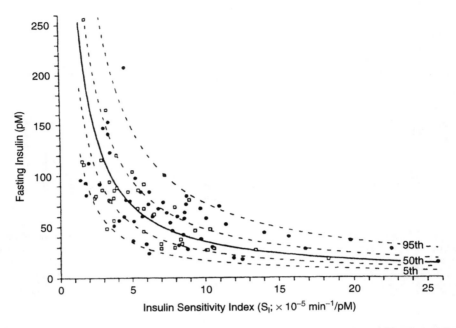

Figure 15.9 Relationship between S_I and fasting insulin in 55 males (●) and 38 females (□). The best fit relationship is represented as a solid line. The 5th, 25th, 75th, and 95th percentiles are represented as dashed lines. The relationship is described by $S_I \times$ fasting insulin = 3.518×10^{-3} ($n = 93$). From [9], with permission from American Diabetes Association, Alexandria, VA.

Unfortunately, when one of these fluxes is included in a compartmental model, estimation of the system parameters becomes ill-conditioned. Instead, an alternate method of estimating V_{max} and K_M is recommended, that involves using the linear compartmental model. For each individual drug dose, $u_x(t)$, the nonlinear flux, $R_{ij}(t)$, may be approximated as $k_{ij}q_j(t)$. Once k_{ij} has been identified for the range of drug doses in the experiments, it may be plotted as a function of the mean mass in compartment j over time, $\bar{q}_j(t)$. If Michaelis–Menten kinetics are present in the system, then k_{ij} will initially be identified as a high value for a low dose/mean mass and eventually decrease to a very low estimated low value for a high dose/mean mass. Further, if $R_{ij}(t) = k_{ij}\bar{q}_j(t)$ is plotted as a function of $\bar{q}_j(t)$ and Michaelis–Menten kinetics are present, these data may then be fit directly to estimate values of V_{max} and K_M with reasonable associated coefficients of variation. This method has been used to investigate the saturability of plasma insulin transport to lymph.

The bilinear relation has been used to describe the dependence of glucose transport on insulin in a remote compartment, $r(t)$. Rather than describing glucose flux as a constant proportional to the glucose concentration, $k_{ij}g(t)$, it is assumed that glucose uptake into tissues is directly dependent on the insulin in this remote compartment, $k_4 r(t)g(t)$. Similarly, it is assumed that glucose production in liver is directly dependent on insulin in the remote compartment, $k_6 r(t)g(t)$. These bilinear relations are critical to the formation of a minimal model of insulin sensitivity that quantifies the influence of decreased peripheral insulin sensitivity on the impairment of the patient's ability to tolerate a standard glucose load.

15.4 REFERENCES

[1]. Carson, E. R., Cobelli, C., and Finkelstein, L. *The Mathematical Modeling of Metabolic and Endocrine Systems: Model Formulation, Identification, and Validation.* John Wiley: New York, 1983.

[2]. Tong, D. D. M. and Metzler, C. M. Mathematical properties of compartment models with Michaelis–Menten type elimination. *Math Biosci, 56,* 293–306, 1980.

[3]. Yang, Y. J., Hope, I. D., Ader, M., and Bergman, R. N. Insulin transport across capillaries is rate limiting for insulin action in dogs. *J Clin Invest, 84,* 1620–1628, 1989.

[4]. Ader, M., Poulin, R. A., Yang, Y. J., and Bergman, R. N. Dose–response relationship between lymph insulin and glucose uptake reveals enhanced insulin sensitivity of peripheral tissues. *Diabetes, 41,* 241–253, 1992.

[5]. Bergman, R. N., Bradley, D. C., and Ader, M. On insulin action in vivo: the single gateway hypothesis. *Adv Exp Med Biol, 334,* 181–198, 1993.

[6]. Steil, G. M., Ader, M., Moore, D. M., Rebrin, K., and Bergman, R. N. Transendothelial insulin transport is not saturable in vivo: No evidence for a receptor-mediated process. *J Clin Invest, 97,* 1497–1503, 1996.

[7]. Bergman, R. N., Ider, Y. Z., Bowden, C. R., and Cobelli, C. Quantitative estimation of insulin sensitivity. *Am J Physiol, 236,* E667–E677, 1979.

[8]. Bergman, R. N. and Bucolo, R. J. Interaction of insulin and glucose in the control of hepatic glucose balance. *Am J Physiol, 227,* 1314–1322, 1974.

[9]. Kahn, S. E., Prigeon, R. L., McCulloch, D. K., Boyko, E. J., Bergman, R. N., Schwartz, M. W., Neifing, J. L., Ward, W. K., Beard, J. C., Palmer, J. P., and Porte, D., Jr. Quantification of the relationship between insulin sensitivity and β-cell function in human subjects: evidence for a hyperbolic function. *Diabetes, 42,* 1663–1672, 1993.

Further Reading

Nonlinear Pharmacokinetics: Shargel, L., and Yu, A. B. C. *Applied Biopharmaceutics and Pharmacokinetics,* 3rd ed. Appleton & Lange: Norwalk, CT, 1993.

15.5 RECOMMENDED EXERCISES

Nonlinear Pharmacokinetics: See Shargel, L., and Yu, A. B. C. *Applied Biopharmaceutics and Pharmacokinetics,* 3rd ed. Appleton & Lange: Norwalk, CT, 1993, Chapter 16, Questions 1, 3, and 4b.

Nonlinear Compartmental Models: See [1], Case study 10.7.1.

16

THE ROLE OF NONLINEAR COMPARTMENTAL MODELS IN DEVELOPMENT OF ANTIOBESITY DRUGS

In this chapter, we discuss the application of a nonlinear compartmental model to the development of antiobesity drugs. **Obesity** is a complex multifactorial chronic disease that occurs in part when mechanisms in the brain regulate body weight to a "set-point" that is associated with excessive accumulation of **adipose** (fat) tissue. According to the National Heart, Lung, and Blood Institute of the National Institutes of Health, obesity is defined as a **body mass index** (BMI) greater than or equal to 30 kg/m^2. BMI, which describes relative weight for height, is significantly correlated with total body fat content [1]. In principle, a drug that could overcome the weak link or links in the brain pathway that prevent the messages conveyed by body weight regulation negative feedback signals from producing an appropriate response in obese individuals would provide an elegant solution to obesity treatment.

16.1 BODY WEIGHT REGULATION

In 1953, Kennedy hypothesized that body adiposity was regulated by **humoral** (transported through body fluids) signals generated in proportion to body fat stores that act in the brain to alter food intake and energy expenditure [2]. In support of this hypothesis, it is known that body fuel stored as adipose tissue remains relatively constant over time, despite short-term (daily) variability in energy intake [3, 4]. Further, changes in body fat content induced by interventions as diverse as dieting [5], behavior modification [6], surgical removal of fat [7], or experimental overfeeding [8] induce compensatory responses that gradually restore adiposity to baseline values.

16.1.1 The Effect of Energy Expenditure

Researchers at Rockefeller University measured 24-hour total energy expenditure in 18 obese subjects and 23 subjects who had never been obese. The subjects were studied at their usual body weight or after losing 10 to 20% of their body weight by underfeeding or gaining 10% by overfeeding. Maintenance of a body weight at least 10% below initial weight was associated with a mean reduction in total energy expenditure of 6 ± 3 kcal per kilogram of fat-free mass per day in subjects who had never been obese ($p < 0.001$) and 8 ± 5 kcal per kilogram per day in obese subjects ($p < 0.001$). Maintenance of a body weight at least 10% above initial weight was associated with a mean increase in total energy expenditure of 9 ± 7 kcal per kilogram per day in subjects who had never been obese ($p < 0.001$) and 8 ± 4 kcal per kilogram per day in obese subjects ($p < 0.001$). After 10% weight loss or gain, the resting energy expenditure between obese and nonobese subjects was significantly different ($p \leq 0.004$). Thus, maintenance of a reduced or elevated body weight is associated with compensatory changes in energy expenditure that oppose the maintenance of a body weight different from the usual weight [9]. The clinical importance of this regulatory system is underscored by the limited success of energy-restricted diets in achieving long-term weight loss.

16.1.2 The Effect of Genetics

This regulatory system possesses a genetic component. In an 84 day study of 12 pairs of young adult male identical twins, the subjects were overfed by 1000 kcal per day, 6 days a week. During the overfeeding period, individual changes in body composition and topography of fat deposition varied considerably. While the mean weight gain was 8.1 kg, the range varied from 4.3 to 13.3 kg. The similarity within each pair in the response to overfeeding was significant ($p < 0.05$) with respect to body weight, percentage of fat, fat mass and estimated subcutaneous fat, with about three times more variance among pairs than within pairs ($r \cong 0.5$). The most likely explanation for the intrapair similarity in the adaptation to long-term overfeeding and for the variations in weight gain and fat distribution between pairs is the involvement of genetic factors [10].

16.1.3 Body Weight Regulation Model

According to the model first proposed by Kennedy, a hormone that acts as a negative feedback signal for body weight regulation is secreted proportional to body fat stores (Figure 16.1). Once the signal reaches the brain, it interacts with **central** (brain) effectors of **energy balance** in the brain **hypothalamus** to oppose the increase in body fat stores. Energy balance refers to the interaction between food intake and energy expenditure. When food intake exceeds energy expenditure, body weight increases. Conversely, when energy expenditure exceeds food intake, body weight decreases. Obesity may occur when the signal is transported to the brain inefficiently or interacts with central effectors ineffectively.

Two candidate feedback signals are **leptin** and insulin. Both are secreted in proportion to body adiposity and act in the brain to reduce food intake and promote weight loss. Numerous **neuropeptides** (proteins in the brain) have been identified as central effectors of energy balance that interact with either or both feedback signals; however, these interactions are beyond the range of our discussion.

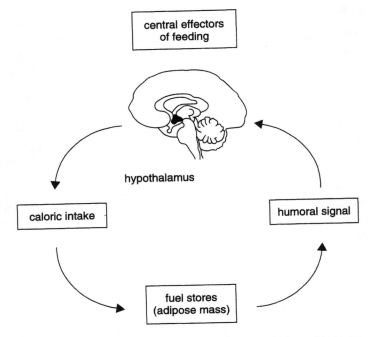

Figure 16.1 Body weight regulation model. The hypothalamus lies at the base of the brain and is shaded.

16.1.4 Leptin

Obesity research exploded in 1995 after leptin was isolated by Jeffrey Friedman's group at Rockefeller University [11]. Leptin is the product of the obesity gene, *ob,* and is produced by adipose cells. It circulates in the blood and enters the brain, where it functions to reduce food intake, reduce plasma glucose and insulin levels, and increase metabolic rate. These actions ultimately lead to a reduction in adipose mass and body weight. Leptin was first isolated in the *ob/ob* mouse, which is extremely obese due to a loss-of-function mutation in the *ob* gene. These mice are deficient in leptin, and dramatically lose body weight when leptin is administered exogeneously. Mice that were injected with 6 μg of mouse leptin per day for 5 days lost $6.8 \pm 1.3\%$ of their initial weight compared to those injected with saline that lost $1.8 \pm 0.4\%$. This difference in weight loss was significant ($p < 0.05$) [12].

In contrast, obese humans possess increased plasma levels of leptin, suggesting that obesity may be due in part to a decreased sensitivity to leptin, rather than a deficiency of leptin [13]. When leptin was administered daily to 60 obese humans with a BMI range of 27.6 to 36.0 kg/m^2 (mean body weight of 89.8 ± 11.4 kg) for 24 weeks, the resulting weight loss fell short of that in *ob/ob* mice. Leptin was administered in one of four doses, including a placebo. The eight subjects receiving the highest dose (0.30 mg/kg) lost a mean weight of 7.1 ± 8.5 kg, but the 12 subjects receiving the placebo lost a mean weight of 1.3 ± 4.9 kg. Further, 10 of the original 18 subjects receiving the highest, and most effective, dose dropped out of the study because of inflammation or **erythema** (reddening of the skin) at the intravenous injection site [14].

16.1.5 Insulin

Insulin was first suggested as a candidate negative feedback signal in 1976 by Steven Woods and Daniel Porte, Jr. of the University of Washington [15]. Porte's group had observed that plasma insulin concentrations varied in proportion to body adiposity [16], and later found that chronic **intracerebroventricular** [direct injection into the cerebral ventricles in the brain, which contain **cerebrospinal fluid** (CSF)] insulin administration in baboons caused a dose-dependent reduction in food intake and body weight over a 20 day period. The dose of 1 μU/kg-day did not reduce food intake, but the larger doses of 10 and 100 μU/kg-day significantly reduced food intake compared to initial intake by 18 and 41%, respectively ($p < 0.05$). Further, the intake during the largest dose was significantly less than that during the intermediate dose ($p < 0.05$). Similarly, the baboons receiving the 100 μU/kg-day dose lost significantly more body weight (2%, $p < 0.05$) than the baboons receiving the lowest [17]. This reduction in food intake and body weight was later reproduced in other species [18, 19, 20]. In support of this hypothesis, it has been shown that the thermic effect of a meal (energy expenditure) is partly dependent on insulin secretion [21], which may be mediated in the **central nervous system** (CNS) and not at a peripheral site [22].

16.2 RECEPTOR-MEDIATED TRANSPORT ACROSS THE BLOOD–BRAIN BARRIER

Regardless of the identities of negative feedback signals for body weight regulation, each must cross the **blood–brain barrier** (BBB). The blood–brain barrier is a membranous barrier that is highly resistant to diffusion and that segregates **brain interstitial fluid** (ISF) from circulating blood. In the periphery, many molecules pass from circulating blood to intersitial fluid by diffusion through the pores between **endothelia,** the cells lining capillaries. However, brain endothelia are endowed with unique anatomic specializations called **tight junctions** that close the **interendothelial** pores and limit diffusion to very small molecules. Instead, large circulating molecules such as peptides may only enter the brain through **transcytosis** (transendothelial transport).

16.2.1 Transcytosis

Transcytosis involves cellular internalization (**endocytosis**) of a given molecule from outside the cell, transport of that substance through the cell, and subsequent secretion (**exocytosis**) of the molecule from the cell opposite the side of entry. It is a highly specific event associated with specific cell surface receptors and their recognition by equally specific **ligands** (the molecules binding to the receptor). The three principal forms of transcytosis are **bulk flow transcytosis, receptor-mediated transcytosis,** and **absorptive-mediated transcytosis.**

In bulk flow transcytosis, transendothelial transport occurs independently of ligand binding to the plasma membrane. Rather, **pinocytotic vesicles** that are freely mobile within the **cytoplasm** of each cell are believed to ferry droplets of fluid across the vascular wall (Figure 16.2). However, because very few of these vesicles exist in brain endothelia, the movement of fluid via bulk flow is minimal.

In receptor-mediated transcytosis, transport is initiated by binding of an extracellular

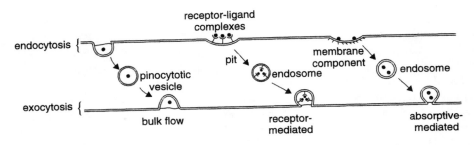

Figure 16.2 Models of bulk flow, receptor-mediated, and absorptive-mediated transcytosis.

ligand to a known receptor protein embedded within the plasma membrane. Following binding, the receptor–ligand complex possesses increased affinity for the **pits,** or invaginations, within the plasma membrane. Upon binding of the complex to a pit, the pit undergoes complete invagination into the cytoplasm and becomes an **endosome** beneath the plasma membrane. The endosome may then move to the opposite pole of the cell, where it fuses with the plasma membrane. A pore is formed in the membrane, enabling the receptor–ligand complex to be secreted from the cell (Figure 16.2). For example, the plasma protein **transferrin,** which binds to metals such as aluminum, enters the brain through this mechanism. It is believed that receptor-mediated transcytosis of aluminum-bearing transferrin is the probable mechanism by which dietary aluminum is deposited in the neuritic plaques of individuals with Alzheimer's disease [23]. Transport across the blood–brain barrier by negative feedback signals is believed to occur by this mechanism.

In absorptive-mediated transcytosis, transport is triggered by an electrostatic interaction between the positively charged portion of an extracellular ligand and a negatively charged plasma membrane surface region. Although the affinity of receptor-mediated endocytosis is much greater than that of absorptive-mediated endocytosis, the capacity of absorptive-mediated endocytosis is far greater. For example, the equilibrium constant, K_D, (recall Chapter 14) of the BBB transferrin receptor for transferrin is 5.6 ± 1.4 nM, whereas the K_D of cationized bovine serum **albumin** binding to bovine brain capillaries is 81 ± 12 nM [24]. Albumin, the most abundant protein in vertebrate blood, is found in egg whites. Upon binding to a membrane component, the component undergoes complete invagination into the cytoplasm and becomes an endosome that moves to the opposite pole of the cell. After the endosome fuses with the plasma membrane, a pore enables the ligand to be secreted from the cell (Figure 16.2).

16.2.2 Transport across the Blood–CSF Barrier

Plasma peptides may also gain access to brain interstitial fluid via the **blood–CSF barrier.** The blood–CSF barrier is composed of the tight junctions of cells between the **choroid plexus,** a single sheet of tissue within the two **lateral ventricles** and **third ventricle** of the brain. The choroid plexus secretes cerebrospinal fluid, the fluid that covers the entire surface of the brain and spinal cord. These tight junctions are slightly more permeable that those of the BBB.

The cerebrospinal fluid is separated from brain interstitial fluid by a single layer of cells lining the ventricles called the **ependyma.** While it is possible for peptides to access brain ISF by receptor-mediated transport across the blood–CSF barrier and through the ependyma, this access is limited to the surface of the brain. Due to the low diffusion coefficients across the ependyma, active drug binding by brain cells, enzymatic inactivation, and high membrane permeability of brain cells, peptide diffusion into deeper brain tissue such as the hypothalamus is extremely limited. Thus, the concentration of peptide at a distance of 1 mm from the ependymal surface approximates a concentration that is less than 0.1% of peptide concentration within the CSF [24].

16.2.3 Nonlinear Compartmental Models of Transport to the Brain

Because brain interstitial fluid is difficult to sample directly, many in vivo studies are based on simultaneous sampling of plasma and CSF. When plasma is used as a forcing function, $P(t)$, one exponential representing a single compartment may be used to model transport to CSF across the blood–CSF barrier [Figure 16.3(b)]. However, if two exponentials are required to fit the data, an extra compartment is present between plasma and CSF. The necessity of an extra compartment implies transport across the blood–brain barrier, through brain interstitital fluid, to CSF [Figure 16.3(c)].

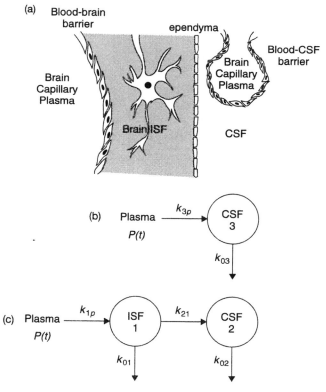

Figure 16.3 Transport from plasma to CSF. (a) Diagram of anatomic structures. (b) One compartment model representing transport from plasma, across blood–CSF barrier, to CSF. (c) Two-compartment model representing transport from plasma, across the blood–brain barrier to brain ISF, and across the ependyma to CSF.

In Figure 16.3b, the rate constant, k_{3p}, represents unidirectional transport across the blood–CSF barrier. In Figure 16.3c, the rate constant, k_{1p}, represents unidirectional transport across the BBB. Both rate constants saturate with increasing plasma concentration. In both models, the rate constant associated with transport from CSF, k_{02} or k_{03}, represents CSF clearance across the **arachnoid villi** to cerebral venous blood. The rate constant associated with transport from ISF, k_{01}, represents the turnover of ISF independent of entering CSF, presumably via metabolism by neurons and **glia**, and/or drainage of ISF into lymph. Because diffusion from CSF to ISF is severely limited, transport between the two compartments is represented unidirectionally as transport from ISF to CSF by rate constant k_{21}. Approximately 20 to 30% of CSF is thought to arise by bulk flow transcytosis of brain ISF across the ependyma.

Using the mass balance principle, transport from plasma to CSF may be described as

$$\dot{c}_3(t) = k_{3p}P(t) - k_{03}c_3(t) \tag{16.1}$$

where $c_i(t)$ is the concentration in compartment i. If transport across the blood–CSF barrier is saturable, k_{3p} will vary with the mean plasma concentration. By the transfer function matrix approach described in Chapter 13, this system is a priori identifiable.

Using the mass balance principle, transport from plasma, through brain interstitial fluid, to CSF may be described as

$$\dot{c}_1(t) = k_{1p}P(t) - (k_{01} + k_{21})c_1(t) \tag{16.2}$$

$$\dot{c}_2(t) = k_{21}c_1(t) - k_{02}c_2(t) \tag{16.3}$$

If transport across the blood–brain barrier is saturable, k_{1p} will vary with the mean plasma concentration. By the transfer function matrix approach, this system is not a priori identifiable. However, Eqs. (16.2) and (16.3) may be arranged to simplify parameter identification. First, let us divide both sides of Eq. (16.2) by k_{1p} and subtract and add the same term:

$$\frac{\dot{c}_1(t)}{k_{1p}} = P(t) - (k_{01} + k_{21} - k_{1p}k_{21} + k_{1p}k_{21})\frac{c_1(t)}{k_{1p}} \tag{16.4}$$

Second, let us rewrite Eq. (16.3) in terms of $c_1(t)/k_{1p}$ as

$$\dot{c}_2(t) = k_{21} \cdot k_{1p}\frac{c_1(t)}{k_{1p}} - k_{02}c_2(t) \tag{16.5}$$

Eqs. (16.4) and (16.5) are the rearranged model of transport from plasma, through brain ISF, to CSF. Because the initial value in the ISF compartment is unknown, the initial values of plasma and CSF are subtracted from their measured dynamic values, such that only dynamic changes in each compartment are modeled. Therefore,

$$P(0) = c_1(0) = c_2(0) = 0 \tag{16.6}$$

The rearranged model is illustrated in Figure 16.4. Using the transfer function matrix approach, this model is a priori identifiable for the parameters $k_{1p}k_{21}$, $(k_{01} + k_{21})$, and k_{02}.

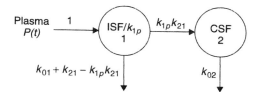

Figure 16.4 Rearranged two compartment model of transport from plasma, through ISF, to CSF. This model is a priori identifiable.

16.2.4 Leptin Transport

Leptin transport across the BBB is believed to be mediated by the leptin receptor, which has been isolated and characterized [25]. The long form of this receptor has been found in the hypothalamus, a site of regulation of feeding behavior. In in vitro studies of isolated human brain capillaries, leptin binding has been observed to be saturable [26]. In a study of 53 human subjects, the CSF leptin levels were correlated to body mass index ($r = 0.43$, $p = 0.001$). However, the CSF to plasma insulin ratio was much lower (5.4 times) in those subjects in the highest plasma insulin quartile compared to those in the lowest plasma insulin quintile, suggesting saturable plasma uptake [27].

16.2.5 Insulin Transport

When Woods and Porte first proposed their hypothesis in 1976, it was not believed that the insulin molecule, which is large in size, could cross the BBB. However, since that time, insulin and its receptor have been identified in the adult mammalian brain [28, 29], and it has been demonstrated that insulin in the brain is derived largely from the circulation [30]. Using **autoradiography,** the process of exposing photographic emulsion by radiation emitted from tissue containing a radionuclide, Porte's group observed that high concentrations of insulin receptors are found in the rat olfactory bulb [31] and in regions of the hypothalamus known to participate in modulation of feeding behavior [32]. Soon thereafter, William Pardridge's group at UCLA provided pharmacokinetic evidence that BBB transcytosis of insulin occurs in developing rabbits [33].

In Section 16.5, we discuss how saturable transport of insulin from plasma to the cerebrospinal fluid in vivo provided the first quantitative evidence in whole animals for facilitated transport of insulin across the blood–brain barrier via the BBB insulin receptor [34].

16.3 PROBLEM SIGNIFICANCE

The number of overweight (BMI of 25 to 29.9) and obese individuals in the United States has increased dramatically since 1960. In the last decade, the percentage of individuals in these categories has risen to 55% of adults age 20 years and older [1]. Although dietary therapy, physical activity, and behavior therapy are the most recommended strategies for weight loss, pharmacotherapy has been approved by the Food and Drug Administration (FDA) as an adjunct to other therapies for patients with a BMI of ≥ 30 with no concomitant risk factors or disease and for patients with a BMI of ≥ 27 with concomitant risk factors or diseases [1].

16.3.1 Obesity Costs

Obesity is associated with increased morbidity and mortality. Weight loss in overweight and obese individuals reduces risk factors for diabetes and cardiovascular disease, including a reduction in blood pressure, reduction in serum triglycerides and increase in high-density lipoprotein cholesterol, and reduction in blood glucose levels [1]. A recent study estimated the economic cost of obesity in the United States in 1995 dollars as $99.2 billion. Approximately $51.64 billion of these dollars were direct medical costs. Direct costs refer to personal health care, hospital care, physician services, allied health services, and medications. These direct costs represented 5.7% of the national health expenditure in the United States [35].

According to a pharmaceutical consulting firm, during the 12 months ending with July, 1999, visits to U.S. office-based physicians for obesity increased by 11% by 7.1 million. During the quarter ending in July, 1999, retail sales of obesity treatments grew to nearly $90 million, a 74% increase from the previous quarter [36]. It is estimated that $33 billion annually is spent on super-low-calorie drinks, exercycles, diet books, and fitness clubs [37].

As estimated in a business report issued by John Wiley & Sons, the potential annual U.S. market for a new antiobesity drug for morbidly obese patients is $750 million. This estimate is based on an assumption of 300,000 morbidly obese people, with 50% receiving a treatment that costs $5000 per year [38].

16.3.2 Pharmacologic Antiobesity Agents

In response to the need to decrease the rate of obesity, antiobesity agents are being developed. The goal of all antiobesity drugs is to induce and maintain a state of negative energy balance until the desired weight loss is achieved [39]. The four general classes of antiobesity drugs are based on this principle. Inhibitors of food intake such as **fenfluramine** (half of the Fen/Phen regimen that acts to inhibit uptake of serotonin, whose approval was withdrawn by the FDA) increase the feeling of fullness and reduce food intake by acting on brain mechanisms. Inhibitors of fat absorption like **Orlistat** reduce food intake through a peripheral, gastrointestinal mechanism of action and do not alter brain chemistry. Enhancers of energy expenditure act through peripheral mechanisms to increase **thermogenesis** (heat generation) without increased physical activity. Stimulators of adipose mobilization like leptin act peripherally to reduce adipose mass without increased physical activity or reduced food intake. Note, however, that weight loss from all of these drug classes may be easily overcome by increased food intake or decreased voluntary physical activity.

As stated above, leptin has not proven to be an effective antiobesity agent. In 1995, the pharmaceutical company Amgen paid a $20 million "signing fee" to Rockefeller University, where the *ob* gene was isolated, for rights to drugs based on this work [37]. More recently, other pharmaceutical companies have created large obesity programs based on drugs that target the leptin receptor and other receptors known to participate in the inhibition of food intake, the enhancement of energy expenditure, or the stimulation of fat mobilization. Interestingly, in a table of potential therapeutic targets (receptors) for new antiobesity drugs collected by an obesity researcher at Hoffmann-La Roche, the insulin receptor was excluded [39].

16.3.3 The Need for Blood–Brain Barrier Studies in Antiobesity Agent Research

Regardless of the receptor targeted, a critical need exists to investigate how a new antiobesity agent will reach this brain receptor from plasma. As stated by leading blood–brain barrier researcher William Pardridge, "few of the new drugs that emanate from the molecular neurosciences will undergo significant transport through the BBB, in the absence of a brain drug delivery strategy. Fundamental research in BBB transport processes provides the platform for CNS drug delivery and CNS drug targeting" [40]. However, he believes that because the majority of neuroscience graduate students study doctorate courses with barely a single lecture on the BBB and less than 0.5% of Society of Neuroscience abstracts are devoted to the BBB, many neuroscientists maintain a persistent indifference to this area of the brain. This indifference is reflected in many pharmaceutical obesity programs, which are solely conducted by peripheral metabolism and central nervous system research groups, whereas the blood–brain barrier is ignored.

16.4 PREVIOUS BLOOD–BRAIN BARRIER INSULIN STUDIES

Previous BBB insulin studies may be categorized as in vitro brain capillary studies, brain uptake studies, and in vivo whole animal studies.

16.4.1 Isolated Brain Capillary Studies

Isolated brain capillaries (Figure 16.5) provide a means of investigating BBB endocytosis. These microvessels are isolated from fresh brain tissue through centrifugation and fil-

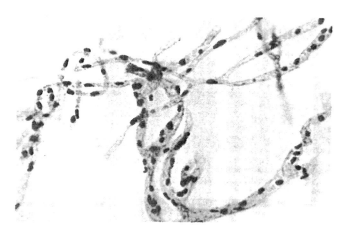

Figure 16.5 Photomicrograph of isolated bovine brain capillaries. The nuclei are stained darkly. The other, lighter colored round objects are red blood cells trapped in the lumen. Reprinted from [41] with permission from the American Diabetes Association, Alexandria, VA.

tering. Although isolated capillaries are depleted of **adenosine** triphosphate (ATP, which provides energy for chemical reactions), they remain metabolically active. Receptors exist on the **luminal** (inner) and **abluminal** membranes of the endothelial cells for mediating endocytosis. However, because energy in the form of ATP is not available, transcytosis is not possible.

Pardridge's laboratory utilized bovine capillaries to detect BBB insulin endocytosis in binding and competitive displacement experiments. In both sets of experiments, brain capillaries were first exposed to radioactive insulin. A concentration of unlabeled insulin was then added. When samples were withdrawn from this mixture, the detected radioactivity was proportional to the unlabeled insulin concentration.

In the binding experiments, capillaries were suspended in an Erlenmeyer flask at a final concentration of 500–800 μg protein/ml. The reaction was initiated by the addition of **[^{125}I]-iodoinsulin** at a final concentration of 0.3 ng/ml. In parallel, a flask was prepared identically except for the addition of 100 μg/ml of unlabeled insulin. Samples were withdrawn at t = 0, 5, 15, 30, 40, and 60 min, and were analyzed to determine insulin concentrations. The experiments were conducted at 22, 30, and 37°C.

To investigate the specificity of binding, competitive displacements experiments were performed in 5 ml plastic tubes containing 50 ml of [^{125}I]-iodoinsulin plus 50 ml of the appropriate dilution of unlabeled insulin (0, 1, 2, 10, 20, 40, 100, 200, 1000, 100000 ng/ml). The incubation was initiated with the addition of 400 ml of capillaries to each tube. Incubations were performed for 45 minutes at room temperature (22°C). Samples were withdrawn and analyzed as described above. The experiments were repeated, with other hormones (**prolactin, growth hormone, thyrotropin,** or **proinsulin**) added in place of insulin.

Over the course of the binding experiment, binding to receptors was rapid, reaching one-half maximal binding in approximately 7 minutes and peaking at 45 minutes at 22°C. Receptor binding is illustrated in Figure 16.6. In the competitive binding experiments, percentage insulin bound was determined for increasing concentrations of insulin. From these data, it was extrapolated that half-maximal displacement occurs at approximately 9 mg/ml. The hormones prolactin, thyrotropin, and growth hormone did not compete with [^{125}I]-iodoinsulin for binding to the receptor. However, proinsulin, the chemical precursor of insulin, competed poorly and was only 4% as active as insulin in displacing the [^{125}I]-iodoinsulin. Through Scatchard analysis of the binding data [42], it was determined that two binding sites exist: a high-affinity, low-capacity site with an **equilibrium constant** of K_D = 2.29 nM and a low-affinity, high-capacity site with K_D = 0.05 nM. As defined in Chapter 14, the equilibrium constant is a measure of equilibrium conditions, when the rate of association of a drug and its receptor equals the rate of dissociation of the drug–receptor complex. Taken together, both experiments provided support for the hypothesis that insulin binds to the BBB insulin receptor [41].

16.4.2 Brain Uptake Studies

More recently, Pardridge's laboratory conducted brain uptake experiments in suckling rabbits. Suckling rabbits were used because the BBB insulin receptor is more active in developing, rather than in adult, animals. To investigate brain uptake, [^{125}I]-iodoinsulin was coinfused with [^{3}H]-albumin and a vehicle solution into the right common **carotid artery** (that transports blood to the brain), at a flow rate of 0.25 ml/min for t = 1, 5, or 10 min. In a second group of rabbits, 100 μg/ml of unlabeled insulin was added to the infusate. The

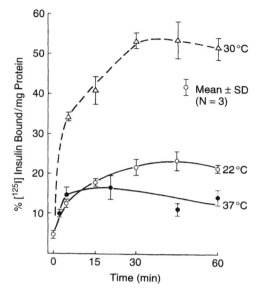

Figure 16.6 Time course of [^{125}I]-iodoinsulin binding to brain capillaries at 22°C, 30°C, and 37°C. Binding is most rapid at 37°C and most active at 30°C. Each point represents the mean ± standard deviation ($n = 3$). Reprinted from [41] with permission from the American Diabetes Association, Alexandria, VA.

blood glucose between both rabbit groups was not significantly different. After the infusion was completed, the animals were decapitated and the brain **hemispheres** were removed. The tissue was homogenized by passage through an 18 g needle and analyzed for the ratio of insulin to albumin uptake.

To investigate transport of insulin from brain capillaries to brain parenchyma, rabbits were exposed to the 10 minute carotid infusion of [^{125}I]-iodoinsulin described above and decapitated. The brain was quickly removed, frozen, and processed for thaw-mount autoradiography. Sections cut 4 mm thick were placed on emulsion-coated slides, exposed at 4°C for one to five months, developed, and examined. Control rabbits were infused with vehicle solution only and decapitated; their brains were processed identically. Finally, to determine if the integrity of insulin is preserved during transport (in the periphery, the insulin receptor degrades insulin), extracts of rabbit brain and blood, obtained after a 10 minute carotid infusion of [^{125}I]-iodoinsulin, were analyzed by **high-performance liquid chromatography** (HPLC, see Chapter 13). The HPLC insulin elution was combined with guinea pig antiinsulin antibody and rabbit antiguinea pig antibody to isolate the [^{125}I]-iodoinsulin fraction. (Recall from Chapter 13 that the double antibody method causes a hormone to precipitate out of solution.)

Relative to tritiated albumin, the brain uptake of [^{125}I]-iodoinsulin was 99.3 ± 5.5%, 110.1 ± 4.3%, and 143.6 ± 7.9%, respectively at 1, 5, and 10 minutes of infusion (Figure 16.7). When unlabeled insulin was added, this uptake decreased significantly ($p < 0.025$). Autoradiography of rabbit brain revealed that insulin, observed as silver grains, was present in the brain parenchyma, as well as in the **pericapillary** (tissue enclosing the capillary) and capillary space. The number of grains per 100 mm^2 over the microvessel, over the parenchyma, and over the control brain were 40.8 ± 3.8, 7.7 ± 0.7, and < 1, respec-

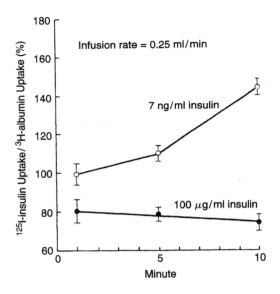

Figure 16.7 Brain uptake of [^{125}I]-iodoinsulin relative to [^{3}H]-albumin as a function of duration of infusion. The open circles represent data from rabbits infused with only 7 ng/ml of the radiolabeled insulin; the closed circles represent data from animals that were infused with an additionl 100 μg/ml of unlabeled insulin. Each point represents the mean and standard error for 4 to 7 animals. Reprinted from [33] with permission from Elsevier Science.

tively. In the HPLC analysis, greater than 90% of the radioactivity in the blood and the majority of the radioactivity in the brain comigrated with the HPLC insulin standard. This elution, compared to other elutions, possessed the greatest **immunoprecipitable radioactivity,** when precipitated with double antibodies.

These experiments demonstrated that insulin is taken up by the brain in a time-dependent, saturable fashion, that insulin crosses from the capillary space to the pericapillary space and brain parenchyma, and that the integrity of insulin is preserved during this transport [33].

16.4.3 Whole Animal Studies

To provide evidence for his insulin as negative feedback signal hypothesis, Michael Schwartz in Daniel Porte, Jr's laboratory developed a method of investigating insulin transport to the canine brain *in vivo*. Using this method, normal adult mongrel male canines were fasted overnight, anesthetized, and placed on a mechanical ventilator. Plasma and CSF samples were acquired during and after an intravenous insulin infusion. Blood (3 ml) was sampled from a cannula inserted into a superficial vein in a forelimb; insulin was infused through a cannula inserted into a superficial vein of a hindlimb. CSF samples (0.4 ml) were obtained from a cannula inserted into the **cisternum magnum,** a natural space between the brain hemispheres. During the infusion, **euglycemia** was maintained by monitoring glucose using a hand-held glucose analyzer, and infusing glucose as necessary to maintain a range of 65 to 105 ml/dl. Plasma and CSF insulin concentrations were measured by radioimmunoassay. Plasma glucose concentrations were measured using the **glucose oxidase method.**

In the first study, either insulin or proinsulin, which has reduced affinity for the insulin receptor, was infused for up to 90 minutes. A significantly lower increment of CSF proinsulin levels ($n = 4$) over 180 minutes was observed, compared to that of insulin ($n = 4$), suggesting that insulin uptake into CSF occurs in part via a mechanism with specificity for insulin [43].

In the second study, either insulin or **inulin** was infused. Inulin is a biologically inert marker of diffusion similar in size to insulin. Infusions were either brief (800–400 mU/kg-min for 5–10 minutes) or sustained (10–15 mU/kg-min for 4 to 12 hours). For the brief infusion, blood was sampled at $t = -15, -5, 2, 4, 6, 8, 10, 12, 15, 20$, and 30 min and thereafter at 30 min intervals for 400 to 500 min. CSF samples were obtained at $t = -15, -5, 6, 15, 30, 60, 90, 120, 150$, and 180 min and thereafter at 60 min intervals until the end of the experiment. For the sustained protocol, blood and CSF samples were obtained at $t = -15, -5, 5, 15$, and 30 min and at 30 to 60 min intervals thereafter for up to 12 hours.

Plasma and CSF insulin or inulin data were fit to the one compartment model of transport across the blood–CSF barrier and to the two-compartment model of transport across the blood–brain barrier. Additionally, two other models were used that are variations of the one-compartment model. In the first variation, k_{3p} was replaced by a Michaelis–Menten term,

$$k_{3p}(t) = \frac{V_{max}}{300 + P(t)} \tag{16.7}$$

where K_M was defined as 300 µU/ml, the value of the insulin receptor dissociation constant, and V_{max} was the maximum velocity [45]. In the second variation, k_{3p} was replaced by the sum of a Michaelis–Menten term and a nonsaturable rate constant, k_x:

$$k_{3p}(t) = \frac{V_{max}}{300 + P(t)} + k_x \tag{16.8}$$

Parameter identification was accomplished with the software program MLAB. Rather than using the **Akaike information criterion** described in Chapter 13, the optimal model was determined by a statistical F-test based on the residual sums of squares from each model.

The time course for plasma and CSF insulin during brief infusion ($n = 5$) at a high dose is shown in Figure 16.8. The two-compartment model fit the high- and low-dose ($n = 4$) data best in terms of smaller residual error, and was significantly different ($p < 0.0001$) from the other three models. In contrast, plasma and CSF inulin data during brief infusion were best fit in terms of smaller residual error by the one compartment model, without variation ($p = 0.02$). The time course for plasma and CSF insulin during sustained infusion for one experiment is shown in Figure 16.9. Although the F-test results implied that the models were significantly different, the residual errors were similar by visual inspection [44].

In this study, the associated coefficient of variation for each identified parameter was never disclosed. The only recorded measure of variation was the standard deviation of a parameter estimate between experiments, which is not particularly useful. However, since both the brief and sustained infusion sampling schedules did not include steady state sam-

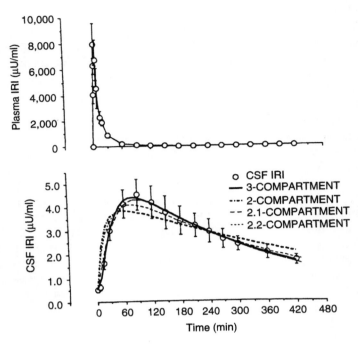

Figure 16.8 Time course of mean (± SEM) plasma and CSF insulin levels during a high-dose brief insulin infusion. The mean infusion rate was 176 ± 23 mU/kg-min over 5 min. The curve fits for various models are shown in the lower panel. The label of "3-compartment" refers to the two-compartment model. The label of "2-compartment" refers to the one-compartment model. The other two labels refer to variations of the one-compartment model. In this article, the plasma forcing function was considered a compartment, even though it was not represented by an exponential during parameter identification. Reprinted from [44] with permission from the American Society for Clinical Investigation, Thoroughfare, NJ; permission conveyed through Copyright Clearance Center, Inc.

ples, the associated coefficients of variation were undoubtedly high, probably much greater than 100%. With the addition of a Michaelis–Menten term in two of the models, parameter identification would be ill-conditioned (recall Chapter 15), causing the associated coefficients of variation to further increase.

Because insulin transport was better represented by a two-compartment model, whereas inulin transport was better represented by a one compartment model, this study provided evidence that plasma insulin is transported to another compartment before being transported to CSF.

16.5 SATURABLE TRANSPORT OF INSULIN FROM PLASMA INTO THE CNS

As discussed in Chapter 15, Michaelis–Menten kinetics may be demonstrated by using a linear compartmental model and showing that the rate constant representing saturable transport decreases with increasing mean concentration. This methodology was not possible in the previous whole animal studies because neither optimal nor practical sampling

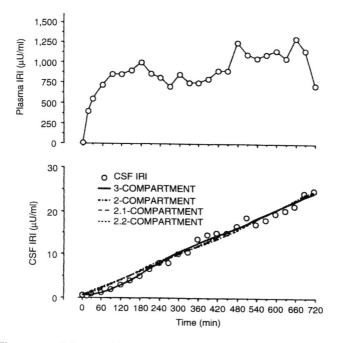

Figure 16.9 Time course of plasma and CSF insulin levels during a sustained insulin infusion. The mean infusion rate was 10 mU/kg-min. The curve fits for various models are shown in the lower panel. The label of "3-compartment" refers to the two-compartment model. The label of "2-compartment" refers to the one-compartment model. The other two labels refer to variations of the one-compartment model. In this article, the plasma forcing function was considered a compartment, even though it was not represented by an exponential during parameter identification. Reprinted from [44] with permission from the American Society for Clinical Investigation, Thoroughfare, NJ; permission conveyed through Copyright Clearance Center, Inc.

was followed, leading to extremely high coefficients of variation. With high coefficients of variation, it would be difficult to demonstrate that a rate constant is stable, much less that it decreases with increasing mean concentration.

16.5.1 A Practical Sampling Schedule

In the current study, the blood and CSF sampling protocols during insulin infusion were substantially modified. In response to a 3 min primed insulin infusion followed by an 87 min insulin infusion at 20% the primed infusion rate, blood samples (1.8 ml) were obtained at $t = -10, -5, 1, 2, 3, 4, 5, 6, 8, 10, 13, 16, 20, 25, 30, 35, 40, 65$ and 90 min, at 5 min intervals for $90 \leq t \leq 150$ min, and at 20 min intervals for $150 \leq t \leq 490$ min. CSF samples (0.4 ml) were obtained at $t = -10, -5, 20, 40, 65, 90, 95, 100, 110, 120, 135$, and 150 min and thereafter at 20 min intervals until 490 min. In the first 3 of 10 experiments, additional blood and CSF samples were taken at 20 min intervals for $510 \leq t \leq 590$ min to insure that steady state values were reached. Except for sampling during the greatest change in the CSF concentration after the infusion was terminated, the number of CSF samples was limited to 3 per hour to insure that CSF production by the choroid plexus would not be modified in response to frequent sampling.

The data were fit to the two-compartment model using the software program SAAM. Typical sets of plasma and CSF insulin curves are shown in Figure 16.10. Because a practical sampling schedule was utilized, the mean coefficient of variation for estimated parameters was 14%. No apparent relationship between either $(k_{21} + k_{01})$ or k_{02} and plasma insulin level was evident. The plasma insulin level utilized for this analysis was P_{ave}, the mean plasma insulin concentration during infusion. The mean value of the rate constants associated with total brain ISF clearance, $(k_{21} + k_{01})$, was 0.011 ± 0.0019 min^{-1}. The mean value of the rate constant associated with CSF clearance, k_{02}, was 0.046 ± 0.021 min^{-1}. However, the value of $k_{1p}k_{21}$, the rate constant associated with insulin uptake from plasma into CSF, varied inversely with plasma insulin levels, being reduced sevenfold from physiologic to supraphysiologic levels (Figure 16.11). Because the sum, $(k_{21} + k_{01})$, was constant, k_{21} and k_{01} were probably also constant. Therefore, the decrease in $k_{1p}k_{21}$ as a function of plasma insulin is mostly likely due solely to the change in k_{1p}.

16.5.2 Estimation of K_M

To estimate K_M, the plasma insulin concentration at which half-maximal transport occurs, the transport velocity was determined. Velocity, $V(P_{ave})$, was calculated by multiplying k_{1p}, the rate constant representing uptake of plasma insulin into brain interstitial fluid, by P_{ave}. As k_{1p} could not be uniquely identified, $V(P_{ave}) \cdot k_{21}$ was used for this calculation. Velocity $\cdot k_{21}$ due to nonspecific uptake is linear, such that at high plasma insulin levels, $V(P_{ave}) \cdot k_{21}$ does not plateau. As this linear component is readily apparent at high plasma insulin levels, velocity due to nonspecific uptake was estimated through graphical analysis as $2 \times 10^{-7} \cdot P_{ave}$. This estimate was subtracted from $V(P_{ave}) \cdot k_{21}$ to isolated $V(P_{ave}) \cdot k_{21}$ due to saturable uptake. Velocity $\cdot k_{21}$ was then plotted as a function of P_{ave} (Figure 16.12). K_M was calculated by fitting the data using a variation of the Michaelis–Menten equation,

$$V(P_{ave}) \cdot k_{21} = \frac{V_{max} \cdot k_{21}}{K_M + P_{ave}} \tag{16.9}$$

The apparent K_M was estimated to be 742 µU/ml (~ 5nM). As discussed in Chapter 15, K_M is a measure of equilibrium conditions, when the rate of association of an enzyme and its substrate equals the rate of dissociation of the enzyme–substrate complex. Therefore, when applied to the BBB insulin receptor, the concepts underlying K_M and the equilibrium constant, K_D, are identical. The estimated value of K_M was similar to the high-affinity K_D of the insulin receptor identified in brain microvasculature by Pardridge's laboratory as $K_D = 2.29$ nM [42].

16.5.3 Independent Estimation of CSF Insulin Clearance

During the parameter estimation process, it was assumed that the larger of the two eigenvalues identified from the system described by Eqs. (16.4) and (16.5) was associated with k_{02}. Similarly, it was assumed that the smaller eigenvalue was associated with $(k_{21} + k_{01})$ and $k_{1p}k_{21}$. To obtain an independent estimate of k_{02}, the rate constant associated with CSF insulin clearance, insulin (750 µU in 750 µl of saline) was infused over 1 min into the cisternum magnum of three dogs that had been fasted overnight, anesthetized, and maintained on mechanical ventilation. Immediately before the infusion was administered,

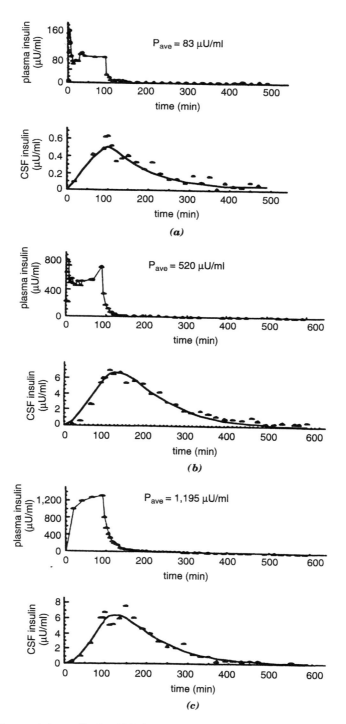

Figure 16.10 Representative studies in which the kinetics of CSF uptake of intravenously infused insulin were determined. Insulin was infused during $0 \leq t \leq 90$ min. Studies shown are representative of low (a), medium (b), and high (c) insulin doses. Upper panels: time course of plasma insulin. Lower panels: time course of CSF insulin and optimal curve fit, from which rate constant parameters were derived. Reprinted from [34] with permission from the American Society for Clinical Investigation, Thoroughfare, NJ; permission conveyed through Copyright Clearance Center, Inc.

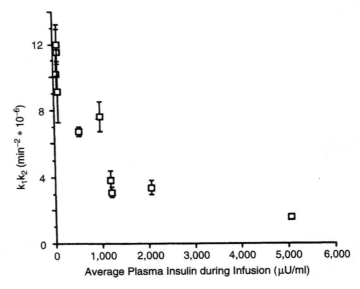

Figure 16.11 Dose–response curve illustrating the relationship between the rate of uptake of plasma insulin into the intermediate compartment and subsequently into CSF. Error bar represents the coefficient of variation associated with the identified parameter value. Here, k_1 refers to k_{1p} and k_2 refers to k_{21}. Reprinted from [34] with permission from the American Society for Clinical Investigation, Thoroughfare, NJ; permission conveyed through Copyright Clearance Center, Inc.

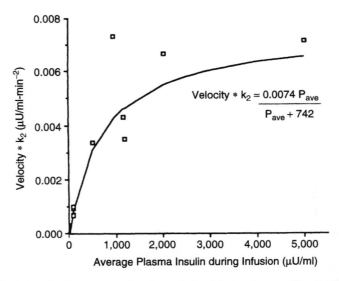

Figure 16.12 Michaelis–Menten curve, based on calculated $k_{1p}k_{21}$, the rate of insulin uptake from plasma, through the brain ISF compartment, and subsequently to CSF, for various insulin doses. Here, k_1 refers to k_{1p} and k_2 refers to k_{21}. Reprinted from [34] with permission from the American Society for Clinical Investigation, Thoroughfare, NJ; permission conveyed through Copyright Clearance Center, Inc.

750 μl of CSF was removed from the cisternum magnum in order to preserve the original CSF volume during the infusion.

CSF samples of 0.4 ml were obtained at $t = -10, -5, 2, 5, 10, 15, 25,$ and 40 min, and thereafter at 20 min intervals until 240 min. The resulting CSF levels were fit to a two-compartment model that had previously been used to model transport kinetics of substances administered intracerebroventricularly. In this model, the first compartment represents CSF and the second represents brain interstitial fluid [46]. The CSF clearance rate was calculated by dividing the estimated bolus concentration administered by the area under the curve of the two exponentials [47].

The mean CSF insulin clearance rate determined from these studies was 0.028 ± 0.0080 min^{-1}, which is within the error of the mean value of k_{02} determined during the intravenous insulin infusion of $k_{02} = 0.046 \pm 0.021$ min^{-1}. Therefore, the larger eigenvalue is associated with insulin clearance from CSF.

16.5.4 Discussion

This study provided evidence that insulin transport from plasma, through brain interstitial fluid, and to CSF is saturable, and is likely facilitated by an insulin-receptor-mediated process. Using this methodology as a foundation for further work, it was later demonstrated that the glucocorticoid dexamethasone impairs plasma insulin uptake, providing evidence that this process may be subject to regulation [48]. Further, it was demonstrated that plasma insulin uptake × basal insulin, measured prior to high fat feeding, is inversely proportional to daily weight gain during high fat feeding, and accounted for 40% of the variance in the data [49]. Because insulin acts in the brain to reduce food intake and body weight, a marked decrease of CNS insulin uptake, which overcompensates for an increase in basal insulin during weight gain, provides a theoretical mechanism by which obesity-prone animals may display marked, progressive weight gain during high-fat feeding.

16.6 SUMMARY

In this chapter, we have discussed the theory of body weight regulation, in which it is hypothesized that a hormone (or group of hormones) is secreted proportional to body fat stores and acts as a negative feedback signal for weight gain. Once the signal reaches the brain, it interacts with central effectors of energy balance in the hypothalamus to oppose the increase in body fat stores. Obesity may occur when the signal is transported to the brain inefficiently or interacts with central effectors ineffectively. Two candidate feedback signals are leptin and insulin. Both are secreted in proportion to body adiposity and act in the brain to reduce food intake and promote weight loss. Numerous neuropeptides have been identified as central effectors of energy balance that interact with either or both feedback signals. Any negative feedback signal must cross the blood–brain barrier to enter brain interstitial fluid through receptor-mediated transcytosis. This transport may be investigated using a nonlinear compartmental model.

With regard to insulin, transport across the BBB was first investigated using isolated bovine brain capillaries. Because isolated capillaries remain metabolically active, they may be used to investigate endocytosis. However, because they lack ATP, they are incapable of transcytosis. In binding experiments, it was shown that insulin binds to BBB insulin receptors rapidly. Through Scatchard analysis of the binding data, it was determined that two binding sites exist: a high-affinity, low-capacity site with an equilibrium constant of $K_D = 2.29$ nM and a low-affinity, high-capacity site with $K_D = 0.05$ nM. Later, BBB insulin transport was investigated by infusing insulin into the carotid artery of suckling rabbits, and analyzing the resulting insulin concentration in brain after decapitation. In this study, it was demonstrated that insulin is taken up by the brain in a time-dependent, saturable fashion, that insulin crosses from the capillary space to the pericapillary space and brain parenchyma, and that the integrity of insulin is preserved during this transport.

BBB insulin transport was also investigated in whole animals in vivo. Plasma and CSF samples were acquired from canines during and after an intravenous insulin infusion. These data were fit to a one- or two-compartment model to determine if brain ISF was involved in the transport process. Because insulin transport is better represented by a two-compartment model, whereas inulin transport is better represented by a one compartment model, this study provided evidence that plasma insulin is transported to another compartment before being transported to CSF.

By modifying the original blood and CSF sampling protocols to accommodate a practical sampling schedule, parameters in the two-compartment model were identified with greater precision. This greater precision enabled rate constant values to be investigated as a function of mean plasma concentration during infusion. It was demonstrated that the rate constant associated with plasma insulin transport is saturable. Further, the apparent K_M from this data was the same order of magnitude as the high-affinity K_D of the BBB insulin receptor. This study provided evidence that insulin transport from plasma, through brain interstitial fluid, and to CSF is likely facilitated by an insulin-receptor-mediated process. Using this methodology as a foundation for further work, it was later demonstrated that that plasma insulin uptake × basal insulin, measured prior to high fat feeding, is inversely proportional to daily weight gain during high-fat feeding, and accounted for 40% of the variance in the data. Because insulin acts in the brain to reduce food intake and body weight, a marked decrease of CNS insulin uptake, which overcompensates for an increase in basal insulin during weight gain, provides a theoretical mechanism by which obesity-prone animals may display marked, progressive weight gain during high-fat feeding.

Because the number of obese individuals in the United States has increased dramatically since 1960, the FDA has approved pharmacotherapy as an adjunct to other therapies for morbidly obese individuals. Several pharmaceutical companies have created large obesity programs based on drugs that target the leptin receptor and other receptors known to participate in the inhibition of food intake, the enhancement of energy expenditure, or the stimulation of fat mobilization. However, these programs are solely conducted by peripheral metabolism and central nervous system research groups, whereas the blood–brain barrier is ignored. Regardless of the receptor targeted, a critical need exists to investigate how a new antiobesity agent will reach this brain receptor from plasma. As stated by leading blood–brain barrier researcher William Pardridge, "few of the new drugs that emanate from the molecular neurosciences will undergo significant transport through the BBB, in the absence of a brain drug delivery strategy." The nonlinear compartmental model approach described provides a new and efficient methodology for investigating this transport system in larger animals. Unlike drug studies performed in mice such as the initial leptin work, these studies in larger animals may provide more insight into how new drugs are metabolized by humans.

16.7 REFERENCES

[1]. National Heart, Lung, and Blood Institute. *Clinical Guidelines on the Identification, Evaluation, and Treatment of Overweight and Obesity in Adults.* National Institutes of Health: Bethesda, MD, 1998.

[2]. Kennedy, G. C. The role of depot fat in the hypothalamic control of food intake in the rat. *Proc R Soc Lond B Biol Sci, 140,* 579–592, 1953.

[3]. Keesey, R. E. A set-point theory of obesity. In Brownell, K. D., ed., *Handbook of Eating Disorders: Physiology, Psychology, and Treatment of Obesity, Anorexia, and Bulimia.* Basic Books: New York, pp. 63–87, 1986.

[4]. Stallone, D. D. and Stunkard, A. J. The regulation of body weight: evidence and clinical implications. *Ann Behav Med, 13,* 220–230, 1991.

[5]. Drenick, E. J. and Johnson, D. Weight reduction by fasting and semistarvation in morbid obesity: long-term follow-up. *Int J Obes, 2,* 123–132, 1978.

[6]. Garner, D. M. and Wooley, S. C. Confronting the failure of behavorial and dietary treaments for obesity. *Clin Psychol Rev, 11,* 729–780, 1991.

[7]. Faust, I. M., Johnson, P. R., and Hirsch, J. Adipose tissue regeneration following lipectomy. *Science, 197,* 391–393, 1977.

[8]. Sims, E. A. H. Experimental obesity, diet-induced thermogenesis and their clinical implications. *Clin Endocrinol Metab, 5,* 377–395, 1976.

[9]. Leibel, R. L., Rosenblum, M., and Hirsch, J. Changes in energy expenditure resulting from altered body weight. *N Engl J Med, 332,* 621–628, 1995.

[10]. Bouchard, C., Tremblay, A., Despres, J., Nadeau, A., Lupien, P. J., Theriault, G., Dussault, J., Moorjani, S., Pinault, S., and Fournier, G. The response to long-term overfeeding in identical twins. *N Engl J Med, 322,* 1477–1482, 1990.

[11]. Zhang, Y., Proenca, R., Maffei, M., Barone, M., Leopold, L., and Friedman, J. M. Positional cloning of the mouse obese gene and its human homologue. *Nature, 372,* 425–432, 1994.

[12]. Campfield, L. A., Smith, F. J., Guisez, Y., Devos, R., and Burn, P. Recombinant mouse OB protein: Evidence for a peripheral signal linking adiposity and central neural networks. *Science, 269,* 546–549, 1995.

[13]. Considine, R. V., Sinha, M. K., Heiman, M. L., Kriauciunas, A., Stephens, T. W., Nyce, M. R., Ohannesian, J. P., Marco, C. C., McKee, L. J., Bauer, T. L., and Caro, J. F. Serum immunoreactive-leptin concentrations in normal-weight and obese humans. *N Engl J Med, 334,* 292–295, 1996.

[14]. Heymsfield, S. B., Greenberg, A. S., Fujiioka, K., Dixon, R. M., Kushner, R., Hunt, T., Lubina, J. A., Patane, J., Self, B., Hunt, P., and McCamish, M. Recombinant leptin for weight loss in obese and lean adults: A randomized, controlled dose-escalation trial. *JAMA, 282,* 1568–1575, 1999.

[15]. Woods, S. C and Porte, D. Jr. Insulin and the set-point regulation of body weight. In Novin, D., Bray, G. A., Wyrwichka, W., eds. *Hunger: Basic mechanisms and clinical implications.* Raven Press, New York, pp. 273–280, 1976.

[16]. Bagdade, J. D., Bierman, E. L., and Porte, D., Jr. The significance of basal insulin levels in the evaluaton of the insulin response to glucose in diabetic and nondiabetic subjects. *J Clin Invest, 46,* 1549–1557, 1967.

[17]. Woods, S. C., Stein, L. J., McKay, L. D., and Porte, D., Jr. Chronic intracerebroventricular infusion of insulin reduces food intake and body weight of baboons. *Nature, 282,* 503–505, 1979.

[18]. Brief, D. J., and Davis, J. D. Reduction of food intake and body weight by chronic intraventricular insulin infusion. *Brain Res Bull, 12,* 571–575, 1984.

[19]. Foster, L. A., Ames, N. K., and Emergy, R. S. Food intake and serum insulin responses to intraventricular infusions of insulin and IGF-1. *Physiol Behav, 50,* 745–749, 1991.

[20]. Florant, G. L., Singler, L., Scheurink, A. J., Park, C. R., Richardson, R. D., and Woods, S. C. Intraventricular insulin reduces food intake and body weight of marmots during the summer feeding period. *Physiol Behav, 49,* 335–338, 1991.

[21]. Rothwell, N. J. and Stock, M. J. Insulin and thermogenesis. *Int J Obes, 12,* 93–102, 1988.

[22]. Menendez, J. A. and Atrens, D. M. Insulin and the paraventricular hypothalamus: Modulation of energy balance. *Brain Res, 555,* 193–201, 1991.

[23]. Pullen, R. G. L., Candy, J. M., Morris, C. M., Taylor, G., Keith, A. B., and Edwardson, J. A. Gallium-67 as a potential marker for aluminum transport in rate brain: Implications for Alzheimers disease. *J Neurochem, 55,* 251–259, 1990.

[24]. Pardridge, W. M. *Peptide Drug Delivery to the Brain.* Raven Press: New York, pp. 117, 189–190, 1991.

[25]. Tartaglia, L. A., Dembski, M., Weng, X., Deng, N., Culpepper, J., Devos, R., Richards, G. J., Campfield, L. A., Clark, F. T., Deeds J., et al. Identification and expression cloning of a leptin receptor, OB-R. *Cell, 83,* 1263–1271, 1995.

[26]. Golden, P. L., Maccagnan, T. J., and Pardridge, W. M. Human blood–brain barrier leptin receptor: binding and endocytosis in isolated human brain microvessels. *J Clin Invest, 99,* 14–18, 1997.

[27]. Schwartz, M. W., Peskind, E., Raskind, M., Boyko, E., and Porte, D., Jr. Cerebrospinal fluid leptin levels: Relationship to plasma levels and to adiposity in humans. *Nat Med, 2,* 589–593, 1996.

[28]. Havrankova, J., Schmechel, D., Roth, J., and Brownstein, M. J. Identification of insulin in the rat brain. *Proc Natl Acad Sci, 76,* 5737–5741, 1978.

[29]. Havrankova, J., Roth, J., and Brownstein, M. Insulin receptors are widely distributed in the central nervous system of the rat. *Nature, 272,* 827–829, 1978.

[30]. Coker, G. T. I., Studelska, D., Harmon, S., Burke, W., and O'Malley, K. L. Analysis of tyrosine hydroxylase and insulin transcripts in human neuroendocrine tissues. *Mol Br Res, 8,* 93–98, 1990.

[31]. Baskin, D. G., Davidson, D. A., Corp, E. S., Lewellen, T., and Graham, M. An inexpensive microcomputer digital imaging system for densitometry: Quantitative autoradiography of insulin receptors with ^{125}I-insulin and LKB ultrofilm. *J Neurosci Meth, 16,* 119–129, 1986.

[32]. Corp, E. S., Woods, S. C., Porte, D., Jr, Dorsa, D. M., Figlewicz, D. P., and Baskin, D. G. Localization of ^{125}I-insulin binding sites in the rat hypothalamus by quantitative autoradiography. *Neurosci Lett, 70,* 17–22, 1986.

[33]. Duffy, K. R. and Pardridge, W. M. Blood-brain barrier transcytosis of insulin in developing rabbits. *Brain Res, 420,* 32–38, 1987.

[34]. Baura, G. D., Foster, D. M., Porte, D., Jr., Kahn, S. E., Bergman, R. N., Cobelli, C., and Schwartz, M. W. Saturable transport of insulin from plasma into the central nervous system of dogs in vivo. *J Clin Invest, 92,* 1824–1830, 1993.

[35]. Wolf, A. M and Colditz, G. A. Current estimates of the economic cost of obesity in the United States. *Obes Res, 6,* 97–106, 1998.

[36]. Scott-Levin website. November 9, 1999. www.scottlevin.com.

[37]. Stipp, D. New weapons in the war on fat. *Fortune, 132,* 64–173, 1995.

[38]. John Wiley & Sons, Press release for the future tech briefing *The Molecular Biology of Obesity: Research Opens New Approaches for Treatment.* November 9, 1999.

[39]. Campfield, L. A., Smith, F. J., and Burn, P. Strategies and potential molecular targets for obesity treatment. *Science, 280,* 1383–1387, 1998.

[40]. Pardridge, W. M., ed. *Introduction to the Blood-Brain Barrier: Methodology, Biology, and Pathology.* Cambridge University Press: New York, p. 7, 1998.

[41]. Frank, H. J. L. and Pardridge, W. M. A direct in vitro demonstration of insulin binding to isolated brain microvessels. *Diabetes, 30,* 757–761, 1981.

[42]. Feldman, H. A. Mathematical theory of complex ligand-binding systems at equilibrium: Some methods for parameter fitting. *Anal Biochem, 48,* 317–338, 1972.

[43]. Schwartz, M. W., Sipols, A., Kahn, S. E., Lattemann, D. F., Taborsky, G. J., Jr., Bergman, R. N., Woods, S. C., and Porte, D., Jr. Kinetics and specificity of insulin uptake from plasma into cerebrospinal fluid. *Am J Physiol, 259,* E378–E383, 1990.

[44]. Schwartz, M. W., Bergman, R. N., Kahn, S. E., Taborsky, G. J., Jr., Fisher, L. D., Sipols, A. J., Woods, S. C., Steil, G. M., and Porte, D., Jr. Evidence for entry of plasma insulin into cerebrospinal fluid through an intermediate compartment in dogs: quantitative aspects and implications for transport. *J Clin Invest, 88,* 1272–1281, 1991.

[45]. Kahn, C. R., Freychet, P., and Roth, J. Quantitative aspects of the insulin receptor interaction in liver plasma membranes. *J Biol Chem, 259,* 2259–2267, 1974.

[46]. Reed, D. J. and Woodbury, D. M. Kinetics of movement of iodide, sucrose, inulin and radioiodinated serum albumin in the central nervous system and cerebrospinal fluid of the rat. *J Physiol (Lond), 169,* 816–850, 1963.

[47]. Gilman, A. G., Rall, T. W., Nies, A. S., and Taylor, P., eds. *Goodman and Gilman's The Pharmacological Basis of Therapeutics,* 8th ed. Pergamon Press: New York, p. 21, 1990.

[48]. Baura, G. D., Foster, D. M., Kaiyala, K., Porte, D. Jr., Kahn, S. E., and Schwartz, M. W. Insulin transport from plasma into the central nervous system is inhibited by dexamethasone in dogs. *Diabetes, 45,* 86–90, 1996.

[49]. Baura, G. D., Vicini, P., Foster, D. M., Porte, D., Jr., and Schwartz, M. W. Central Nervous System Insulin Uptake Predicts the Tendency to Obesity In Dogs in Vivo. *Manuscript in preparation.*

[50]. Schwartz, M. W. University of Washington and Puget Sound Veterans Affairs Medical Center. Research data, 1989.

[51]. SAAM II User Manual. SAAM Institute: Seattle, WA. 1997.

Further Reading

Leptin: Campfield, L. A. and Smith, F. J. Overview: neurobiology of OB protein (leptin). *Proc Nutr Soc, 57,* 429–440, 1998.

Insulin: Schwartz, M. W., Figlewicz, D. P., Baskin, D. G., Woods, S. C., and Porte, D., Jr. Insulin in the brain: a hormonal regulator of energy balance. *Endocrine Rev, 13,* 387–414, 1992.

Obesity: Bjorntorp, P. and Brodoff, B. N., eds. *Obesity.* Lippincott: Philadelphia, 1992.

16.8 CENTRAL INSULIN KINETICS EXERCISES

Engineer H at Startup Company (a small biotech company) has been asked to investigate insulin transport in dogs across the blood–brain barrier. He decides to infuse various doses of insulin peripherally for 150 minutes, and to sample blood and cerebrospinal fluid. The dosage during the first 3 min of the infusion was five times greater than the dosage during the remaining 147 min of the infusion. During the infusion, euglycemia was maintained. If the rate constant representing transport across the BBB decreases as the dose increases, these results will provide evidence that saturation occurs. We recreate his analysis below.

16.8.1 Clinical Data

The plasma and insulin data in Figures 16.13 and 16.14 were obtained from the same dog by Engineer G during two studies. In the first low-dose study, the average plasma level during the infusion was 73 μU/ml. In the second high-dose study, the average plasma level during the infusion was 1914 μU/ml. The measurement errors in the insulin radioimmunoassays were assumed to be 10%.

These data were generously provided by Dr. Michael Schwartz of the University of Washington and VA Puget Sound Health Care System in Seattle, Washington [50].

This exercise will be conducted using a demo version of the program SAAMII [51]. The demo version is identical to the original version, except that file saving and printing functions have been disabled.

16.8.2 SAAMII Instructions

To download a demo version of SAAMII, go to the internet site *http://www.saam.com/download/download.htm*. Read the Test Drive License Agreement,

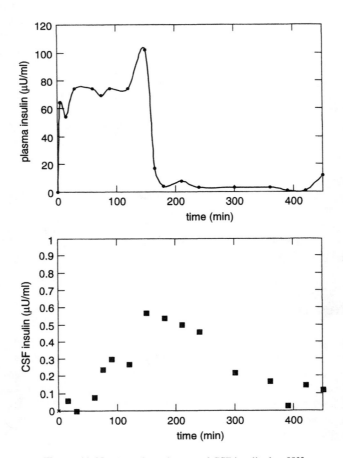

Figure 16.13 Low-dose plasma and CSF insulin data [50].

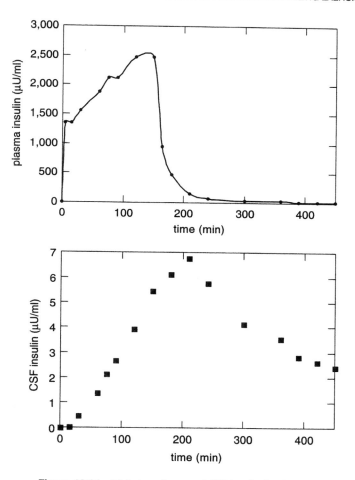

Figure 16.14 High-dose plasma and CSF insulin data [50].

and specify your acceptance. Enter your name, address, and e-mail address. Read the download directions, and double-click the *Download* tab. Run *s2demov12.exe* to uncompress program files. These files are saved in the *saam* directory.

Please move the *csf1.stu* and *csf2.stu* files from *ftp://ftp.ieee.org/uploads/press/baura* to the *saamIIvi.2demo* directory. From this directory, start the *Compwin.exe* program. Using the menu, *File/Open csf1.stu*. The compartmental model will appear. Click the *Experiment* tab in the Toolbox. The plasma insulin forcing function is represented by compartment 1; brain ISF insulin is represented by compartment 2. CSF insulin is represented by compartment 3 and is sampled. To view the data values, use the menu to *Show/Data*. When finished, close the *Data* window to enable other functions. To view the CSF insulin plot, use the menu to *Show/Plot*. The first time you use the plot function, you will have to select "s1:csf."

16.8.3 Exercises

1. Observe the initial values of the parameters by selecting *Show/Parameters* from the menu. Please note that $(k_{21} + k_{01})$ in the chapter text is represented by $k(0, 2)$ in the software, k_{02} is represented by $k(0,3)$, and $k_{1p}k_{21}$ is represented by $k(3, 2)$. Solve for Eqs. (16.4) and (16.5) by selecting *Compute/Solve* from the menu. How well do these initial values fit the data (check the plot)?
2. Iterate the fit by selecting *Compute/Fit* from the menu. (a) Has the fit improved? (b) Are the residuals random?
3. Evaluate the statistics by selecting *Show/Statistics* from the menu. (a) What is the coefficient of variation associated with each identified parameter? (b) What is the Akaike information criterion?
4. Repeat steps 1 to 3 for *csf2.stu*. Note that the upper limit of $k_{1p}k_{21}$ has been doubled to account for the increased insulin infusion. With the original initial parameters, SAAMII is unable to fit the data. (a) Try three other sets of other initial values by editing and saving the initial values in the *Show/Parameters* window. (b) Are you able to iterate?
5. (a) Is the sampling schedule a practical sampling schedule? (b) How does this affect parameter identification? (c) Is it possible to demonstrate saturation?

IV

SYSTEM THEORY IMPLEMENTATION

In Parts I through III of this textbook, we discussed various system theory technologies and their application to medical instruments. In Part IV, we discuss more practical implementation issues. Specifically, in Chapter 17 we describe how these techniques are implemented through discussions of data types, digital signal processing (DSP) processors, embedded systems, and the Food and Drug Administration medical device software requirements. These topics were chosen because they are not commonly reviewed in the standard biomedical engineering curriculum.

In Chapter 18, the history of system theory implementation in low-cost medical monitoring (~ $5000 per instrument) is summarized. Although digital signal processing traces its roots to the Cooley and Tukey fast fourier transform journal article published in 1965,* digital signal processing has only appeared in low-cost industrial monitors since 1995. Control systems have yet to be widely accepted in low-cost monitoring/diagnosis systems. It is shown that many of the recent leaps in instrumentation technology can be traced to the implementation of system theory, and that the time is ripe for more system theory advances.

---

*Cooley, J. W. and Tukey, J. W. An algorithm for the machine computation of complex Fourier Series. *Math Comp, 19,* 297–301, 1965.

17

ALGORITHM IMPLEMENTATION

In Parts I through III, we discussed the theory behind various system theory technologies and the application of these theories to medical instruments. In this chapter, we describe how these technologies are implemented through discussions of **data types, digital signal processors** (DSP processors), **embedded systems,** and the Food and Drug Administration (FDA) medical device software requirements. These topics were chosen because they are not commonly reviewed in the standard biomedical engineering curriculum.

Historically, to conserve processing time, power, and cost, system theory algorithms were coded using the data types of **integer** or **fixed-point** arithmetic. As costs decreased, **floating-point** arithmetic began to be utilized to obtain a wider range of the numbers represented. Due to further drops in cost, DSP processors have become a viable option in medical instrumentation design. These specialized microprocessors support repetitive, numerically intensive tasks. If present, the DSP processor is part of an embedded real-time system that contains the hardware and software that enable a medical instrument to perform its tasks. The software within an embedded system is scrutinized in great detail as part of the medical device approval process by FDA.

17.1 DATA TYPES

Data may be represented using the fixed-point, or integer format. Although floating-point is the representation with the largest range, it is also more expensive and slower than the fixed-point and integer formats.

17.1.1 Floating-Point Representation

A binary floating-point number, X_2, is represented as the product of two signed numbers, the **mantissa,** or fraction, M, and the **exponent,** E:

$$X_2 = M \cdot r^E \tag{17.1}$$

where r is the **radix**, or base, of the system. The mantissa determines the accuracy of the numbers, whereas the exponent determines the range of the numbers. Floating-point numbers with exponents in the range $0 < E < 255$ are said to be **normalized.** In this section, we discuss **single-precision** implementation, which requires 32 bits. Double-precision implementation requires 64 bits.

Using the normalized format for a 32 bit floating-point binary number, which agrees with the ANSI/IEEE 754–1985 Standard for binary floating-point arithmetic [1], bits 0 through 22 represent the mantissa, bits 23 through 30 represent the exponent, and bit 31 is the sign bit. When the sign bit is 0, the number is positive; when the sign bit is 1, the number is negative. The **radix point** lies between bits 22 and 23 to separate the mantissa and exponent fields (Figure 17.1).

Using this standard, the decimal equivalent, X, for a floating-point number $E \cdot M$ is calculated as

$$X = (-1)^s (1 + M) 2^{E-127} \tag{17.2}$$

where s = sign bit. Because the exponent is stored in excess 127 code, the allowable range of representable exponents is $-126 = 00000001_2$ through $+126 = 11111110_2$.

For example, the floating-point number $100001111.0110\ldots 0000_2$ is equal to the decimal value 2.65×10^{-34}. The exponent is $00001111_2 = 15$, and the mantissa is $0.0110\ldots_2 = \frac{1}{4} + \frac{1}{8} = 0.375$. Using Eq. (17.2) yields

$$X = (-1)^1 (1 + 0.375) 2^{15-127} = -1 \times 1.375 \times 2^{-112} = 2.65 \times 10^{-34} \tag{17.3}$$

17.1.2 Fixed-Point Representation

In contrast to floating-point, a binary fixed-point number is represented as a **two's complement** fraction. In this representation, the binary point is to the right of the most significant bit, which is also the sign bit. Each fixed-point number lies in the range from -1 to $(1 - 2^{-B})$, where B is the number of bits used to represent the number. Two's complement positive numbers are in the natural binary form. A negative number is formed from the corresponding positive number by complementing all the bits of the positive number and then adding 1 least significant bit. For example, the two's complement representation of -3 is obtained from $3 = 0011_2$ as $1100_2 + 0001_2 = 1101_2$.

In the common **Q15 format** that uses 16 bits, the most significant bit is the sign bit and the other 15 bits are fractional bits. Since a maximum of 2^B different numbers can be rep-

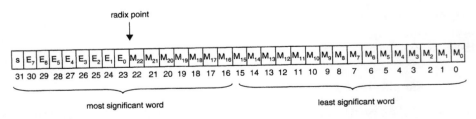

Figure 17.1 ANSI/IEEE 754-1985 Standard for binary floating-point arithmetic.

resented, with adjacent numbers separated by approximately 2^{-B}, the accuracy of fixed-point representation is $\pm 0.5 \times 2^{-B}$. For a decimal fraction consisting of d digits, its accuracy is $\pm 0.5 \times 10^{-d}$. To retain the same accuracy for the two representations, B must be chosen such that

$$0.5 \times 10^{-d} = 0.5 \times 2^{-B} \tag{17.4}$$

$$B = d \log_2 10 \cong 3.3d \text{ bits} \tag{17.5}$$

Using Eq. (17.5), the accuracy of 15 fractional bits is equal to the accuracy of 4.5 decimal digits.

17.1.3 Integer Representation

Integers are most commonly represented using two's complement representation, without using one bit explicitly to represent the sign. Each two's complement integer lies in the range from $-0.5\ 2^B$ to $(0.5\ 2^B - 1)$.

17.1.4 Implementation

The choice of data type depends on factors such as accuracy, execution time, cost, and power. In floating-point arithmetic, roundoff errors occur in both additions and multiplications, whereas in fixed-point and integer arithmetic, roundoff errors are only possible in multiplication. However, overflow occurs in integer and fixed-point arithmetic, but not as often in floating-point arithmetic because of its very wide **dynamic range** (ratio between largest and smallest numbers that can be represented).

Although floating-point arithmetic is preferred for its wider dynamic range, its operations require greater execution intervals. Floating-point processors also are more expensive and require more power. Cost is a constraint because physiologic monitors are often bundled with their medical disposables during an initial hospital sale, with the monitor essentially given for free. Using this marketing strategy, profit is made through continous disposables sales. Power is a constraint because monitors must often be portable and able to run on battery power for several hours.

For example, the commonly used Intel 80386 DX 33 MHz processor requires 61 ns for integer additions, 364–758 ns for integer multiplications, 375–906 ns for floating-point additions, and 594–1000 ns for floating-point multiplications. The fixed-point processor and floating-point math coprocessor use 5 V supply voltages. Typically, 150 mA is required for the processor, and 95 mA is required for the coprocessor [2, 3]. This processor was introduced by Intel in 1985. During this era, integer arithmetic was often utilized and included frequency-selective filters (see Chapter 1), based on powers of two that could be implemented as bit shifts.

17.2 DIGITAL SIGNAL PROCESSORS

As prices have dropped, the use of programmable digital signal processors has become a viable option in medical instrumentation design. A DSP processor is a specialized microprocessor that supports repetitive, numerically intensive tasks. Rather than the traditional

Von Neumann architecture that accesses a single bank of memory through a single set of buses, DSP processors use a variation of the **Harvard architecture,** in which two separate memory spaces are accessed by two separate buses (Figure 17.2). This architecture allows the processor to fetch an instruction while simultaneously fetching operands for the instruction or storing the result of the previous instruction in memory. The processor may use fixed- or floating-point arithmetic.

For example, a **multiply–accumulate operation** (MAC) is usually performed in a single instruction cycle. This fast operation is useful in algorithms involving a vector product, such as frequency selective filtering, correlation, or Fourier transforms. In addition, DSP processors generally provide enough bits in their **accumulator registers** (registers for intermediate and final results of arithmetic operations) to accommodate growth of the accumulated product without overflow.

17.2.1 Specialized Features

DSP processors commonly feature a specialized **addressing mode.** Dedicated **address generation units** allow arithmetic processing to proceed at maximum speed, and allow specification of multiple operands in one small instruction word. Because many algorithms involve repetitive computations, most processors provide special support for **efficient looping.** Most processors also incorporate one or more serial or parallel **input/output** (I/O) interfaces and specialized mechanisms such as **direct memory access** (data transfer without processor involvement) to allow low-cost, high-performance I/O. Finally, power consumption has been reduced by using **complementary metal-oxide silicon** (CMOS) technology which reduces the supply voltage from 5.0 to 3.3 V. Many processors incorporate power management features to provide control over the processor's master clock or over the processor parts that receive the clock signal. CMOS energy consumption is linearly proportional to clock frequency and increases with the number of active parts.

17.2.2 Pricing

DSP processor pricing is a function of circuit complexity, performance, and packaging. The representative unit prices of several processors purchased in quantities of 1000 parts are shown in Table 17.1. Each cited processor uses inexpensive plastic quad flat packaging. Note that the prices decreased dramatically between 1995 and 2000 as new architectures and higher clock speeds increased performance. Increased performance has been driven by communications applications such as cellular telephony and personal computer

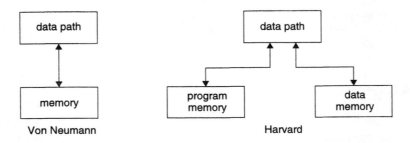

Figure 17.2 Data buses required for various architectures.

Table 17.1 Price of selected fixed-point DSP processors in quantities of 1000 parts with plastic quad flat packaging. 1995 pricing was obtained from [4]. 2000 pricing was obtained from [5]

Manufacturer	Processor	Instruction clock speed (MIPS)	Voltage (V)	Unit Price ($) June, 1995	January, 2000
Analog Devices	ADSP–21062	40	5	249.00	102.00
Motorola	DSP560002	40	5	38.20	15.90
Texas Instruments	TMS320C31	25	5	54.10	36.74

modems. By 1997, the world market for DSP processors had tripled in three years to $3 billion. Annual increases of 30–40% to some $14 billion by the end of 2002 have been predicted by Forward Concepts, a Tempe, Arizona marketing research firm [6].

17.2.3 Implementation

Processor selection requires consideration of processor speed, cost, ease of use, performance, time to market, and on-chip peripherals. DSP processors typically have irregular instruction sets and specialized features that complicate programming. Therefore, good development tools such as assemblers, linkers, simulators, in-circuit emulators, and debuggers dramatically affect the productivity of DSP software developers. The choice of floating- versus fixed-point arithmetic improves developer productivity, but increases chip cost and power consumption. Typical word sizes are 16 bits for fixed-point and 32 bits for floating-point.

17.3 EMBEDDED SYSTEMS

The DSP processor is an optionally utilized part of a real-time embedded system. An embedded system contains computer hardware, such as a general purpose microprocessor; computer software, such as system theory algorithms; and possibly additional mechanical or other parts that function to control the behavior of an **intelligent product.** Input for this control is obtained from the **man–machine interface** of the embedded system.

17.3.1 Programming Environment

Many people equate the development of an embedded system with the coding of software. However, before the software for an embedded system may be effectively designed, the programming environment should first be established. This environment contains three components: the computer hardware, the software development process, and the hardware interface.

First, the specification and prototype of computer hardware provides a platform for the software. The hardware includes the processor, **memory,** and **peripherals.** Memory devices function to store and retrieve data and code. The three types of memory devices are **random-access** (RAM), **read-only** (ROM), and **hybrid** devices. Hybrid devices enable the programmer to overwrite data in a ROM-like device. Peripherals coordinate the interaction with the outside world (**I/O = input/output**) or perform a specific hardware function.

388 ALGORITHM IMPLEMENTATION

Second, the establishment of a software development process provides a foundation for the coding. Embedded software development tools enable **source files** to be compiled into an **object file**, object files to be linked together to produce a single file called a **relocatable program**, and physical memory addresses to be assigned to the relative offsets with the relocatable program. The output of this process is an executable file that is ready to be run on the embedded system and tested using some type of software **debugger**.

Finally, the design of **device drivers** and selection of an **operating system** provide the interface between software and hardware. A device driver interacts directly with the peripheral **control** and **status registers**. In a good embedded system design, the device driver modules are the only pieces of software in the system that read and write that particular devices' registers directly. In this way, the state of the hardware is more accurately tracked and software changes that result from hardware changes are localized to the device driver. The operating system schedules the execution of and maintains information about the state of software tasks, enabling **multitasking**. Although it is possible to create an operating system, it is more desirable to purchase a commercial operating system, as it will have been extensively tested. The choice in commercial systems involves a compromise between functionality, performance, and price (including royalties).

For example, let us establish a programming environment for a variation of the classic Hello, World! example, as first described by Michael Barr [7]. We wish to code an executable program that prints the string "Hello, World" at 10 second intervals to a dumb terminal, and toggles a red LED at a rate of 10 Hz.

We choose the low-cost, high-speed Target 188EB board from Arcom Control Systems (Kansas City, MO) as the hardware for this project. This board contains:

1. An Intel 80188EB processor (25 MHz)
2. 128K RAM
3. 128K ROM
4. Two RS232-compatible serial ports
5. Three programmable timer/counters
6. A remote debugging adapter containing two additional RS232-compatible serial ports

Free development tools and utilities are included with this board that allow development of the software in C/C++ using Borland's C++ compiler (Inprise Corporation, Scotts Valley, CA). Additionally, a debug monitor preinstalled in the onboard ROM enables Borland's Turbo Debugger to find and fix bugs easily. Arcom also includes a library of hardware interface routines for manipulating the onboard hardware.

With this hardware and software in place, we write the following device driver modules: an LED driver, a timer driver, and a serial port driver. The serial port driver provides the interface to the dumb terminal. We choose the free ADEOS operating system (Netrino, Baltimore, MD) for this application. This programming environment is illustrated in Figure 17.3 [7].

17.3.2 Embedded System Characteristics

Once a programming environment has been established, the embedded system software can be coded. The development of embedded system software is influenced by at least eight characteristics:

1. Design for mass production. This goal requires emphasis on efficient memory and processor utilization.
2. Static structure. The computer system is embedded in an intelligent product of given functionality and rigid structure. This known a priori static environment can be analyzed during the design to simplify the software, increase robustness, and improve the efficiency of the embedded system.
3. Man–machine interface. The interface must be easy to operate and require minimal training. This interface is mandated by the FDA.
4. Minimization of electrical or mechanical subsystem. This minimization reduces the manufacturing and increases reliability.
5. Functionality determined by software in read-only memory. The quality standards for this software are high, as this software may not be modified after release, or may be intermittently upgraded if an **erasable programmable read-only memory** (EPROM) or **flash memory** has been used.
6. Maintenance strategy. An excellent diagnostic interface and a self-evident maintenance strategy are important for maintenance in the field.
7. Ability to communicate. Standards are evolving for communication between devices in the hospital environment [8].
8. Safety and efficacy. For medical devices, the embedded system must minimize risk to the patient.

17.4 FDA REVIEW OF MEDICAL DEVICE SOFTWARE

The software within the embedded system is scrutinized in great detail as part of the review process by FDA. The **premarket submissions** for medical devices to FDA are categorized as two types. If a new type of device will be marketed, a **premarket application** (PMA) is submitted. If the device is "substantially equivalent" to a legally marketed device that is not subject to a PMA, a **premarket notification** called a **510(k)** is submitted. A 510(k) typically takes 3 months to receive approval, whereas the time to approval for a PMA is variable and typically much longer than 3 months. For example, the Gensia combination drug/product closed-loop system, combining a new drug with a new device (see Chapter 14) was approved after 44 months.

We will restrict this discussion to the general principles of software validation for a premarket notification 510(k). As defined by FDA, design validation is the process of

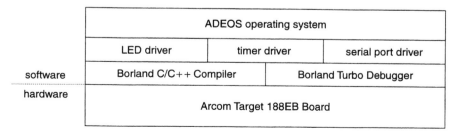

Figure 17.3 Programming environment for software example.

"establishing by objective evidence that device specifications conform with user needs and intended use(s)" [9]. The detailed requirements for the full 510(k) may be found in public FDA documentation [10]. The 13 components required for software documentation are:

1. Level of concern
2. Software description
3. Device features controlled by software
4. Device hazard analysis
5. Software requirements specification
6. Architecture design chart
7. Design specification
8. Traceability analysis
9. Development
10. Validation, verification, and testing
11. Revision level history
12. Unresolved anomalies
13. Release version number

17.4.1 Level of Concern

As defined by FDA, **level of concern** refers to "an estimate of the severity of injury that a device could permit or inflict (directly or indirectly) on a patient or operator as a result of latent failures, design flaws, or using the medical device software" [9]. The three levels of concern are major, moderate, and minor levels. A major level of concern indicates that device failure or latent flaws could result in death or serious injury of the patient and/or operator. A moderate level of concern indicates that device failure or latent flaws could result in nonserious injury of the patient and/or operator. A minor level of concern indicates that device failure or latent flaws would not be expected to result in any injury of the patient and/or operator. The level of concern may be determined through the flow chart in Figure 17.4. The identified level of concern for a medical device directly determines the amount of software documentation required.

17.4.2 Software Description

The software description contains a comprehensive overview of the device features that are controlled by software. Additionally, this description contains the intended operational environment, including the programming language, hardware platform, operating system, and use of off-the-shelf components.

17.4.3 Device Features Controlled by Software

An overview of device features controlled by software describes the role of the software in the device, including functionality, user interface, and user control. It also contains a description of redundant subsystems that act to improve device safety.

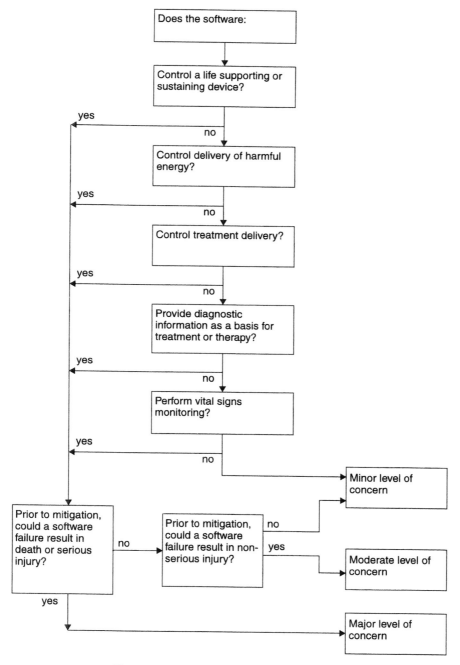

Figure 17.4 Level of concern flow chart. From [9].

17.4.4 Device Hazard Analysis

A device hazard analysis takes into account all device hazards associated with its intended use, hardware, and software. This information is typically presented in tabular form, and includes the following items: the hazardous event, level of concern of the hazard, cause(s) of the hazard, method of control, corrective measures taken, testing to demonstrate correct implementation.

17.4.5 Software Requirements Specification

The software requirements specification typically includes functional, performance, interface, design, and developmental requirements. Typical requirements include hardware requirements, programming language and program size, interface requirements, and software performance and functional requirements. For a minor concern device, it is usually sufficient to provide only functional requirements.

17.4.6 Architecture Design Chart

The architecture design chart depicts the partitioning of the software into its functional subsystems. For moderate and major concern devices, the chart should be more detailed, including a description of the role each functional module plays in fulfilling the software requirements.

17.4.7 Design Specification

The design specification is a description of the function and implementation of the software. It should include documentation of the acceptance of the software requirements specification, development standards and programming standards, hazard analysis, systems documentation, hardware used, parameters to be measured or recorded, logic, data structures and data flow diagrams, definition of variables, error and alarm messages, supporting software, communications links, and security measures. This section is not required for devices of minor concern.

17.4.8 Traceability Analysis

A traceability analysis or matrix links requirements, design specifications, hazards, and validation. It acts as a map for providing the links necessary to locate information. This section is not required for devices of minor concern.

17.4.9 Development

The software development life cycle plans are summarized in 2–4 pages in the submissions of moderate or major concern devices. Additionally, for major concern devices, an annotated list of the control/baseline documents generated during the software development process and configuration management and maintenance plan are included. This section is not required for devices of minor concern.

17.4.10 Validation, Verification, and Testing

As defined by FDA, software **verification** involves a systematic application of various analyses, evaluations, assurances, and testing of the software and its supporting documentation at each stage of the software development process, to assure that all the requirements specified for that stage have been fulfilled. Software **validation** uses similar analytical techniques and assures that the finished device is appropriate for its intended use and will be reliable and safe. For a minor level device, a software functional test plan with pass/fail criteria, data, and an analysis of the results is sufficient for submission. For a moderate level device, the verification activities and system level test protocols are also required. Further, for a major level device, the unit and integration test protocols are also required.

17.4.11 Revision Level History

The revision history log documents all major changes to the software during its development cycle. This section is not required for devices of minor concern.

17.4.12 Unresolved Anomalies

A list of all unresolved software anomalies includes the following items: the problem, the impact on performance, and any plans for correcting the problem. This section is not required for devices of minor concern.

17.4.13 Release Version Number

The release version number and date are based on the software that will be included in the marketed device.

17.4.14 System Theory Considerations

Within its software guidance documentation, FDA specifically acknowledges the special considerations inherent in using artificial neural networks and closed-loop control in medical device software [9]. A neural network should be tested sufficiently, separately from the training stage, to insure that the "performance remains as expected and relevant data is extracted appropriately." In the case of closed-loop control, safety and testability should be maximized to minimize risk to the patient and/or operator.

17.5 SUMMARY

In this chapter, we have discussed how the various system theory technologies of Chapters 1 to 16 are implemented in medical instruments. Historically, to conserve processing time, power, and cost, system theory algorithms were coded using the data types of integer or fixed-point arithmetic. As costs decreased, floating-point arithmetic began to be utilized to obtain a wider range of the numbers represented. Due to further drops in cost, DSP processors have become a viable option in medical instrumentation design. These specialized microprocessors support repetitive, numerically intensive tasks. Rather than the traditional Von Neumann architecture that accesses a single bank of memory through a single set of buses, DSP processors use a variation of the Harvard architecture, in which two separate memory spaces are accessed by two separate buses. This architecture allows the processor to fetch an instruction while simultaneously fetching operands for the instruction or storing the result of the previous instruction in memory. The processor may use fixed- or floating-point arithmetic.

The DSP processor is an optionally utilized part of an embedded real-time system that contains the hardware and software that enable a medical instrument to perform its tasks. An embedded system contains computer hardware, such as a general purpose microprocessor; computer software, such as system theory algorithms; and possibly additional mechanical or other parts that function to control the behavior of an intelligent product. Input for this control is obtained from the man–machine interface of the embedded system. Before the software for an embedded system may be effectively designed, the programming environment (computer hardware, software development process, hardware interface) should be first established.

The software within an embedded system is scrutinized in great detail as part of the medical device approval process by FDA. The amount of software documentation required is determined by the identified level of concern. The 13 components required for software documentation are: level of concern; software description; device features controlled by software; device hazard analysis; software requirements specification; architecture design chart; design specification; traceability analysis; development; validation, verification and testing; revision level history; unresolved anomalies; and release version number.

17.6 REFERENCES

[1]. IEEE. *IEEE Standard for Binary Floating-Point Arithmetic.* ANSI/IEEE Standard 754-1985.

[2]. *Intel 386™ DX Microprocessor 32-bit CHMOS Microprocessor with Integrated Memory Management.* Rev. 11. Intel Corporation: Santa Clara, CA, 1995.

[3]. *Intel 387 DX Math Co-Processor.* Rev. 5. Intel Corporation: Santa Clara, CA, 1995.

[4]. Lapsley, P., Bier, J., Shoham, A., and Lee, E. A. *DSP Processor Fundamentals: Architectures and Features.* IEEE Press: New York, 1997.

[5]. Avnet Electronics Marketing website. January 11, 2000. www.em.avnet.com.

[6]. Stevens, J. DSPs in communications. *IEEE Spectrum, 35,* 39–46, 1998.

[7]. Barr, M. *Programming Embedded Systems in C and C++.* O'Reilly: Sebastopol, CA, 1999.

[8]. Kopetz, H. *Real-Time Systems: Design Principles for Distributed Embedded Applications.* Kluwer: Boston, 1997.

[9]. FDA Center for Devices and Radiological Health Office of Device Evaluation. *Guidance for FDA Reviewers and Industry: Guidance for the Content of Premarket Submissions for Software Contained in Medical Devices.* CDRH: Washington, DC, 1998.

[10]. Rice, L. L. and Lowery, A. *Premarket Notification 510(k): Regulatory Requirements for Medical Devices.* HHS Publication FDA 95-4158. CDHR: Washington, DC, 1995.

18

THE NEED FOR MORE SYSTEM THEORY IN LOW-COST MEDICAL MONITORING

In the challenging environment of mergers and acquisitions in the medical device industry, it is difficult to innovate new products. Studies have shown that the threat of future downsizing negatively impacts the willingness of employees to take risks and the degree to which performance is motivated by a desire to do a job well. Despite this, technology innovation in low-cost monitoring instrumentation has continued during this period by incorporating system theory. System theory is the transdisciplinary study of the synthesis and design of systems, and the analysis of their performance. It is possible to continue this level of innovation by hiring recent biomedical and electrical engineering graduates who are trained in system theory.

18.1 FUTURE EMPLOYMENT FOR BIOMEDICAL ENGINEERING GRADUATE STUDENTS

The future employment for biomedical engineering graduate students is uncertain. According to the National Science Foundation (NSF), the percentage of PhDs who go on to do postdoctoral work in science and engineering rose from 25% for the pre–1965 cohort to 41% of the 1992 to 1994 group. Additionally, the median time served increased from 20 to 29 months, a 45% increase [1] that may be related to the difficulty of obtaining a full-time job. Surveys by university postdoctoral associations determined that future job placement and postdoctoral salary levels outweigh health and safety as major concerns [2].

Based on NSF surveys, in 1973, 52% of scientists and engineers with PhDs from American universities were engaged in basic research or teaching activities. However, in

1991, only 37% were in such positions [3]. It is possible that the newly created (September, 2000) National Institute of Biomedical Imaging and Engineering will increase academic hiring of new Biomedical Engineering PhDs, but these results remain to be seen. Although many PhDs are driven to alternative employment in business and industry, only 1% of those answering a recent NSF survey were given advice by their mentors aimed mainly at obtaining posts in industry, government, or the nonprofit sector [1].

18.2 THE LOSS OF INNOVATION IN THE MEDICAL DEVICE INDUSTRY

One industry in which graduating biomedical and electrical engineers may seek employment is the medical device industry. The medical device industry is relatively small, consisting of approximately 6000 companies covering 50 clinical specialties. Only about 100 of these companies produce annual revenues over $100 million. Approximately 72% of these manufacturers employ fewer than 50 people each [4]. According to a 1998 survey by the U.S. Department of Labor, 15,094 engineers were employed in this industry in 1998, and employment is forecasted to increase to 20,131 engineers by 2008 [5].

Medical instrumentation manufacturers are a subset of the medical device manufacturers, distinguished by manufacture of devices that incorporate embedded or PC-based systems for use in diagnostic and patient monitoring applications. For example, CardioDynamics manufactures an impedance cardiography module that measures noninvasive cardiac output within a GE Medical Systems Solar modular monitor. Both the module and the monitor are embedded systems [6]. Alternatively, Medical Graphics Corporation manufactures a spirometer for measurement of lung volumes. The spirometer consists of Windows-based software, a PC card, and a pneumotach for conversion of airflow to a differential pressure [7]. The innovations of medical device manufacturers have changed the practice of medicine. As estimated by the National Institutes of Health (NIH), about 25 million Americans benefit from devices such as heart valves, implantable pumps, pacemakers, defibrillators, and stents [4].

18.2.1 Cost Containment

The medical device industry has been greatly affected by the consolidations among providers (hospitals) and payers (insurance companies), and changes in Medicare, due to increased health care costs. In the early 1990s, the United States became the first country in which rising health care costs exceeded 10% of the gross national product. In response to these escalating costs, hospital and purchasing groups formed to take advantage of economies of scale. By 1996, three organizations—the Columbia Health Care System, the Voluntary Hospital Association of America, and Premier—possessed the purchasing power of over 70% of the hospital beds in the United States [8]. During the same period, hospitals began to compete for managed care contracts with health maintenance organizations (HMOs) and preferred provider organizations (PPOs), which cut costs while presumably improving patient outcomes. From 1991 to 1996, HMO membership rose from 40 to 55 million. Simultaneously, in-patient days decreased from approximately 2900 per 1000 population to 900 per 1000 population [9, 10]. In 1997, Medicare reform cut Medicare spending growth by $115 billion for the next five years and altered the payment structures of Medicare itself. Medicare accounts for some 20% of all U.S. personal health expenditure, and represents the nation's single largest purchaser of health care services [11].

18.2.2 Mergers and Acquisitions

In response to these consolidations in Medicare providers and reduced payment structures, the medical device industry began its own consolidation through mergers and acquisitions. Consolidation allowed device companies to lower costs and offer providers "one-stop shopping" through broadened product lines, and price concessions through increased volume. From 1991 to 1994, the average number of mergers and acquisitions per year in the device industry was 50, with an average annual value of $1.8 billion. By 1995, the number of acquisitions rose to 88, with a total value of $66.4 billion [12].

For example, the former startup company Nellcor, which popularized the use of pulse oximetry, went public in 1987 with stock worth $250 million [13]. Nellcor then purchased ventilator manufacturer Puritan-Bennett for $457 million in 1995, in order to extend its range of product offerings in the respiratory care market [14]. Subsequently, imaging and specialty pharmaceuticals manufacturer Mallinckrodt purchased Nellcor Puritan Bennett for $1.9 billion in 1997, in order to expand Mallinckrodt's higher-margin critical care products operation. Although Mallinckrodt's annual sales growth in its respiratory group was 7–8%, its imaging products business declined more than expected, resulting in a reduced share price in late 1998 [15]. In late 2000, Mallinckrodt was acquired by manufacturing conglomerate Tyco International Ltd. for $3.24 billion. Tyco, whose businesses range from medical supplies to fire detection equipment, assumed nearly $1 billion in Mallinckrodt debt. With this acquisition, Tyco became a leader in bulk analgesic pharmaceuticals and a leader in the respiratory care business [16].

18.2.3 Merger and Acquisition Byproducts

Although device company consolidations potentially result in more solutions to customer needs, efficiencies of scale, and more global distribution channels, negative effects also occur. If a company went into debt to facilitate the consolidation, expendable costs such as personnel and research and development (R&D) are contained. For example, when Eli Lilly sold IVAC Corporation, the original inventor of controllable drug infusion and the digital thermometer, to River Medical in 1994, $178.5 million of the $185 million purchase price (96%) for IVAC, was borrowed. Within a year, 30% of the work force was laid off, there was a "sharp curtailment of product development," and IVAC's large corporate campus was sold [17]. Layoffs may also result from duplication of services between the two parent companies. For instance, 18 months after IVAC was acquired by River Medical, the combined company was acquired by IMED Corporation, another drug infusion company. Within one year, 14% of the combined 1600 employees were laid off [18, 19].

This downsizing negatively impacts innovation. According to a recent study by Wharton researchers Dougherty and Bowman, downsizing seems to interfere with the informal relationships that innovators utilize to win support and resources for new products. This support and these resources normally assist in meshing innovative activities with those of the firm as a whole [20]. More recently, Bommer and Jalajas surveyed 150 R&D engineers from 15 major corporations. They determined that the threat of future downsizing negatively impacted the willingness to take risks, to make suggestions, and the degree to which performance was motivated by a desire to do the job well [21]. The effect of downsizing on innovation provides a rationale for the results of a study by Massachusetts consultancy Mitchell Consultants. Mitchell determined that although shares of downsizing firms outperform the stock market during the six months after a restructuring is announced, they lag the market after three years [22].

18.3 LOW-COST MEDICAL MONITORING AND SYSTEM THEORY

Ironically, technology innovations occurred in medical instrumentation companies during this major period of mergers and acquisition. These innovations occurred in low-cost medical monitoring instruments (< $5000) through utilization of system theory techniques. The primary techniques that are discussed throughout this textbook are digital signal processing and control theory.

18.3.1 Digital Signal Processing

Excluding frequency-selective filters, digital signal processing (DSP) techniques have been utilized in commercial medical instrumentation such as electrocardiographs and pacemakers since the 1980s. More recently, the IVAC Signature Edition® Pump, which is now manufactured by ALARIS Medical Systems, uses a **pseudorandom binary sequence** to encode its flow waveforms (Figure 18.1). The resulting pressure waveforms, having been transmitted through the resistive medium of the patient's vein, catheter, and infusion tubing, are then decoded by cross-correlation [23, 24]. Due to this coding method, which is essentially a bandpass filter with cutoff frequencies selected by the system, resistance waveforms are minimally affected by motion artifact and blood pressure pulsations. Accurate resistance monitoring at low flow rates, such as 1 ml/hr, enables much faster detection of downstream occlusions than does pressure monitoring. For patients such as neonates, who require such low flow rates, the time during which critical medication is not administered and the risk of a bolus infusion are minimized.

The most commercially publicized digital signal processing application is the Massimo SET algorithm. This algorithm, which received FDA approval in 1998, is incorporated into dozens of patient monitors (Figure 18.2), and minimizes erroneous readings in **pulse oximeters** due to motion or poor peripheral perfusion. During pulse oximetry, tissue oxygenation is estimated by monitoring oxygen saturation in arterial blood (S_pO_2). Ironically, the basis of the noise compensation is an old DSP technology that traces its roots to Bernard Widrow's work in 1960: **adaptive filtering** [25]. The novelty of the SET algorithm is that the reference noise source required for a variation of adaptive filtering is derived from two noisy signals with different wavelengths [26]. In a validation study of 10 healthy volunteers using Nellcor N-200, Nellcor N-3000, and Masimo prototype pulse oximeters, the subjects were subjected to various concentrations of oxygen and motion. Including periods of **dropout** when the oximeter provided no display data, the Masimo oximeter possessed the highest sensitivity and specificity [27].

18.3 LOW-COST MEDICAL MONITORING AND SYSTEM THEORY

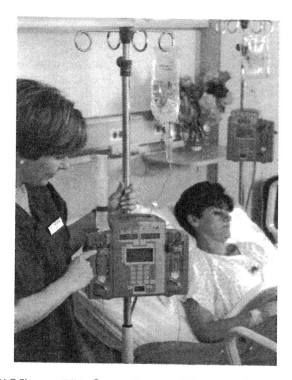

Figure 18.1 IVAC Signature Edition® pump. Courtesy of ALARIS Medical Systems, San Diego, CA.

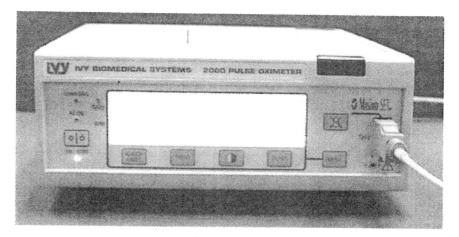

Figure 18.2 Ivy 2000 Pulse Oximeter with Masimo SET adaptive noise cancellation algorithm. Courtesy of Masimo Corporation, Irvine, CA.

18.3.2 Control

Strictly speaking, closed-loop drug delivery systems do not meet the criterion of low cost. However, significant instruments are included for completeness in discussing system theory applications.

The first commercial closed-loop drug delivery system was marketed by Hook Lane Nyetimber in England in 1982. This instrument controlled the infusion of oxytocin for initiation of labor using intrauterine pressure, measured every 15 minutes, as a feedback signal. The control algorithm was a **proportional–integral–derivative controller.** This device was able to initiate labor with 20% of the total dose of oxytocin when compared with manual intervention [28].

More recently, Gensia Automedics developed the GenESA system for **pharmacologic stress testing** that was based on proportional–intergral control of their drug arbutamine (Figure 18.3). In 1997, it was predicted that the pharmacologic stress test market had the potential to achieve annual sales of "$100 million in three to five years," based on the percentage of cardiac patients who are unable to take a conventional exercise stress test for diagnosis of coronary heart disease [30]. In a multicenter study of 210 patients with symptoms and angiographic evidence of coronary artery disease, the patients underwent both GenESA and exercise stress tests in randomly assigned order, with 20 hours to 14 days between tests. The sensitivity for detecting ischemia by either **angina** (severe chest pain) or a change in the ECG **ST segment** (≥ 0.1 mV horizontal or downsloping depression between the S and T waves) was 84% using arbutamine versus 75% using exercise ($p \leq 0.014$) [31].

These control and DSP applications, and others that the author has investigated, are described in detail in the even-numbered application chapters of this textbook.

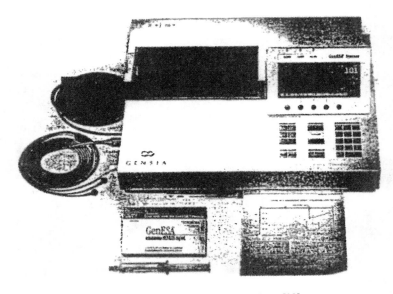

Figure 18.3 GenESA System. From [29].

18.4 ADDRESSING THE NEED FOR INNOVATION IN A COST-CONSCIOUS ENVIRONMENT

In this section, some personal recommendations are given, based on the observations described above.

18.4.1 Increased Information and Guidance During Graduate School

During graduate school, students need better guidance in understanding the variety of possible career options. Research can be conducted in other environments besides academia. According to the quadrant model of scientific research, first postulated by Donald Stokes, research may be categorized by quest for fundamental understanding and considerations of use. The four combinations of these two categories result in four quadrants (Figure 18.4). Pure basic research, which is conducted in academia, involves a search for mechanism but not utility. Stokes cites Niels Bohr's quest of a model atomic structure as an example of pure basic research. In contrast, user-inspired basic research, which is conducted in industry, searches for both mechanism and utility. Here, Stokes cites Louis Pasteur's discoveries of microbiological processes that let to the "pasteurization" of milk and immunization of patients from disease. Certainly, pure applied research is also conducted in industry, as exemplified by Thomas Edison's experiments towards commercially prof-

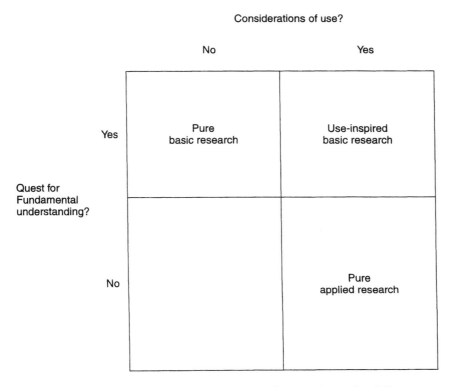

Figure 18.4 Stokes' Quadrant Model of Scientific Research. Based on [32].

itable electric lighting without deeper scientific understanding. However, wildly inefficient scavenger hunts for solutions can hardly be justified economically in today's cost-conscious environment. In the fourth quadrant, research is inspired by particular phenomena without a search for mechanism or utility [32].

I feel strongly that I should see my research benefiting patients during my lifetime. Similarly, Floyd Bloom, former editor of *Science,* stated that "I want to know more about the brain in my lifetime, and doing it the old-fashioned way isn't going to get us there. I think the private company's ability to discover will be far greater than anything I can do here (at Scripps) with my own two hands" [33].

While working at AT&T Bell Laboratories, I observed first hand that industrial research may equal or even surpass academic research. This is not common in the medical device industry, but it is possible. As more researchers with solid academic training enter this industry, the research standards will increase. Unfortunately, foundational medical device issues are not taught in graduate school and are not presented in academic journals or conferences. Although it is common knowledge in industry, students never learn that four of the well-funded topics of patient monitoring are continuous, noninvasive monitoring of cardiac output, blood pressure, glucose, and depth of anesthesia. Further, in many of his editorial columns, Panos Papamichalis, the 2000 President of the IEEE Signal Processing Society (SPS), requested that DSPers who work on industrial applications write articles or present at conferences, as this "would create significant interest and a better level of understanding of the technology" [34]. The genesis of this textbook was a conversation I had with an academic friend at the 1996 International Conference of Acoustics, Speech, and Signal Processing, asking why no applications were presented. He replied that no one has access to real data. This textbook is an attempt to provide a foundation for system theory within medical instrumentation research.

18.4.2 Increased System Theory Awareness in Medical Instrumentation

With simple exceptions such as frequency-selective filters and linear control systems, system theory has primarily been taught at the graduate level [35]. Since the majority of engineers employed in medical instrumentation possess only bachelor's degrees, they are unfamiliar with the utility of system theory. I cannot count the number of times that fellow engineers assumed I design Butterworth filters or empirically derived algorithms. While Masimo demonstrated that DSP in their pulse oximeter can be marketed as a technical advantage, it is unfortunate that the basis of their technology, which was published in 1960, only became well recognized in this industry in the 1990s. The Signature Edition® pump and Edwards Life Sciences Vigilance monitor, which also utilizes the pseudorandom binary sequence [36], were never marketed as possessing superior DSP.

Now that Masimo has made our marketing departments aware of DSP, marketing and engineering managers are becoming more open to incorporation of DSP and control technologies. However, no easy-reference guide with DSP examples exists. A search of the abstracts of articles published in the last 10 years by an industry publication, *Medical Device and Diagnostic Industry,* revealed that system theory has never been profiled, whereas other technologies such as virtual reality, implantable chips, and blood substitutes received much attention. This textbook will provide one reference source, but others need to be written. Further, commercial instruments that do possess DSP or control theory should tout these technologies as marketing features.

18.4.3 Smart R&D

Biomedical and electrical engineers who are finishing their MS and PhD degrees and possess a working knowledge of system theory should be recruited to work in medical instrumentation. As David and Jim Matheson observed in their Harvard Business School Press textbook, *The Smart Organization: Creating Value Through Strategic R&D*, "creativity involves nothing more than looking at a problem with fresh eyes, seeing in it something others have missed, and refusing to accept the apparent solution" [37]. While engineers who have worked in downsized environments are probably unwilling to take R&D risks in new areas, new graduates who are trained in system theory would naturally gravitate toward system theory. As demonstrated above, many of the problems present in physiologic monitoring signals can be solved efficiently with standard DSP and control research techniques. The novelty of these approaches lies in matching the optimum technique with a particular application. A recent article on the state of U.S. healthcare in the *Harvard Business Review* advised our industry to "invest less money in high-end, complex technologies and more in technologies that simplify complex problems" [38].

Ideally, the marketing department provides the knowledge of particular customer demands that, when met, can be translated into competitive advantages. Researchers need to work with marketing to focus their resources on the most pressing customer needs. For example, at CardioDynamics, marketing creates the priority list for research. As the Director of Research and Development, I discuss the most logical series of steps to address these needs with the Director of Marketing. These discussions lead to a revised priority list. Similarly, Ashok Ganguly, former chairman of Unilever Company in India, wrote in his textbook, *Business-Driven Research and Development: Managing Knowledge to Create Wealth*, that "the choice of a manageable number of priorities, and the assessment of risks and rewards, probably are the most important defining steps which distinguish a consistently successful corporation. In more philosophical terms, *less is more* (sic) probably provides an ideal holistic definition of such management style and practice" [39]. As illustrated throughout this textbook, innovative technologies by themselves cannot guarantee successful products.

In summary, it is difficult to innovate new products in this challenging environment of mergers and acquisitions in the medical instrumentation industry. Despite this, continued innovation is achievable through implementation of system theory. More recent medical instrumentation innovations are based on system theory. By hiring new biomedical and electrical engineers trained in system theory, further innovation is possible.

18.5 REFERENCES

[1]. Schmidt, K. Will the job market ever get better? *Science, 285*, 1517–1519, 1999.

[2]. Ferber, D. Getting to the front of the bus. *Science, 285*, 1514–1517, 1999.

[3]. National Academy of Sciences. *Reshaping the Graduate Education of Scientists and Engineers*. National Academy Press: Washington, DC, 1995.

[4]. Marwick, C. Device manufacturers consider constraints. *JAMA, 283*, 1410–1412, 2000.

[5]. National Industry–Occupation Employment Matrix. U.S. Department of Labor Bureau of Labor Statistics. *stats.bls.gov/empoils.htm*.

[6]. CardioDynamics International Corporation website. August 11, 2001 update. www.cdic.com.

[7]. Medical Graphics Corporation website. July 31, 2001 update. www.medgraph.com.
[8]. Castelini, C. Today's health-care industry focuses on cost containment. *MD&DI*, Sept. 56, 2000.
[9]. Lewis, T. S. Managed care—Evolution or revolution? *Med Dev Tech*, 7, 24–26, 1996.
[10]. Kordash, J. Can the managed care revolution in the US prosper in Europe? In *Proceedings of Managed Care Conference*. London, UK. June, 1996.
[11]. Mannen, T. Voyaging into the new world of Medicare reform. *MD&DI*, Nov. 30, 1997.
[12]. Graf, A. J. Challenges to competitiveness in a changing medical device market. *MD&DI*, Jan. 50, 1997.
[13]. Nesheim, J. L. *High Tech Start Up*. John L. Neshiem, Saratoga, CA, 1997.
[14]. Fisher, L. M. Mallinckrodt to buy maker of medical gear for $1.8 billion. *New York Times*, *146*, D1, D5, July 24, 1997.
[15]. Miller, J. P. "New" Mallinckrodt finds makeovers aren't miracles: After acquisition and good profit and stock-price news, doubters remain. *Wall Street Journal*, *131*, B4, June 2, 1999.
[16]. Tyco buying Mallinckrodt for $3.2 billion. *New York Times*, *4446*, C2, June 29, 2000.
[17]. Rose, C. D. IVAC says cuts were needed to compete. *San Diego Union Tribune*, *2196*, I1, February 4, 1996.
[18]. Kupper, T. Alaris share jump despite wider losses. *San Diego Union Tribune*, *2198*, C2, August 14, 1997.
[19]. Advanced Medical Inc. work force to be cut 14% following IVAC purchase. *Wall Street Journal*, *392*, B4, December 11, 1996.
[20]. Dougherty, D. and Bowman, E. H. The effects of organizational downsizing on product innovation. *Calif Mgt Rev*, *37*, 28–45, 1995.
[21]. Bommer, M. and Jalajas, D. S. The threat of organizational downsizing on the innovative propensity of R&D professionals. *R&D Mgt*, *29*, 26–34, 1999.
[22]. Unthinking shrinking: Research shows that downsizing damages information relationships in a company that are important to innovation. *Economist*, *336*, 70, 1995.
[23]. Voss, G. I., Butterfield, R. D., Baura, G. D., and Barnes, C. W. *Fluid flow impedance monitoring system*. U.S. Patent 5,609,576. Mar. 11, 1997.
[24]. Butterfield, R. D. and Farquhar, A. *Fluid flow resistance monitoring system*. U.S. Patent 5,803,917. Sept. 8, 1998.
[25]. Widrow, B. and Hoff, M., Jr. Adaptive switching circuits. *IRE WESCON Conv Rec, Part 4*, 96–104, 1960.
[26]. Diab, M. K., Kiani-Azarbayjany, E., Elfadel, I. M., McCarthy, R. J., Weber, W. M., and Smith, R. A. Signal processing apparatus. *U.S. patent 5,632,272*. May 27, 1997.
[27]. Barker, S. J., and Shah, N. K. The effects of motion on the performance of pulse oximeters in volunteers (revised publication). *Anesth*, *86*, 101–108, 1997.
[28]. Modular labour-management system. *J Med Eng Technol*, *6*, 89–92, 1982.
[29]. Gensia, Inc., GenESA System. *FDA Application Number P940001*. September 12, 1997.
[30]. FDA allows use of device in diagnosing heart disease. *Wall Street Journal*, *401*, B4, September 15, 1997.
[31]. Dennis, C. A., Pool, P. E., Perrins, E. J., Mohiuddin, S. M., Sklar, J., Kostuk, W. J., Muller, D. W. M., and Starling, M. R. Stress testing with closed-loop arbutamine as an alternative to exercise. *J Am Coll Cardiol*, *26*, 1151–1158, 1995.
[32]. Stokes, D. E. *Pasteur's Quadrant: Basic Science and Technological Innovation*. Brookings Institution Press: Washington, DC, 1997.
[33]. Hammond, H. K. Entrepreneurial scientists overcome conflict issues. *San Diego Union Tribune*, *9*, H6, October 29, 2000.

[34]. Papamichalis, P. E. Advances in "Techno-Speak" or Shrinking the Digital Divide. *IEEE Sig Proc Mag, 17,* 6–7, 2000.

[35]. McClellan, J. H., Schafer, R. W., and Yoder, M. A. A changing role for DSP education. *IEEE Sig Proc Mag, 15,* 17–18, 1998.

[36]. McKown, R., Yelderman, M., and Quinn, M. Method and apparatus for continuously measuring volumetric flow. *U.S. Patent 5,146,414.* Sept. 8, 1992.

[37]. Matheson, D. and Matheson, J. *The Smart Organization: Creating Value Through Strategic R&D.* Harvard Business School Press: Boston, 1998, p. 41.

[38]. Christensen, C. M., Bohmer, R., and Kenagy, J. Will disruptive innovations cure health care? *Harv Bus Rev, 78,* 102–112, 2000.

[39]. Ganguly, A. *Business-Driven Research and Development: Managing Knowledge to Create Wealth.* Purdue University Press: West Lafayette, IN, p. 27, 1999.

GLOSSARY

α-amino acid—protein component containing a primary carbon ion, attached to an amide group, carboxyl group, hydrogen ion, and side chain

absolutely summable—property of impulse response whereby summation of all impulse response values is finite

absorbance—also known as optical density, the product of the extinction coefficient, concentration, and path length

absorption—the rate at which a drug leaves its site of administration

accumulator register—register for intermediate and final results of arithmetic operations

action potential—cell membrane potential change that occurs during nerve or cardiac impulse propagation

activation function—function that determines the output of a modeled neuron from a weighted sum of inputs, typically a hyperbolic tangent

acute—referring to short-lived

adaptive comb filter—filter with adjustable structure in which specified periodic frequencies are minimized

adaptive filter—filter with a structure that is adjustable in such a way that its performance improves through contact with its environment

adaptive linear neuron (ADALINE)—single-neuron model developed by Widrow and used for adaptive filtering. A multiple-neuron model is called Many ADALINE, or MADALINE

adaptive noise canceller—adaptive filter in which the reference noise source is filtered and subtracted from a primary input containing the signal and noise to eliminate the noise by cancellation

adaptive predictor—type of adaptive line enhancer in which the input, composed of signal plus periodic noise, is delayed and used to create the filtered reference input. The output of this filter is the filtered reference input.

adaptive resonance theory (ART) network—unsupervised neural network based on ability of the input and a stored category to "resonate" when they are sufficiently similar. When an input pattern is not sufficiently similar to any existing category, a new category is formed, using a previously uncommitted output neuron

adaptive self-tuning filter—type of adaptive line enhancer in which the input, composed of signal plus periodic noise, is delayed and used to create the filtered reference input. The output of this filter is the periodic noise

address generation unit—specialized addressing mode of DSP processor that allows arithmetic processing to proceed at maximum speed and allows specification of multiple operands in one small instruction word

adenocarcinoma—classification of malignant invasive tumor composed of endocervical, endometrial, or extrauterine cells

adenosine triphosphate (ATP)—molecule found in all living organisms that is the main immediate source of usable energy for the activities of the cells

adipose—fat tissue

administration set—tubing used during intravenous drug administration

adsorption—accumulation by the surface of a solid or liquid of the atoms, ions, or molecules of a gas or other liquid

affine class—time-scale distributions that include the continuous wavelet transform

affinity—attraction or force between particles that causes them to combine

afterload—sum of all loads against which cardiac fibers must shorten during systole

agonist—drug molecule that interacts with the receptor and elicits a maximal pharmacologic response

air embolism—entry of air into patient's circulatory system

albumin—class of simple water-soluble proteins found in tissues such as egg white

aliasing—distortion that occurs when a continuous signal that is restricted to frequency components below the Nyquist frequency is then sampled at a frequency less than double the Nyquist frequency

allergic reaction—exaggerated and sometimes harmful reaction to external substances

alveolus—tiny thin-walled sac in lungs filled with small blood vessels

ambient light—surrounding light

amide—ion with structure NH_3^-

anabolism—processes concerned primarily with the assembly of complex organic molecules

anatomical dead space—volume of conducting airways, from trachea to bronchioles, in which gas is not exchanged

anemia—inadequate hemoglobin synthesis

anemic hypoxia—diminished oxygen capacity of blood due to blood loss, inadequate hemoglobin synthesis, formation of methemoglobin, or carbon monoxide poisoning

anesthesia—absence of physical sensation in part or all of the body, typically a reversible condition that is induced using anesthetic drugs

angina—severe chest pain

antagonist—drug that elicits no response, but reversibly or irreversibly inhibits the receptor interaction of a second agent

antibody—any of several kinds of normally occurring protein molecules that are produced in the body of cells called lymphocytes and that act primarily as a defense against invasion by foreign substances

antigen—foreign substance that has entered the body

apparent volume of distribution—the proportionality constant relating drug concentration in blood or plasma to the amount of drug in the body

applanation—application of sufficient pressure by a sensor over an artery of interest

arachnoid villi—microscopic projections that extend into venous channels and provide CSF–vascular interfaces.

area under curve (AUC)—in kinetic modeling, the total area under the drug concentration in blood or plasma versus time curve

arrhythmia—abnormal heart rhythm

arterial hypoxia—inadequate oxygen in tissue due to decrease in ventilation–perfusion ratio

arterial oxygen saturation—functional arterial oxygen saturation is the ratio of oxyhemoglobin concentration to total concentration of arterial hemoglobin available for reversible oxygen binding. Fractional arterial oxygen saturation refers to the ratio of oxyhemoglobin concentration to oxyhemoglobin and deoxyhemoglobin concentrations. When MetHb and COHb are not present, fractional and functional saturations are equal.

arterial tonometry—noninvasive blood pressure measurement by applying sufficient pressure to a sensor over an artery such as the radial artery, which possesses sufficient bony support. With sufficient pressure, it is assumed that the transmural pressure equals zero and the external pressure equals the internal pressure.

artery—blood vessel that transports blood away from heart. The largest arteries are the aorta, which transports blood from left ventricle, and pulmonary artery, which transports blood from right ventricle. The radial artery at the wrist transports blood to the hand; the carotid artery transports blood to the brain. A small artery is called an arteriole.

artificial intelligence—computer science specialty that focuses on creating machines that can engage in behaviors that humans consider intelligent

artificial neural network (ANN)—mathematical model of system operator for MIMO system, so-named because original models of this type, such as the Hopfield network, were used to model human brain processing

assay—quantitative or qualitative analysis of a substance to determine its components

atrial pacing—pacing through a catheter inserted into the esophagus, which is in the proximity to the left atrium

atrium—either of the two upper chambers (left, right) of the heart that receives blood from the veins and forces it into a ventricle

atypical glandular cells of undetermined significance (AGUS)—classification of cellular changes in glandular cells exceeding those expected in a benign reactive or reparative reaction, yet not abnormal enough to be clearly neoplasic

atypical squamous cells of undetermined significance (ASCUS)—unusual cells that are not abnormal enough to be classified as dysplasia

auscultation—noninvasive measurement of blood pressure, during which a clinician inflates an external cuff around the tissue over the brachial artery and listens for specific Korotkoff sounds during slow cuff deflation to determine the systolic and diastolic pressures

autocorrelation—calculation of the agreement between neighboring data observations in a time sequence

automatic external defibrillator (AED)—device that recognizes and treats rapid ventricular arrhythmias in patients with cardiac arrest without requiring interpretation of the rhythm by medical personnel

autonomic nervous system—one of the two main divisions of the nervous system that supplies impulses to the body's heart muscles, smooth muscles, and glands. Two antagonistic divisions make up the autonomic nervous system: the sympathetic division that stimulates the heart, dilates the bronchi, contracts the arteries, and inhibits the digestive system, preparing the organism for physical action; and the parasympathetic, or craniosacral, division, which has the opposite effects, and prepares the organism for feeding, digestion, and rest. Anesthetics such as thiopental reduce sympathetic nervous activity.

Autonomous Intelligent Cruise Control (AICC) system—intelligent cruise controller that maintains speed and distance from preceding vehicles in a passenger car

autoradiography—process of exposing photographic emulsion by radiation emitted from tissue containing a radionuclide

autoregressive moving average exogenous input model (ARMAX)—linear difference equation by which the current output value is calculated based on past output values, past and current input values, and past and current disturbance values. Variations such as AR, MA, and ARX are possible.

back propagation—optimization method for determining multilayer feedforward neural network coefficients, in which the system error is propagated back from the outputs of the network to the inputs

backward shift operator—representation of delay of one sample in a signal

bandlimited—referring to a restriction in frequencies below the Nyquist frequency

bandpass filter—frequency-selective filter that minimizes all frequencies outside a specified range

bandstop filter—frequency-selective filter that maximizes all frequencies except those within a certain band or range

Beer–Lambert law—physical relationship in which transmitted light resulting from incident light decreases exponentially as a function of the extinction coefficient of an absorbing substance at a specific wavelength, concentration, and path length

benign—referring to a noncancerous change

bilinear relation—relationship in which flux is proportional to the product of two masses

bilinear—referring to a calculation in which a signal enters twice

bioavailability—extent to which a drug reaches its site of action

bivalent—referring to two states, such as true and false

blood gas analyzer—instrument that quantifies gas concentrations within a blood sample

blood pressure (BP)—pressure changes generated during cardiac cycle and transported from the left ventricle to the systemic circulation and from the right ventricle to the pulmonary circulation

blood–brain barrier (BBB)—membranous barrier that is highly resistant to diffusion and that segregates brain ISF from circulating blood through tight junctions in brain endothelia

blood–CSF barrier—tight junctions of cells between the choroid plexus that separate the blood from the CSF

body mass index (BMI)—description of relative weight for height, which is significantly correlated with total body fat content

Briggs–Haldane analysis—theory that when an enzyme and substrate interact to form an enzyme–substrate complex, the complex dissociates faster as more complex is formed

bronchus—branch of trachea. The tubular extension of a bronchus is a bronchiole.

bundle branch block—cardiac condition that occurs when right and left ventricles do not simultaneously depolarize

bundle of His—group of cardiac muscle fibers that span the connective tissue wall separating the atria from the ventricles

Butterworth filter—filter with magnitude response that is maximally flat in the passband

calibration curve—curve that adjusts the quantitative output of measuring device to a known standard

cancer—multifactorial disease characterized by cell growth free of the restraints that regulate normal cells

canonical form—standard state-space representation of a transfer function

capillary bed—dense network of narrow blood vessels with extremely thin walls that allows gas exchange between blood and cells

carboxyhemoglobin—molecule in which carbon monoxide displaces oxygen from hemoglobin, making hemoglobin unavailable for oxygen transport

carboxyl—ion with structure COO^-

carcinoma—cancer cell, usually grouped with like cells to form a lump or mass called a malignant tumor

carcinoma-in-situ—abnormal cell growth during which entire cervical lining thickness becomes disordered, but the abnormal cells have not spread below the lining

cardiac bypass—surgical procedure during which an extra vein from leg or chest wall is used to "bypass" or detour around a blocked artery. The vein is grafted above and below the point of blockage, allowing blood to flow freely around the affected area.

cardiac index—cardiac output normalized by body surface area

cardiac output—total blood flow

cardinality—total number

cassette—infusion pump disposable component that displaces reservoir fluid

catabolism—processes related to degradation of complex substances with subsequent generation of energy

catecholamine—any of a group of amines derived from catechol that has an important physiologic effect. The catecholamines dopamine, dobutamine, and epinephrine possess vasodilative and vasoconstrictive properties.

catheter embolism—decrease in blood pressure with pain along the vein that occurs when piece of catheter breaks off and floats freely in the vessel

catheter—needle inserted into the vein for drug administration or pressure measurement

causality—property of linear time-invariant system whereby current output sequence value depends only on past and current input sequence values

central compartment—in compartmental modeling, representation of highly perfused organs and tissues such as the liver and kidney, in rapid distribution equilibrium with the blood

central nervous system (CNS)—portion of the vertebrate nervous system consisting of the brain and spinal cord

central neuron—brain cell consisting of a soma (cell body with its nucleus), projections called dendrites, and an axon that projects from the soma and communicates to the dendrite of another neuron through a synapse

centroid—center of area

cerebral cortex—extensive outer layer of gray matter of the cerebral hemispheres, largely responsible for higher brain functions

cerebrospinal fluid (CSF)—fluid that covers the entire surface of the brain and spinal cord, secreted by the choroid plexus

cervix—lower portion of uterus that is a narrow muscular canal connecting the body of the uterus to the vagina. Outer surface is lined with epithelial flat cells called squamous cells. Canal of cervix is lined with epithelial tall cells called columnar cells. These two cell types meet at squamo–columnar junction, also called the transformation zone.

characteristic function—for time-frequency distributions, the product of the kernel and symmetric ambiguity function

Chebychev filter—filter with magnitude response that is equiripple in the passband and monotonic in the stopband

cholesterol—white crystalline substance in animal tissues and various foods, normally synthesized by the liver

choroid plexus—single sheet of tissue within two lateral ventricles and third ventricle of the brain that secretes cerebrospinal fluid

chronotropic—referring to an effect in speed or rate

circulatory overload—infusion of fluids at a rate greater than patient's system can accommodate

cisternum magnum—a natural space between the brain hemispheres

Clark electrode—electrode used to measure arterial oxygenation in which current produced by negative electrode in a blood sample is directly proportional to availability of oxygen molecules at electrode tip

clearance rate—ratio of the intravenous dose to the total area under the drug concentration in blood or plasma versus the time curve

closed-loop drug delivery (CLDD)—technique during which a drug is administered from a drug delivery device such as an infusion pump, while the drug infusion rate is continuously adjusted by a closed-loop algorithm or controller that is based on control theory techniques. The drug sodium nitroprusside, which changes blood pressure, was extensively investigated as an agent for use in CLDD.

coefficient of variation (CV)—also known as fractional standard deviation, the ratio of error standard deviation to mean error

Cohen's class—time-frequency distributions, including the spectrogram and Wigner distribution, that can be generated using the kernel method devised by Cohen

colposcopy—examination of the vagina and cervix under magnification, after application of 5% acetic acid. For unknown reasons, many HPV-associated lesions demonstrate whitening after this application.

combinatorial optimization—minimization of a performance function in which the solution involves multiple steps

compartment—a well-mixed and kinetically homogeneous amount of material that is not necessarily a physiologic space or well-delimited physical volume

compartmental model—type of kinetic model based on the law of conservation of mass, described using ordinary differential equations that are based on compartments

competitive learning—method for determining weight values of the Kohonen network, based on minimizing the Euclidean distance

complementary metal–oxide silicon (CMOS)—chip fabrication technology that reduces supply voltage from 5.0 to 3.3 V

complete heart block—condition in which conduction from atria to ventricles is completely interrupted

compliance—flexibility of infusion tubing that may cause large fluctuations in flow rate, typically modeled as an electrical capacitor in circuit models

conditional mean—calculation of the average frequency at a particular time, or average time at a particular frequency, for a time-frequency distribution

conditional standard deviation—calculation of the frequency standard deviation at a particular time, or time standard deviation at a particular frequency, for a time-frequency distribution

continuous noninvasive blood pressure (CNIBP)—blood pressure monitoring technique during which systolic, mean, and diastolic blood pressure values are updated at each beat, using a noninvasive apparatus

contractility—estimation of maximum velocity of contraction of cardiac muscle fibers

control—regulation of the output of a process using the mechanism of feedback to the desired output. A controller is the implementation of a control process.

convective transport—bulk flow transport

convex—referring to outward curving

convolution sum—common description of output calculation for linear time-invariant system, composed of convolution of input with impulse response

convolution theorem—also known as dual domain theory of duality; the statement that convolution in the time domain translates to multiplication in the frequency domain

cooximeter—instrument that uses four wavelengths to determine deoxyhemoglobin, oxyhemoglobin, carboxyhemoglobin, and methemoglobin in a single blood sample

coronary artery disease (CAD)—result of fatty build-up on the inner walls of the arteries that nourish the heart. These build-ups can narrow the arteries and thus restrict the normal flow of oxygen-rich blood, or can actually block the flow of blood altogether.

covariance—mean squared error between estimated and observed signals

crisp set—conventional set containing real numbers within a specified range

critical care—treatment of the sickest patients

cross power density spectrum—Fourier transform of cross-correlation function

cross term—interference term that may be present in calculation of time-frequency distribution

cross-correlation—calculation of the agreement between two signals

Dalton's law—physical relationship in which each gas in a mixture exerts a partial pressure proportional to its share of the total volume

damped sinusoid waveform—traditional external defibrillation waveform that is shaped by resistor, capacitor, and inductor

data type—representation of data as integer, fixed-point, or floating-point within a computer

de novo—referring to from the start

debugger—software development tool used to test and find bugs in embedded software

defibrillation threshold—minimum shock for successful defibrillation, typically measured as total voltage or energy required for 50% successful defibrillation

defibrillation—method during which a strong current of short duration is administered to convert rapid twitching of the ventricles to a slower rhythm that allows the heart to pump blood. If external, current is applied across the thorax via paddles or electrode adhesive pads. If internal, current is applied through electrodes in contact with heart tissue.

deoxyribose nucleic acid (DNA)—genetic material of all cellular organisms and most viruses that carries the information needed to direct protein synthesis and replication

depolarization—increase in potential in a cell membrane

depot—site of drug administration

design of experiments (DOE) factorial design—statistical technique that enables user to determine statistically significant factors that affect a specified response

GLOSSARY

desired signal—signal that does not possess noise

detection error—error in Pap smear reading that occurs when a few dozen abnormal cells are not identified among hundreds of thousands of normal cells present in a well-taken smear

determinism—property by which each value of a sequence is uniquely determined by a mathematical expression, table of data, or rule of some type

device driver—driver that interacts directly with peripheral control and status registers; typically the only piece of software in the system that reads and writes a particular devices' registers directly

diastole—portion of the cardiac cycle during which the ventricles fill with blood, characterized by minimum blood pressure

dicrotic notch—characteristic deflection in a blood pressure waveform that occurs as blood pressure is transported from the left ventricle to the aorta, smaller arteries, and arterioles, due to damping from reflections in the arterial tree

diffusion—movement of substance along a decreasing concentration gradient

digital signal processor (DSP processor)—specialized microprocessor that supports repetitive, numerically intensive tasks

digital—referring to a finger

dilation—imaging technique involving inversion of binary pixels based on mathematical manipulations, which is stopped before individual objects begin to touch

direct memory access (DMA)—data transfer without processor involvement

direct search method—optimization method, including the simplex search, that requires only calculation of current values of the performance function

direct-form realization—block diagram representation of ARMAX model in which delays are represented by system function z^{-1}

discrete time—sampled time

disjunctive interpretation—fuzzy property in which outputs must be combined by union to approximate the compatibility relation

dissociation curve—graphical plot of saturation as a function of gas partial pressure

distribution—transport of drug to various sites in the body

disturbance sequence—noise sequence

dominant pacemaker—cardiac cells capable of autoexcitation that are present in the sinoatrial (SA) node

Doppler ultrasound—measurement technique used to measure blood flow, based on the Doppler shift

dose-dependent kinetics—deviations from linear pharmacokinetics at higher drug doses

downsampling—process whereby the sampling rate of a signal is reduced to the desired sample rate

downstream occlusion—flow obstruction below the infusion device

drip chamber—plastic tube for visually verifying flow

dropout—period during which medical instrument does not provide display data

drug—therapeutic agent

dynamic range—ratio between largest and smallest numbers that can be represented

dysplasia—abnormal cell growth, which is a precursor to cancer

echocardiography—ultrasound for detection of wall-motion abnormalities

effect compartment—target site for a drug

efficient looping—specialized feature of DSP processor for repetitive computation

eigenvalue—root of a differential equation

electrocardiogram (ECG)—measured surface biopotentials originating from electrical activity of the heart

electrocautery—device used during surgery to cut and stop bleeding by applying a radio-frequency spark between a probe and tissue

electrocorticogram (EcoG)—nonstationary signal taken from electrodes that are implanted over lobes of the cerebral cortex after surgical incision of the skull

electrogastrogram (EGG)—recording of gastric activity from muscular electric activity obtained by placing electrodes on the abdomen

elimination—removal of drug from the body

elution—removal of one substance from another, usually an adsorbed material from an adsorbent

embedded system—combination of computer hardware and software, and possibly peripherals, that are designed to perform a specific function

empirical—referring to a reliance on observation and experiment

endocervical curettage—scraping of the inner uterine lining

endosome—sac beneath the plasma membrane

enzyme—catalyst that promotes the enormous variety of reactions that channel metabolism into essential pathways

ependyma—single layer of cells lining the ventricles that separates the CSF from brain ISF

epicardial first heart sound—sound that occurs during initial cardiac contraction, with the sound recorded from inner layer of the sac surrounding the heart

epithelia—interconnected sheets of cells that work in concert to secrete mucus

equalization—adjustment of the amplitude of an electronic signal equiripple—equally spaced ripple that varies between the value 1 ± a small number

erosion—imaging technique involving isolation of all objects within an image that are greater than or equal to the smallest size of a pathological cell nucleus

erythema—reddening of the skin

erythrocyte—red blood cell

euglycemia—maintenance of glucose within the normal range

even symmetry—property by which $f(k) = f(-k)$

exclusive-or (XOR) function—binary function in which the output is true if the two inputs are not identical

expected value ($E\{\ \}$)—mean of a function

expiration—act of breathing out

expiratory tidal volume—dead space tidal volume plus alveolar space tidal volume

extension tubing—infusion tubing option that extends length

extinction coefficient—property of absorbing substance that is dependent on wavelength

extravasation—infusion of a medication capable of causing a blister or tissue destruction into tissue surrounding a blood vessel

extravascular tissue—tissue surrounding a blood vessel

fast Fourier transform (FFT)—digital calculation of Fourier transform

fibrillation—erratic heart rhythm, caused by continuous, rapid discharges from either areas of atria (atrial fibrillation) or ventricles (ventricular fibrillation). One cause of atrial fibrillation is rheumatic heart disease.

Fick method—estimation of cardiac output made by determining change of oxygen concentration in pulmonary circulation

fiducial point—deflection in electrocardiogram or impedance cardiography waveform that represents event during cardiac cycle

filter—infusion tubing option that removes particulate matter suspended in the infusion solution

finite impulse reponse (FIR) filter—filter described by past and present values of an input sequence

finite support—for a joint distribution, it is ideal that the distribution equals zero for frequencies outside a specified frequency range if the Fourier transform of the signal is zero outside this frequency range, and that it equals zero for times outside a specified time range if the signal is zero outside this time range. If each requirement is met, then the distribution has weak finite frequency or time support. If the distribution equals zero when the signal or spectrum is zero, the distribution has strong finite support.

flexibility—possibility of using a chosen model structure to describe most of different system dynamics that can be expected in an application

floating point—data type in which a number is represented as the product of two signed numbers: the mantissa (fraction) and exponent of a radix. Single-precision implementation requires 32 bits; double-precision implementation requires 64 bits. The radix point separates bits of the mantissa and exponent fields.

fluid viscosity—property of a fluid that tends to prevent it from flowing when subjected to an applied force. High-viscosity fluids resist flow, whereas low-viscosity fluids flow easily

Food and Drug Administration (FDA)—agency of the United States Department of Health and Human Services that monitors the safety of medical devices

forcing function—in compartmental modeling, replacement of the normal equations associated with a compartment of interest by a specified set of equations, thus "forcing" the rest of the system to utilize this function

function approximation—modeling and prediction applications

fuzzy conditional statement—rule composed of an antecedent and consequent

fuzzy input—ordered pair containing a linguistic label and degree of membership

fuzzy logic—form of logic in which variables can have degrees of truthfulness or falsehood represented by a range of values between 1 (true) and 0 (false)

fuzzy model—system description based on fuzzy logic, including the linguistic fuzzy, fuzzy relational, and Takagi–Sugeno models

fuzzy partition—family of fuzzy membership functions

fuzzy set—mathematical model of the vagueness present in our natural language with which we describe phenomena that do not possess sharply defined boundaries

gain—logarithmic magnitude, typically calculated in decibels (dB), as 20 times the $\log_{10}$ magnitude

Gauss–Newton iterative scheme—practical approach to determining compartmental modeling parameters involving calculation of the gradient

generalized linear phase—property of function in which phase response is linear

glia—network of supporting tissue and fibers that nourishes nerve cells within the brain and spinal cord

glottis—opening to the windpipe

glucagon—hormone secreted by the pancreas when glucose levels fall several hours after a meal, promoting both glucose release from intracellular stores and glucose synthesis

gluconeogenesis—glucose synthesis

glucose intolerance—inability to process blood sugar

glucose oxidase method—blood glucose measurement technique based on this enzyme's catalytic effect to oxidize o-dianisidine, which can be quantified as an increase in absorbance at 460 nm

goodness of fit—model validity measure usually determined by Akaike's final prediction error criterion or information criterion, in which the number of model parameters is weighed against the number of data points and error

gradient method—optimization method that requires calculation of accurate values of at least the first, and sometimes the second, derivative of the performance function

gradient—vector consisting of partial derivatives of any function with respect to each coefficient

gravity flow regulation—regulation of flow in an infusion device by gravity

group delay—negative derivative of phase response

half-life—time for the amount of drug in the body to decrease by 50%

harmonic—integer multiple frequency of a fundamental frequency

Harvard architecture—microprocessor access of two separate memories by two separate buses

hematoma—leaking of blood into surrounding tissue that may occur during intravenous insertion if catheter pierces the back of a vein

heme—molecule within each hemoglobin protein chain containing an iron ion that reversibly binds to an O_2 or CO molecule

hemisphere—right or left side of the brain

hemoglobin (Hb)—also known as deoxyhemoglobin; a molecule containing four protein chains, each containing a heme molecule

high-performance liquid chromatography (HPLC)—technique during which a column is packed with a material that can selectively adsorb molecules on the basis of some difference in their chemical structure, leading to measurements of substance concentrations within the fluid poured through the column

high-resonance pole—solution of characteristic equation in the frequency domain that has a long time constant

high-grade squamous intraepithelial lesion (HSIL)—classification associated with moderate or severe dysplasia or carcinoma-in-situ, with cells usually deficient in nucleoli

highpass filter—frequency-selective filter that minimizes all frequencies below a specified frequency

hormone—peptide or steroid secreted by one tissue and conveyed by the blood stream to another to effect physiologic activity

human papillomavirus (HPV)—commonly manifested as warts, this virus infects the squamous epithelia of the skin and mucuous membranes

humoral—referring to transportation through body fluids

hydrophobic interaction—association of nonpolar molecules in aqueous solution

hyper-, hypotension—high or low blood pressure

hyper-, hypothermia—high or low temperature

hyper-, hypovolumia—high or low volume

hypothalamus—part of the brain that regulates emotions, hormone production, and the autonomic nervous system

hypoxemia—insufficient oxygen in blood

hysterectomy—surgical removal of uterus. A simple hysterectomy involves the uterus only; a radical hysterectomy involves the uterus and adjacent tissues.
ictal—referring to a seizure
immune response—in reaction to a foreign substance, such as a protein invading the tissues of a higher vertebrate, an organism defends itself by synthesizing and attaching an antibody to the invading substance. Specific cells then destroy the marked substance
impedance cardiography (ICG)—also known as thoracic bioimpedance or impedance plethysmography; a voltage measurement across the thorax in response to applied constant current. Features in the voltage waveform are then used to estimate stroke volume. The first commercial ICG product was the Minnesota Impedance cardiograph.
impulse response—a single-input, single-output system operator
impulse—pulse of unbounded amplitude and zero duration, with an integral of 1
in vitro—outside of an organism
in vivo—within an organism
indicator-dilution method—measurement of cardiac output obtained by applying a detectable indicator upstream in the circulation and detecting it downstream to determine the flow rate by which it was mixed
infarction—tissue death from obstruction of local blood supply. Myocardial infarction is typically called a "heart attack."
infiltration—infusion of an intravenous solution or medication into tissue surrounding a blood vessel
infinite impulse response (IIR) filter—filter described by autoregressive moving average model with constraint that number of feedforward coefficients is less than or equal to number of feedback coefficients
infusion—intravenous administration modeled as a unit step function
injection—intravenous administration modeled as an impulse
inotropic—referring to an muscular contraction effect
input sequence—discrete samples of an input
inspiration—act of breathing in
instantaneous frequency—calculation of the average frequency at a particular for a time-frequency distribution
instantaneous power—absolute value of a squared signal
insulin resistance—inefficient processing of insulin
integral of time multiplied by absolute error (ITAE) function—common performance function calculated as the integral of the product of time and the absolute error
integrated optical density (IOD) screen—sum of the pixel grey values for each object
interstitial fluid (ISF)—fluid between cells
intraarterial blood pressure (IAP)—blood pressure measured invasively with 20 gauge catheter in artery of interest, typically the radial artery (in the wrist)
intracerebroventricular—referring to direct injection into the cerebral ventricles in the brain
intramuscular administration—drug administration within a muscle
intravenous administration (IV)—drug administration within a vein
intravenous glucose tolerance test (IVGTT)—measurement of insulin and glucose plasma samples in response to an intravenous glucose injection
inulin—biologically inert marker of diffusion similar in molecular size to insulin

ion channel—ion-selective pore in the cell membrane

ischemic hypoxia—restriction of organ perfusion leading to inadequate oxygen delivery to cells

isovolumetric contraction—period of onset of ventricular systole, during which intraventricular pressure rises, causing immediate closure of AV valves

isovolumetric relaxation period—diastole period during which valves are closed

joint energy density—also known as a probability distribution and joint distribution; a density that is a function of both time and frequency

Kalman filter—recursive filter described by state equations

kinetic—referring to changes over time

knowledge base—fuzzy control rule base for fuzzy inference and data for fuzzification and defuzzification membership functions

Kohonen network—unsupervised neural network that possesses a structure similar to a single-layer feedforward network

Korotkoff sounds—sounds generated by partial occlusion of the brachial artery that can be used to determine systolic and diastolic blood pressure

labeling—identification of a molecule for measurements of concentration. With exogenous labeling, an isotope is introduced outside of the body. With endogenous labeling, a labeled precursor of the molecule of interest is used to label the tracer, such as infusion of a radiolabeled amino acid to label a protein.

laminar—referring to flow that is nonturbulent

Laplace frequency—s, which by definition is equal to $j\Omega$

law of mass action—physical observation that a reaction reaches the same position of equilibrium (at a given temperature), no matter what the starting conditions

lead—differential voltage recording of biopotentials generated by heart, defined by Einthoven

leadscrew—pump screw that produces constant linear advancement of syringe plunger

learning algorithm—type of parameter identification that may be supervised or unsupervised. With supervised learning, a model is trained by presentation of input and output pairs. With unsupervised learning, a model automatically classifies input data according to the distribution of input data and their relations

least mean squares (LMS)—minimization of the mean squared error between an approximation and the true signal

least squares estimate—vector that minimizes the mean squared error performance function

left ventricular ejection time (LVET)—systolic ejection time interval from opening of aortic valve to closing of aortic valve

leptin—hormone isolated from the *ob/ob* mouse that causes a negative feedback signal for obesity

level of concern—as defined by FDA, an estimate of the severity of injury that a device could inflict on a patient or operator as a result of latent failures or design flaws

level of coverage—minimum value in a fuzzy partition

ligand—molecule binding to a receptor

light emitting diode (LED)—semiconductor that produces light when current passes through it

linear peristaltic mechanism—infusion pump mechanism in which a portion of administration set tubing is positioned in a linear channel against a rigid backing plate, while an array of cam-driven actuators sequentially occlude tubing starting with the section nearest reservoir, forcing fluid toward patient with a sinusoidal wave, or peristaltic, motion

linear programming—optimization technique by which a nonlinear function is approximated using Taylor's expansion and solved in a recursive fashion

linear separation—differentiation between two linearly separable sets of patterns

linear time-invariant (LTI) system—system that possesses superposition and that does not depend on the time origin

linearity—property of superposition

Lineweaver–Burk plot—linear plot obtained by taking reciprocals on both sides of the Michaelis–Menten equation

linguistic fuzzy model—also known as Mamdani model, this original model developed by Mamdani involves three steps: (1) transformation of crisp inputs into fuzzy inputs using fuzzification, (2) rule base inference to map fuzzy inputs into fuzzy outputs, and (3) transformation of fuzzy outputs into crisp outputs by defuzzification

lipid bilayer—double layer of regularly arranged phospholipid molecules that is widely accepted as forming the basic structure of cell membranes and other biological membranes

local infection—contamination, usually bacterial, at break in skin

low-grade squamous intraepithelial lesion (LSIL)—classification associated with hollow cells, called koilocytes, that possess atypical nuclei and/or mild dysplasia

lowpass filter—frequency-selective filter that minimizes frequencies above a specified cutoff frequency

Lukasiewicz intersection—fuzzy mathematical operator that results in taking the minimum of two degrees of membership

luminal—referring to inner

lymph—fluid containing white cells, chiefly lymphocytes, that is drained from tissue spaces by the vessels of the lymphatic system

magnitude—amplitude portion of frequency response

Mamdani max–min inference—use of Zadeh intersection for rule-based inference and the centroid for defuzzification

marginal conditions—ideal conditions under which the sum of all distributions for all frequencies at a specified time should describe the instantaneous power, and the sum of the distributions for all times at a specified frequency should describe the power density spectrum

mass spectroscopy—concentration measurement technique during which the difference in mass-to-charge ratios of ionized atoms or molecules is exploited to separate them from each other

maximum likelihood—probability distribution function of the observations conditioned on a parameter vector

McCullough–Pitts model—simple model of a brain neuron as a binary threshold unit, by which a neuron computes a weighted sum of its inputs from other units and outputs a "1" or "0" according to whether this sum is above or below a certain threshold

mean arterial pressure (MAP)—integral of a pressure waveform over one cardiac cycle

mechanical flow regulator—fluid reservoir positioned above a patient, with a mechanical clamp around the attached tubing

median filter—filter in which the inputs are sorted in increasing order, and the middle value is used as the filter output

membership function—measurement of the degree of similarity of an element to the fuzzy set. The maximum degree of membership is termed the height. If the height equals 1, the primary fuzzy set is called a normalized fuzzy set. The support of a primary fuzzy set is a crisp set that contains all the elements that possess nonzero degrees of membership. The core of a primary fuzzy set in a universal set is a crisp set that contains all elements that possess degrees of membership equal to one.

memory—computer storage that includes changeable random-access memory (RAM), fixed read-only memory (ROM), and hybrid memory. Hybrid devices enable the programmer to overwrite data in a ROM-like device, and include erasable, programmable read-only memory (EPROM) and flash memory.

metabolism—totality of chemical reactions that occur in animal cells

metastasis—process by which cancer spreads as cells from a tumor break away and travel to other part of body, where they can continue to grow

methemoglobin—hemoglobin mutation that does not bind oxygen

method of steepest descent—also known as Cauchy's method; a gradient method in which the next iteration is based on the steepest change in the value of the performance function

methylene blue—physiologic dye

Michaelis–Menten dynamics—relationship in enzyme kinetics with which, at high substrate concentrations, a limiting velocity is approached because the enzyme molecules become saturated

minimal model of insulin sensitivity—nonlinear compartmental model that quantifies the influence of decreased peripheral insulin sensitivity on the impairment of the patient's ability to tolerate a standard glucose load

model identifiability—possibility of obtaining unique estimates of all the unknown model parameters. Theoretical identifiability is termed a priori identifiability. Practical identifiability is termed a posteriori identifiability. A model may be uniquely identifiable, nonuniquely identifiable, or nonidentifiable.

modulation—process of changing the amplitude, frequency, or phase of a transmitted signal using an encoder. When the modulated signal is received, it is returned to its original state using a corresponding decoder.

modulo-2—referring to a binary system

monotonic—referring to a consistent increase over time

motion artifact—noise created within a physiologic signal from external motion

multilayer feedforward network—most common neural network architecture, in which inputs are fully connected to neurons of one or more hidden layers, which are then fully connected to the neurons of an output layer

multiple input, multiple output (MIMO) system—system described by system operator that has many inputs and outputs

multiply–accumulate (MAC) operation—microprocessor operation involving multiplication and addition

myocardial ischemia—insufficient blood flow to heart tissue

myoelectrical—referring to electrical impulses from muscles

neonatal—referring to a premature infant

neoplasia—new cell growth

neuropeptide—protein in the brain

neurotransmitter—substance present in the presynaptic terminal of a neuron that is released by stimulation

Newtonian fluid—fluid whose stress–rate-of-strain relationship is linear, following Newton's law

noncompartmental model—type of kinetic model that considers the input and output relationships of a system accessed through a single accessible compartment, usually the plasma

nondegenerate reparameterization—transformation to forms such as the canonical form

nonlinear least squares estimation—system identification technique used with compartmental models. Since the model is nonlinear in terms of its parameters, the performance function is not of quadratic form as a function of the parameters, and an iterative approach to a solution, rather than explicit analytical solution, is required.

normalization—conformance to some standard reference, such as body surface area, by division of that reference

normoxemia—room air

number-of-templates property—adaptive resonance theory network property that the number of output neurons, or templates, required is smaller than the number of total input patterns for certain values of network parameters

Nyquist frequency—bandwidth of a sampled signal

Nyquist rate—a sampling rate that is twice the Nyquist frequency

obesity—complex multifactorial chronic disease that occurs in part when mechanisms in the brain regulate body weight to a "set-point" that is associated with excessive accumulation of fat tissue

object file—file that is output when a source file is compiled

occupation theory—theory that a signal drug molecule interacts with a receptor with a single binding site to produce a pharmacologic response

ocular tonometry—measurement of intraocular pressure by applying sensor to the cornea until its central area is flattened. This flattening indicates that the circumferential stresses in the corneal wall have been removed and that the internal and external pressures are equal.

Ohm's law—assuming that impedance is purely resistive; calculation of voltage as the product of the current and resistance

operating system—system that schedules execution of and maintains information about the state of software tasks, enabling multitasking

optimal sampling schedule (OSS)—time schedule for sampling substance of interest, with which the maximum precision of model parameter estimates is obtained, under the practical constraints of sample acquisition

optimization—method used to estimate the best solution of a performance function

oral ingestion—administration of drug through the digestive tract

orthogonality—property of two vectors when their inner vector is zero

oscillometry—noninvasive measurement of blood pressure, during which an external cuff is inflated around the tissue over the brachial artery and slowly deflated. Specific changes in the amplitude of pressure oscillations correspond to the systolic, mean, and diastolic blood pressures.

output sequence—discrete samples of an output signal

oxidation—chemical process requiring oxygen as an input

oxyhemoglobin (HbO_2)—hemoglobin bound to oxygen

P wave—deflection in electrocardiogram that represents depolarization and simultaneous contraction of both atria

Papanicolaou (Pap) smear—procedure during which cervical cells are collected during a pelvic exam and preserved on a glass slide for cancer classification
parenteral administration—drug administration by method other than oral ingestion
Parkinsonism—neurological disorder due to abnormally low dopamine levels in the substantia nigra region of the brain
parsimony principle—principle that the simpler of two possible model structures will on average result in better accuracy
partial agonist—drug that elicits a partial, lower than maximal, response
passband—region within which the magnitude of the frequency response must approximate unity
pattern classification—application in which various features are consistently identified
peptide—protein that is formed by a bond between the amino and carboxyl groups of two amino acids, with water as a byproduct
perceptron—model first described by Rosenblatt that incorporates supervised learning into a model of the neuron
performance function—function used as goal during the optimization process
perfusion—blood supply
pericapillary—tissue enclosing the capillary
peripheral compartment—in compartmental modeling, presentation of poorly perfused tissues and fluids of the central compartment and the poorly perfused or less readily accessible tissues
peripheral—external device that coordinates the interaction of a computer with the outside world or that performs a specific hardware function
perturbation—excitation of a system
pharmacodynamics—relationship between drug concentrations at the site of action and subsequent pharmacologic response
pharmacokinetics—transport of drugs to the site of action
pharmacologic stress testing— use of a drug to simulate exercise, for the purpose of inducing ischemia in areas where blood flow is insufficient
phase response—angular portion of frequency response
phlebitis—inflammation of vein from infusion fluid
photodetector—instrument that measures transmitted light, often a photodiode
pinocytotic vesicle—in bulk flow transcytosis, a small sac that is freely mobile within the cell cytoplasm that ferries droplets of fluid across the vascular wall
piston—in a large-volume pump, a disk that interfaces to a cassette containing fluid and moves to cause the displacement of fluid
pit—invagination within the plasma membrane
plasma—fluid portion of blood and lymph
plasticity—ability of a neural network to react to new data
plunger—portion of a syringe that moves to displace fluid
Poiseuille's equation—also known as Hagen–Poiseuille's equation, an expression for flow through a uniform tube based on radius, pressure, length, and fluid viscosity
pole—root of denominator or characteristic equation of a transfer function
polypeptide—long chain of amino acids
positional catheter—thin needle inserted into a vein that has been displaced so that the tip wedges against the internal lining of the vein wall, restricting fluid flow
positron emission tomography—technique for measuring the concentrations of positron-

emitting radioisotopes within a three-dimensional object by the use of external measurements of the radiation

possibility—degree of membership in a fuzzy set

postsynaptic membrane—receptor site on a neuron for neurotransmitters that have been released from another neuron

power density spectrum—absolute value of the squared Fourier transform of a signal

prediction error method—system identification that is based on use of a prediction error performance function

prefiltering—use of lowpass filtering for limiting the bandwidth of an analog filter before digitization, in order to avoid aliasing

preload—passive load on heart muscle that establishes initial muscle length of cardiac fibers before contraction

premarket submission—submission to FDA requesting that a medical device be marketed. If a device is "substantially equivalent" to a legally marketed device, a premarket notification called a 510(k) is submitted. If a new type of device will be marketed, a premarket application (PMA) is submitted.

pressure—force of circulating fluid upon the walls of its container; for example, blood pressure in an artery

presynaptic terminal—portion of neuron within which an action potential causes proteins called neurotransmitters to be released

primary input—in an adaptive filter, the input that contains the signal and noise

primary structure—the amino acid sequence of a protein

primitive polynomial—alternative sequence representation of a pseudorandom binary sequence

principal component analysis—also known as the Karhunen–Loeve transformation; an optimal transformation that enables decorrelation of the input variables

probability density function—function of a continuous variable such that the integral of the function over a specific region yields the probability that its value will fall within the region

product intersection—fuzzy mathematical operator that results in taking the product of two degrees of membership

proportional–integral–derivative (PID) controller—method used in classical control theory based on a transfer function containing a proportional gain, integral term, and derivative term. A variation is the proportional–integral controller.

protein—complex molecule composed of amino acids linked by peptide bonds

pseudorandom binary sequence (PRBS)—sequence of 1s and 0s that simulates a random sequence in which the number of 1s is greater by +1 than the number of 0s

pulmonary artery (PA) catheter—thin tube passed through skin into central vein until the tip reaches the pulmonary artery

pulmonary circulation—movement of blood from right ventricle to left atrium

pulse oximetry—measurement in which tissue oxygenation is estimated by monitoring oxygen saturation in arterial blood

pulse pressure—difference between systolic and diastolic pressures in an artery of interest

pulsed-field gel electrophoresis (PFGE)—movement of charged particles in a colloid or suspension when a pulsed electric field is applied to them

Purkinje fiber—one of a network of specialized cardiac muscle fibers that rapidly transmit impulses from the atrioventricular node to the ventricles

QRS complex—electrocardiogram deflections composed of Q, R, and S waves that represent ventricular depolarization and beginning of ventricular contraction
quadratic equation—any second-degree polynomial equation
radial basis function network—artificial neural network in which the weights in the feedforward network are replaced by Gaussian functions at each hidden neuron
radiation therapy—treatment of cancer with high-energy rays such as x-rays
radioimmunoassay (RIA)—hormone concentration measurement technique involving combining a hormone or other protein of unknown concentration with a fixed amount of radiactive antigen. A fixed amount of antibody is then added, which binds to both radio- and unlabeled antigen. When the antigen and antibody molecules bind, they are precipitated out of solution. The bound radioactive antigen is then detected, and is inversely proportional to the unlabeled antigen present.
radionuclide perfusion imaging—use of radioisotope for detection of blood flow defects
range limiting—referring to integration to finite limits
rate theory—theory that the pharmacologic response is not dependent on the drug–receptor complex concentration but depends on the rate of association of the drug and receptor
receptor—molecule on a cell surface that binds with a specific molecule, antigen, hormone, or antibody, that is a precursor to a specific action. For example, adrenergic receptors in smooth muscle and heart bind with multiple drugs, resulting in contraction or relaxation of the muscle. For the β_1-adrenergic receptor, isoproterenol and arbutamine are agonists and metoprololis is an antagonist.
rectangular window—window with value 1 within range $\{0, M\}$, and value 0 otherwise
recurrent network—artificial neural network in which data moves in both directions, giving rise to feedback
recursive—referring to the repeated application of a function to its own values
reference noise source—in an adaptive filter, a second noise source that is correlated to the noise contained in the input signal
refractory period—time after which a nerve or muscle cell has received a stimulus. During an absolute refractory period, a myocardial cell is unexcitable. During a relative refractory period, a myocardial cell may be reexcited
regression vector—vector containing past input and output values
relocatable program—single file consisting of object files that have been linked together
Remez exchange algorithm—algorithm that iterates between solution of estimated filter coefficients and desired frequency response until a filter is found that meets the desired response with the lowest possible number of filter coefficients
repolarization—decrease in potential in a cell membrane
reservoir—container for fluid or drug that is to be infused into the body
residual—difference between estimated and observed signals
resistance—impulse response between flow and pressure, created within the vasculature
respiration frequency—number of breaths per unit time
respiration—process during which oxygen and carbon dioxide are exchanged between the cells and their surroundings
resting potential—baseline voltage of cell membrane
robust control system—system that exhibits desired performance in the presence of significant plant uncertainty
roller clamp—fluid flow regulator

sampling error—error in Pap smear reading that occurs when abnormal cells are not placed on smear

sampling theorem—theory stating that samples of a continuous bandlimited signal that have been acquired frequently enough are sufficient to represent the signal exactly

saturation—occurrence when every enzyme molecule is occupied by substrate and is carrying out a catalytic step

scale—physical attribute representing compression

scalogram—squared modulus of a wavelet transform

secondary pacemaker—cells in atrioventricular node that may initiate a heartbeat if the dominant pacemaker cells do not

secondary structure—in a protein, the regular folding due to hydrogen bonds between amino and carboxyl groups that leads to local regions of helical structure

sensitivity—correct detection percentage of an alarm condition

sepsis—infection that has entered the patient's circulation

servo—feedback mechanism that consists of a sensing element, amplifier, and controller

settling time—time by which a system settles to within 2% of its final value

shift—change in time or frequency

signal-to-noise ratio (SNR)—ratio of the strength of a signal of interest to unwanted interference

single input, single output (SISO) model—system with scalar input, disturbance, and output

somatostatin—hormone produced in the hypothalamus that inhibits the release of growth hormone and suppresses endogenous insulin release

source file—original computer file that has not been compiled

specific gravity—ratio of the density of a substance to the reference density of water

specificity—correct detection percentage of a nonalarm condition

spectrogram—short-time Fourier transform (STFT)

speed shock—reaction when a substance, unfamiliar to the body, is rapidly infused into the circulation

sphygmomanometer—measurement device with an inflatable bladder within a cuff, a rubber bulb for cuff inflation, and a gauge to determine pressure readings

spike—hard plastic tube with a sharp point for penetrating reservoir

spirometer—device that measures volume of air during inspiration and expiration

spread-spectrum—modulation technique which spreads the spectrum of a signal by using a very wideband-spreading signal, and has properties facilitating demodulation of the transmitted signal by the intended receiver

squamous cell carcinoma—malignant invasive tumor or squamous cells

ST segment—short, relatively isoelectric segment between S and T waves of an electrocardiogram, representing the plateau before ventricular repolarization

stability—property of linear time-invariant system whereby every bounded input sequence produces a bounded output sequence

stackup—signal property that occurs when the time decay of a perturbed signal response back to its original value is longer than its period

start-up—new company funded with venture capital money

stereoselectivity—property of one optical isomer, the mirror image of a molecule, being generally 10 to 100 times more powerful than the other optical isomer

stethoscope—medical instrument used for listening to breathing, heartbeats, and other sounds made by the body

stochastic signal—member of an ensemble of signals that is characterized by a set of probability density functions; a random signal

stopband—region in which magnitude response must approximate zero

strictly positive real (SPR)—transfer function in which the real part is always greater than zero

stroke volume—ejection volume from left ventricle during systole, which is typically ~ 60 ml

subcutaneous administration—drug administration through implantation under the skin

substrate—substance acted upon by an enzyme during a biochemical reaction

sudden cardiac arrest (SCA)—cessation of heartbeat; may be caused by ventricular fibrillation

superposition—property that if input x_1 produces output y_1 and input x_2 produces output y_2, then input $(x_1 + x_2)$ produces output $(y_1 + y_2)$

suprasternal notch—key cardiothoracic landmark just above the junction of the clavicles (collarbones)

surface tension—property of liquids caused by the interaction of molecules at or near the surface that tend to cohere and contract the surface into the smallest possible area

syringe pump—pump in which fluid is held in a syringe for small-volume fluid administration

system identification—process of determination of dynamic models from experimental data

system operator—mathematical representation of the relationship between the inputs and outputs of a system

system output power—power of the system error

system—group of interacting components

systemic circulation—movement of blood from left ventricle through body, excluding lungs, to right atrium

systemic vascular resistance (SVR)—peripheral resistance of systemic circulation

systole—contraction portion of cardiac cycle during which blood is expelled into large arteries, characterized by peak blood pressure

tertiary structure—in a protein, the specific compact structure, based on folds in helically coiled regions, for the entire polypeptide chain

The Bethesda System (TBS)—classification of cervical cells, by the natural history of HPV and cervical dysplasia

thermodilution—estimation of cardiac output by injection of iced saline through a pulmonary artery catheter into right atrium and measurement of temperature changes downstream by a catheter thermistor in pulmonary artery. Cardiac output is calculated by the Stewart–Hamilton equation, a calculation in which the temperature change is proportional to blood flow.

thermogenesis—heat generation

thin-layer cytology—more precise Pap smear sample obtained by rinsing sample brush in vial containing buffered alcohol preservative solution and then, under microprocessor control, removing blood, mucous, and nondiagnostic debris from sample. The sample is then drawn through a filter to collect a thin, even layer of diagnostic cellular material, which is then transferred to a glass slide.

thorax—chest cavity

thrombosis—clot

tidal volume—volume of each breath

tight junction—unique anatomic specialization in brain endothelia that closes the interendothelial pores and limits diffusion to very small molecules

time to alarm (TTA)—for an infusion pump, the time by which pressure rises to a preset threshold in the tubing section, proportional to the tubing compliance and inversely proportional to the flow rate

time-frequency representation—two-dimensional mapping of the fraction of the energy of a one-dimensional signal at time and frequency. Common representations include the spectrogram, Wigner–Ville (or Wigner) distribution, Choi–Williams distribution, and Zhao–Atlas–Marks distribution

time-invariance—system property whereby delay of the input sequence causes a corresponding shift in the output sequence

time-scale representation—two-dimensional mapping of the fraction of the energy of a one-dimensional signal at time and scale

time-shifting property—property under which a delay in the time domain translates to an exponent of z in the z domain

tissue hypoxia—condition in which the supply of oxygen to a tissue is inadequate to meet its needs

topology preservation—validity measure of goodness of fit for a Kohonen network, in which the quality of training is monitored by quantifying the accuracy of various maps in preserving the network relations of the input space

tracee—endogenous molecule under study when a labeled molecule is introduced into the body

tracer—labeled molecule introduced into the body

trachea—windpipe

transcytosis—transendothelial transport involving cellular internalization, or endocytosis, of a given molecule from outside the cell; transport of that substance through the cell; and subsequent secretion, or exocytosis, of the moecule from the cell opposite the side of entry. The three forms are bulk flow transcytosis, receptor-mediated transcytosis, and absorptive-mediated transcytosis.

transfer function—Fourier transform of system impulse response

transmural—referring to both sides of the arterial wall

transthoracic impedance—impedance across thorax that consists of impedance from extratissue sources, including defibrillator, leads and electrodes; impedance from tissue sources, which include intracardiac and extracardiac tissue; and impedance from interface between electrode and tissue

truncated exponential waveform—defibrillation waveform shaped only by resistor and capacitor that may be monophasic or biphasic. Biphasic truncated exponential (BTE) is defined by total duration, phase duration (percentage of total duration occupied by first shock phase), and tilt (ratio percentage of difference between initial and terminal second phase voltage to initial voltage).

two's complement—natural binary form in which a negative number is formed from the corresponding positive number by complementing all the bits of the positive number and then adding one least significant bit. Q15 format uses 16 bits, with the most significant bit as sign bit and other 15 bits as fractional bits.

uncertainty principle—principle that the product of the standard deviations, with respect to time and frequency, of a signal must be greater than or equal to ½

unit step function—function with amplitude of 1 that begins at $t = 0$

upper limit of vulnerability hypothesis—hypothesized mechanism of defibrillation by which a defibrillation shock must halt conduction fronts propagating through tissue by directly exciting or prolonging refractoriness of myocardium just in front of these conduction fronts, and must prevent new conduction fronts at the border of the directly excited region that reinitiate fibrillation

upstream occlusion—flow obstruction above the infusion device

validation—as defined by FDA, process of establishing by objective evidence that device specifications conform with user needs and intended use

valve—membranous flap covering the inlets and outlets of ventricles and bases of larger arteries. The mitral valve is in left ventricle; the tricuspid valve is in right ventricle. These together are called atrioventricular (AV) valves and prevent regurgitation of blood into atria during ventricular systole. Aortic and pulmonary valves at bases of the larger arteries prevent regurgitation into ventricles during diastole.

Van der Walls interactions—transient dipole bonds induced in neighboring nonpolar groups

vasoactive—referring to a drug that enhances heart muscle contractility

vasoconstriction—decreased blood vessel radius. Epinephrine causes arterioles in skeletal muscle to dilate at low concentrations and vasconstrict with larger doses

vasodilation—increased blood vessel radius

vein—blood vessel that transports blood to the heart. The largest veins are the vena cava, which transports blood to right atrium, and the pulmonary vein, which transports blood to left atrium. The metacarpal vein, located at back of hand, is often used for drug infusion. A peripheral vein runs in neurovascular complexes with arteries and nerves. A small vein is called a venule.

venipuncture—puncture of a vein

venous hypoxia—inadequate oxygen to veins

ventilation—portion of pulmonary respiration during which oxygen is transported from outside air by bulk flow transport to alveoli of the lung

ventricle—either of the two lower chambers (left, right) of the heart that receives blood from an atrium and ejects it into a large artery

venture capital—investment money for new technologies in which risk for loss and potential for profit are considerations

verification—as defined by FDA, a systematic application of various analyses, evaluation, assurances, and testing at each stage of the development process to assure that all the requirements specified for that stage have been fulfilled

vesicant—medication capable of causing a blister or tissue damage

vital sign—traditionally, an external indicator that a patient has not lapsed into cardiac arrest. The four official vital signs are heart rate, temperature, respiration, and blood pressure. Pulse oximetry is often considered the fifth vital sign.

volume clamp method—also known as method of Penaz; noninvasive measurement of blood pressure during which light from an LED is passed through a finger and detected while a cuff placed over the same finger is continuously adjusted to maintain a constant transmural pressure across the finger arteries. It is assumed that an approximately linear relationship exists between the intraarterial pressure and detected light.

volumetric infusion pump—pump that displaces constant flow, rather than relying on hydrostatic pressure to produce flow, for drug infusion

voluntary apnea—self-induced cessation of breathing

Von Neumann architecture—microprocessor access of single bank of memory through a single set of buses

wavelength—distance between two peaks in a wave, used to differentiate portions of light

wavelet transform—time-scale distribution that may be calculated through convolution of a wavelet with a lowpass (scaling) or highpass (wavelet) filter. When a scaling filter is used, the resulting wavelet transform is composed of approximation coefficients. When a wavelet filter is used, the resulting wavelet transform is composed of detail coefficients. One example is the Haar wavelet transform.

wavelet—function with a mean of zero

white noise—sequence of independent and identically distributed random variables with zero mean

wide-sense stationarity—signal property whereby signal possesses a constant mean and does not vary with time

window method—method for designing FIR filter involving truncation of an ideal frequency response

windowing theorem—dual of the convolution theorem that states that multiplication in the time domain is equivalent to convolution in the frequency domain

working myocardium—main mass of the heart that performs the mechanical work of pumping

Zadeh intersection—fuzzy mathematical operator that results in taking the minimum of two degrees of membership

zero—root of numerator of a transfer function

INDEX

Abe, Shigeo, 203, 209, 216
Absolute summability, 5
Adaptive filter, 46–65
   adaptive noise cancellation, 46
      example, 60–61
      proof, 46–48
      pulse oximetry, 78–83
   adaptive predictor, 61, 62
   adaptive self-tuning filter, 61–64
   IIR filter, 52–60
      HARF, 54–57
      limitations, 52–53 ,60
      SHARF, 57–60
   least mean squares algorithm, 50–52, 58–60
      FIR filter example, 52
      FIR versus IIR filter performance, 58–60
      optimization, 51
      performance function, 50–51
   primary input, 46
   reference noise source, 46
   requirements, 46
Aliasing, 9
   Nyquist frequency, 9
   Nyquist rate, 9
   sampling theorem, 9
Anatomy
   cardiac output, 112–115
   central neuron, 195–196
   cervix, 218–219
   electrocardiogram, 163–165
   respiration, 67–70
Aoyagi, Takuo, 75
ARMAX model, 6, 141
   backward shift operator, 142
   feedback coefficients, 141
   feedforward coefficients, 141
   instrumental variable method, 150–152
   prediction error method, 145–150
      assumption of white noise, 148–150
      derivation, 146–147
      example, 147–148
      performance function, 145–146
   recursive least squares algorithm, 152–157
      derivation, 152–155
      example, 156–157
      forgetting factor, 155–156
      initial conditions, 155
   regression vector, 144
   variation models, 7–8
      AR model, 142–143
      ARX model, 143, 174–176, 280
      FIR model, 143
      IIR model, 144
      MA model, 143
Artificial neural network, 195
   applications, 195
Autocorrelation, 6

Baura, Gail
   central insulin transport, 369–373, 376–377
   defibrillation waveform optimization, 174–187, 191
   fuzzy control, 274–285, 289
   impedance cardiography fiducial point detection,128–133, 136
   infusion pump resistance monitoring, 40–43
Bergman, Richard, 344, 348, 353
Bernstein, Donald, 123, 128, 135

Bezdek, Jim, 241
Blood pressure, 262
    auscultation, 264–265
        Korotkoff sounds, 264
        limitations, 265
    continuous noninvasive blood pressure, *See* Continuous noninvasive blood pressure
    diastolic, 263
    dicrotic notch, 263
    intraarterial measurement, 43–44, 264
        complications, 264
    mean arterial pressure, 263
    oscillometry, 265–266
        limitations, 266
    systolic, 263
Body weight regulation, 354–357
    energy expenditure, 355
    genetics, 355
    insulin, *See* Central insulin transport
    model, 355
        feedback signals, 355–357
        hypothalamus, 355
Bundle branch block, 129

Cardiac arrhythmia, 166
    atrial fibrillation, 44, 136
    ventricular fibrillation, 163, 166–167
Cardiac cycle, 112–114
    diastole, 112
        filling period, 114, 262
        isovolumetric relaxation period, 114, 262
    systole, 113
        ejection period, 113, 262
        isovolumetric contraction period, 113, 262
Cardiac electrical conduction, 163
    action potential, 163–165
        depolarization, 164
        refractory period, 164
        repolarization, 164
        resting potential, 163–164
    spread of excitation, 165
        atrioventricular node, 165
        bundle of His, 165
        pacemakers, 165
        Purkinje fibers, 165
        sinoatrial node, 165
Cardiac output, 112
    afterload, 115
    cardiac index, 115
    contractility, 115
    measurements, 116–122
        Doppler ultrasound, 121–122

        Fick method, 118
        impedance cardiography, *See* Impedance cardiography
        indicator-dilution methods, 116
        thermodilution, 116–118
    ventricular preload, 115
Cardiac valves, 113
Carpenter, Gail, 210, 217
Causality, 4
Central insulin transport, 357–374
    autoradiographic studies, 361
    estimation of CSF insulin clearance, 370, 373
    evidence for saturable transport, 368–373
    discussion, 373
    K_M estimation, 370, 372
    isolated brain capillary studies, 363–366
    nonlinear compartmental model, 359–361, 370–372
    practical sampling schedule, 369
    transport across blood-brain barrier, 358–359
    whole animal studies, 366–368
Cervical cancer screening system, 229–234
    AutoPap system
        fuzzy model, 229–231
        versus PAPNET comparison, 234
    PAPNET neural network, 231–234
        processing, 231–233
        training data, 231
        validation, 233–234
Cervical cancer, 219–227
    treatment, 219
        chemotherapy, 219
        hysterectomy, 219
        radiation therapy, 219
Clark electrode, 71–72
    blood gas analyzer, 71–72
Closed-loop drug delivery, 316
    limitations, 316
    pharmacologic stress test, *See* Gensia pharmacologic stress test
    system(s), 320
        control supervisor, 324–325
        historical, 327
Cobelli, Claudio, 309, 314, 348, 353
Cohen, Leon, 91, 100, 110
Continuous noninvasive blood pressure, 262, 266
    arterial tonometry, 269–273
        assumptions, 269
        early VitalWave processing, 272–273

NCAT, 269–271
Vasotrac, 271–272
clinical significance, 266–267
Finapres, 267–268
limitations, 267
method of Penaz, 267
VitalWave blood pressure processing, *See* VitalWave processing
Convolution
sum, 4
theorem, 5
Dual domain theory of duality, 5
Corporations
Blood pressure monitors
Colin, 269
Critikon, 266
Medwave, 269, 271–274
Nellcor, 269
Ohmeda, 267
Tensys Medical Systems, 284
TNO Biomedical Instruments, 267–268
VitalWave Corporation, 273–274, 276, 284
Cardiac output monitors
Baxter Edwards, 117–118, 123
BoMed Medical Manufacturing, 123, 125
CardioDynamics International Corporation, 123, 128–129
Interflo Medical, 118
Lawrence Medical Systems, 121
Renaissance Technologies, 123, 128, 131–133
Wantagh Incorporated, 123
Cervical cancer screening systems
Autocyte, 228
Cytyc Corporation, 227
Neopath, 227–229
Neuromedical Systems, 227–229
Tripath Imaging, 228, 231, 234
Closed loop drug delivery systems
Esaote Biomedica, 327
Gensia Automedics, 325, 328
Hook Lane Nyetimber, 327
IVAC Corporation, 327
Miles Labs, 327
Defibrillators
Agilent Technologies, 169, 172
Cardiotronics Systems, 174
Heartstream, 170, 172
Medtronic Physio Control, 172
Zoll Medical, 172

Exercise Data
SAAM Institute, 338, 378
VitalWave Corporation, 86
Welch Allyn Incorporated, 192–193
Infusion pumps
Alaris Medical Systems/IVAC Corporation, 35–42
Baxter, 32, 35, 37
Obesity
Amgen, 362
Hoffmann-La Roche, 363
Oxygen monitors
Biox, 75
Criticare, 77
Hewlett-Packard, 75
Instrumentation Laboratory, 71
Mallinkrodt, 78
Masimo, 77–83
Minolta, 75
Nellcor, 75, 76–78, 80, 82
Nihon Kohden, 75
Ohmeda, 75, 78
Puritan-Bennett, 78
Cross power density spectrum, 6
Cross-correlation, 6
Cybenko, George, 202, 215

Data types
fixed-point, 384–385
Q15 format, 384–385
two's complement, 384–385
floating-point, 383–384
double-precision, 384
exponent, 383
mantissa, 383
radix, 384
single-precision, 384
integer, 385
tradeoffs, 385
Design of experiments, 40
Detection
sensitivity
Pap smear, 224
pulse oximetry, 82
specificity
Pap smear, 224
pulse oximetry, 82
Determinism, 5
Diab, Mohamed, 77, 85
Digital signal processor, 385–387
architecture, 386
Harvard, 386

Digital signal processor *(continued)*
    Von Neumann, 386
    implementation, 387
    multiply-accumulate operation, 386
    pricing, 386–387
    specialized features, 386
Direct-form realization, 8
Discrete time, 3
Doppler ultrasound, 121–122, 272, 274
Downsampling, 10
Drug administration
    absorption, 29
    bioavailability, 29
    catheter, 30
    infusion, 30
    injection, 30
    oral ingestion, 29
    parenteral administration, 29
        intramuscular, 29
        intravenous, 29
    venipuncture, 30
Drug infusion
    administration set, 30
    complications
        bolus, 36
        catheter embolism, 31
        extravasation, 31
        hematoma, 31
        infiltration, 31
        local infection, 31
        phlebitis, 31
        positional catheters, 31
        short half-life drugs, 36
        upstream and downstream occlusion, 31
    controller, 33
    delivery system, 32
    mechanical flow regulator, 32–33
    pump flow profiles, 34
    syringe pump, 33, 35
        hardware, 33–34
    time to alarm, 36
    volumetric infusion pump, 33
        cassette, 33
        linear peristaltic mechanism, 33
Drugs
    arbutamine, 317, 319, 326–327
    atropine, 326
    cisplatin, 219
    dipyridamole, 325–326
    dobutamine, 325–326
    dopamine, 36, 296–297
        kinetics, 303

epinephrine, 264
insulin
    central transport, 357–374
    minimal model, 348–351
    peripheral sampling schedule, 308–311
    peripheral transport, 344–347
isoproterenol, 319, 325
leptin, 356, 361
metoprolol, 319
pharmacologic antiobesity agents, 362–363
probenecid, 317
sodium nitroprusside, 316
somatostatin, 346
thiopental, 264

Eckerle, Joseph, 269, 289
Editorial, 395–405
Einthoven, Willem, 166
Electrocardiogram, 76–77, 163
    leads, 166
    P wave, 166
    QRS complex, 166
        detection, 106–107
    ST segment, 166
Electrocorticogram, 96
Electrogastrogram, 87, 89
Embedded system, 387
    characteristics, 388–389
    man-machine interface, 387
    programming environment, 387–389
        driver, 388
        memory, 387–389
        multitasking, 388
        operating system, 388
        peripherals, 387
        register, 388
        software development process, 388
Endothelium, 357
Epicardial first heart sound, 89–90
Epithelium, 218
Exercises
    Matlab, 43
        cardiac output, 237–238
        central insulin model, 377–380
        digital thermometry, 192–194
        infusion pump occlusion detection, 289–291
        intraarterial blood pressure, 43–45
        noninvasive blood pressure, 85–86
        peripheral insulin model, 337–339
        QRS detection, 136–137
    reference,

adaptive filter, 64
artificial neural networks, 217
frequency-selective filters, 28
fuzzy logic and control, 261
linear compartmental model, 315
linear system identification, 152
nonlinear compartmental model, 353
pseudorandom binary sequence, 28
time-frequency and time-scale analysis, 111
Expected value, 46
External defibrillation, 163
  hardware, 167–168
    automatic external defibrillator, 171–172
  pads, 167
  waveform(s), 167–171
    biphasic truncated exponential, 170–171
    damped sinusoid, 167–169
    monophasic truncated exponential, 169–170
    optimization using Cardiotronics method, 185–188

FDA
  premarket submission, 389
    510(k), 389
    premarket application (PMA), 328, 389
  software documentation, 390
    architecture design chart, 392
    controlled device features, 390–391
    design specification, 392
    development, 392
    device hazard analysis, 392
    level of concern, 390–391
    release version number, 393
    revision level history, 393
    software description, 390
    software requirements specification, 392
    system theory considerations, 393
    traceability analysis, 392
    unresolved anomalies, 393
    validation, verification, and testing, 393
Finite response filter design, 20–24
  window method, 20–23
Fluid(s)
  cerebrospinal fluid, 357
  flow, 30–31
    Hagen–Poiseulle's equation, 30, 263
    Newtonian fluid, 30
    Ohm's law, 30
    viscosity, 30

  interstitial fluid, 345
  lymph, 344–345
  plasma, 295
Founders with academic backgrounds
  Huntsman, Lee, 121
  Nelson, Alan, 229
  Yelderman, Mark, 118
Fourier transform pair, 5
Frequency-selective filter, 9
  bandpass, 9
  bandstop, 9
  gain, 12
  highpass, 9
  lowpass, 9
  magnitude, 12
  passband, 12
  phase, 12
  stopband, 12
Friedman, Jeffrey, 356, 375
Fuzzy
  linguistic fuzzy model, *See* Linguistic fuzzy model
  logic, 239
    membership function, 240
    set, 239–240
  relational model, 239

Gas transport
  alveolar diffusion, 66
    Dalton's law, 68
    lung perfusion, 68
    partial pressure, 68
  blood gas transport, 69–70
    dissociation curve, 69–71
    functional arterial oxygen saturation, 69
    hemoglobin, 69
  tissue respiration, 70
    hypoxia, 70–71
  ventilation, 66–68
    expiration, 67
    inspiration, 67
    respiration frequency, 67
    tidal volume, 67–68
Geddes, Leslie, 173, 184, 190
Gensia pharmacologic stress test
  clinical
    significance, 325, 328
    testing phases, 328
  data acquisition and processing, 331–332
  pharmacodynamic model, 329–331
  PI control algorithm, 332–333
  validation, 333–324

physiologic targets, 324
Grossberg, Stephen, 210, 217

Hemoglobin, 69
  dysfunctional, 70–71, 76
Human papillomavirus, 219–220
  screening, 220
Huntsman, Lee, 121, 135
Hyperstability, 53
  strictly positive real transfer function, 53–54

Ideker, Raymond, 187, 191
Impedance cardiography, 112
  CardioDynamics wavelet processing,
      128–133
    data acquisition and processing, 128–133
    validation, 129–131
  Drexel spectrogram processing, 124–127
    data acquisition and processing, 124–126
    validation, 125
  impedance plethysmography, 119
  Minnesota impedance cardiograph, 121
  original Kubicek theory, 119–120, 127–128
  spectrogram versus wavelet processing,
      131–133
  Sramek-Bernstein equation, 122–123
  thoracic bioimpedance, 119
Impulse reponse, 4
  filter
    Butterworth, 14
    Chebyshev, 14
    design, 13–20
Input sequence, 2
Internal defibrillation, 163, 167

Kerber, Richard, 173, 184, 186, 190
Kiani, Joe, 77, 85
Kinetic model, 299, 340
  bilinear relation, 340, 347–351
    example, 348–351
  linear compartmental model, 300–303
    central compartment, 302
    definition, 300–302
    example, 303
    forcing function, 302
    nonlinear least squares estimation,
        306–308
    peripheral compartment, 30, 302
    properties, 302
  Michaelis–Menten dynamics, 340–347
    central insulin transport, 368–373
    definition, 341–343

    example, 344–347
    parameter estimation, 343
  noncompartmental model, 299–301
    apparent volume of distribution, 300
    AUC, 299
    diffusion, 29
    distribution, 30, 300
    elimination, 30, 300
    half-life, 300
  sampling schedule, 308–312
    optimal, 299, 308–311
    practical, 311–312
Kohonen, Teuvo, 206, 217
Kubicek, W. G., 19, 135

Larry as a sweetie, vii, xvii
Law of mass action, 69
Left ventricular ejection time, 114
Linear time-invariant system, 4, 141
Linearity, 4
Linguistic fuzzy model, 239
  cervical cancer screening, 229–231
  defuzzification, 240, 246
    centroid, 248
    example, 248
    maximum, 248
    mean of maximum, 248
  fuzzification, 240, 242
    example, 243–244
    fuzzy partition, 242–243
    membership function, 242
  fuzzy control, 251–252
    example, 252–255
  fuzzy pattern recognition, 255–257
    empirical classification, 255–259
    nearest prototype classification, 255, 257
  history, 240–242
  knowledge base, 240, 248–249
    neural network tuning, 249–251
  Mamdani model, 240
  rule base inference, 240, 244–246
    example, 245–246
    fuzzy logic, 245
    rules, 245
Lloyd, Jack, 75

Mamdani, Ebrahim, 240, 260
Market forces
  cervical cancer screening, 227–228
  continuous noninvasive blood pressure, 267
  external defibrillation, 171
  impedance cardiography, 123

infusion pumps, 37
obesity, 362
pharmacologic stress test, 328
pulse oximetry, 78
McCullough, Warren, 196, 215
Median filter, 42
Metabolism, 66, 295
  anabolism, 66
  catabolism, 66
  respiration, 66
Milllikan, Glenn, 75
Minimal model of insulin sensitivity, 348–351
  derivation, 348–350
  validation, 351
Model identifiability, 145, 304–306
  practical identifiability, 305–306
  theoretical identifiability, 304–305
    transfer function matrix approach, 304–305
Model structure choice, 144–145
  parsimony principle, 144
  performance function, 144
Model validation, 157, 306, 312–313
  coefficient of variation, 157–158, 307
  correlation matrix, 212
  fuzzy models, 251
  goodness of fit, 158
    Akaike Information criterion, 158, 312
    final prediction error, 158
  linear example, 158–159
  number of templates, 213
  plausibility, 158, 313
  residual statistics, 158, 313
  topology preservation, 213
Multiple input, multiple output system, 195
Myocardial
  infarction, 115
  ischemia, 167, 325

Nelson, Alan, 229
Neonate, 33, 36
Neuron model, 196–198
  ADALINE, 197
    MADALINE, 197
  McCullough–Pitts model, 196–197
    activation function, 196
    threshold, 196
  perceptron, 197–198
    learning rate, 197
    linear separation, 197
New, William, 75

Newgard, P. M., 269, 288
Nicolai, Ludwig, 75
Noise
  sequence, 3
  sources
    cervical cancer screening, 224
    drug infusion, 39–40
    impedance cardiography, 121
    maternal heartbeat, 60–61
    pulse oximetry, 76, 80, 82
  white, 7, 141

Obesity, 354
  clinical significance, 361–362
  NIH body mass index, 354
  pharmacologic antiobesity agents, 362–363
    need for blood-brain barrier studies, 363
Optimization, 48–49
  direct search methods, 48
    simplex search, 48
  gradient methods, 48
    method of steepest descent, 49, 201
  performance function, 48

Pap smear, 218
  cell classification, 221–224
    adenocarcinoma, 224, 227
    AGUS, 224, 227
    ASCUS, 221, 223
    HSIL, 224–225
    LSIL, 223
    the Bethesda system (TBS), 221
  limitations, 224
  origin, 220
Papanicolaou, George, 220
Pardridge, William, 361, 363–364, 376–377
Pedrycz, Witold, 240, 260
Penaz, Jan, 267, 288
Pharmacodynamics, 317
  drug receptor theory, 317–319
    agonists and antagonists, 318–319
    occupation theory, 317–319
    rate theory, 319
  drug receptors, 317
    drug-receptor interaction, 317
    protein types, 317
    feedback signals, 319–320
Pharmacokinetics, 316–317
  effect compartment, 316
Philip, James, 37, 43
PID controller. 320–324
  definition, 321

PID controller *(continued)*
   robust control, 321–324
      example, 323–324
Pitts, Walter, 196, 215
Poiseuille's equation, 30, 263
Porte, Jr., Daniel, 351, 353, 357, 361, 366, 375–377, 366
Power density spectrum, 6
Prefilter, 9–12
   blood pressure monitoring, 283
   impedance cardiography, 124, 128
   pulse oximetry, 80, 81
   resistance monitoring, 40
   transthoracic impedance estimation, 176
Pressman, G. L., 269, 288
Protein, 295–299
   functions, 295
   measurement
      high-performance liquid chromatography, 297
      positron emission tomography, 297
      radioimmunoassay, 297–298
      tracers, 298–299
   structure, 295–297
      amino acid, 295
      peptide bond, 295
      primary, 296
      secondary, 296
      tertiary, 296–297
Pseudorandom binary sequence, 24
   continuous thermodilution, 117–118
   filter design, 24–26, 38–42
   resistance monitoring, 38–42
Pulmonary artery catheterization, 116
   complications, 122
      infarction, 122
      sepsis, 122
      thrombosis, 122
Pulmonary circulation, 112, 115
   left atrium, 69
   metacarpal vein, 39
   pulmonary artery, 112
   pulmonary vein, 112
   right ventricle, 70
Pulse oximetry, 72–84
   Beer–Lambert law, 72
      absorbance, 72–73
      extinction coefficient, 72
      incident light, 72
      path length, 72
      transmitted light, 72
      wavelength, 72
   calibration curve, 74–76
   ear oximeter, 75
   empirical processing, 76–77
   hardware, 72–74, 81
      cooximeter, 73
      light emitter diode, 74
      photodetector, 72, 74
   Kalman filter, 77–78
   limitations, 76
      dysfunctional hemoglobins, 76
      electrocautery, 76
      motion artifact, 76
      poor peripheral perfusion, 76
   Masimo adaptive noise cancellation, 78–83
      derivation, 78–79
      implementation, 83
      reference noise extraction, 79–80
      validation study, 80 ,82
   Nellcor, 75, 77–78
   ratio, 73–75
   standard of care, 75–76

Ramsey, Michael, 266, 287
Resistance
   infusion system, 31
      catheter resistance, 31
      IVAC pump, 35, 38–42
      Philip studies, 37–38
      pseudorandom binary sequence, 38–42
   systemic vascular, 115
   transthoracic, 173–174
Rosenblatt, Frank, 197, 215

Scale, 87
Schade, Christy, 327, 336
Schwartz, Michael, 366, 376–378
Shaw, Robert, 75
Single input, single output model, 3, 141
Spread-spectrum
   demodulation, 24
   modulation, 24
Sramek, Bohumir, 122–123, 135
Stability, 5
Stackup, 26
Stochastic
   probability density function, 5
   process, 5
   signal, 5, 148
Stroke volume, 113
Sudden cardiac arrest, 166
Sugeno, Michio, 240, 260
Sugimachi, Masaru, 44, 110, 136, 161, 237

Sun, Hun, 123, 135
Superposition, 4
Supervised neural network, 195
   multilayer feedforward network, 199–203
      activation function, 201–202
      back propagation, 199–201
      cervical cancer screening, 231–234
      definition, 199–200
      example, 203–205
      fuzzy model tuning, 249–251
      necessary number of hidden layers, 202
      selection of input variables, 202–203
   radial basis function network, 203–204
   recurrent network, 203–204
System
   identification, 141
   operator, 3
Systemic
   circulation, 112, 114–115
      aorta, 43, 112
      arteriole, 70
      capillaries, 66, 115
      left ventricle, 69
      radial artery, 32, 44, 85
      right atrium, 70
      vena cava, 112
      venule, 70
      vascular resistance, 263–264
         autonomic nervous control, 264

Takagi, Hideyuiki, 240, 260
Takagi-Sugeno model, 239
ThinPrep 2000 processor, 228–229
Thorax, 67, 119
Time-frequency representation, 87–104
   Choi-Williams Distribution, 101–102
   impedance cardiography, 124–127
   kernel method, 100–103
      Cohen's class, 100
      properties, 100–101
   properties, 91–94
      characteristic function, 94
      global means, 92
      local means, 92–93
      marginals, 91
      scaling invariance, 93
      time and frequency shift invariance, 93
      total energy, 92
      uncertainty principle, 94
      weak and strong finite support, 94
   Spectrogram (short-time Fourier transform), 91, 94–97
      derivation, 95
      example, 96
      properties, 95–96
   Wigner distribution, 97–99
      blood pressure example, 274–275
      derivation, 98
      example, 99
      properties, 98–99
      pseudo Wigner distribution, 99
   Zhao–Atlas–Marks Distribution, 102–104
Time-scale representation, 87, 104–109
   affine class, 104
   continuous wavelet transform, 105
   dyadic discrete wavelet transform, 105, 107–109
      approximation coefficients, 107–108
      blood pressure example, 279–280
      detail coefficients, 107–109
      impedance cardiography, 128–133
   properties, 104–105
   scalogram, 105–107
      example, 106–107
Time-shifting property, 8
Transcytosis, 357–358
Transfer Function, 5
   aortic-radial artery, 147–149, 156–157
   pole, 9
   zero, 9
Transthoracic impedance, 173–178
   Cardiotronics impedance estimation, 174–187
      capacitance estimation, 176–179
      resistance estimation, 174–176
      validation, 178–185
   components, 173
   Geddes' pure resistance model, 173
   resistance and reactance in dogs, 173–174

Unsupervised neural network, 195
   adaptive resonance theory network, 210–212
      definition, 210–211
      implementation, 211–212
      stability-plasticity dilemma, 210
   Kohonen network, 205–213
      competitive learning, 206, 207–208
      definition, 206–207
      example, 208–210

Venture capital, 78
Vital signs, 78

VitalWave blood pressure processing, 273–284
    data acquisition and processing, 283
    diastolic pressure estimation, 279–280
    fuzzy control, 282–283
    hypothesized mechanism, 276–279
    MAP estimation, 274–279
    orthogonal signal, 274
    systolic pressure estimation, 280–282
    validation, 283–286

Wang, Xiang, 123, 135
Wesseling, Karel, 267, 288
Wide-sense stationarity, 6, 46
Widrow, Bernard, 49, 53, 60, 65, 198, 215, 327
Wood, Earl, 75
Woods, Stephen, 357, 361, 375

Yalow, Rosalind, 298
Yelderman, Mark, 118, 135

Zadeh, Lotfi, 239, 260

About the Author

Gail Dawn Baura received a BSEE from Loyola Marymount University in 1984, and an MSEE and MSBME from Drexel University in 1987. She received a PhD in Bioengineering from the University of Washington in 1993. Between these graduate degrees, Gail worked as a loop transmission systems engineer at AT&T Bell Laboratories. Since graduation, she has served in a variety of research positions at IVAC Corporation, Cardiotronics Systems, Alaris Medical Systems, and VitalWave Corporation. Gail is currently the Director of Research and Development at CardioDynamics. Her Research interests are the application of system theory to patient monitoring and insulin metabolism.

R 857 .S47 B38 2002
Baura, Gail D.
System theory and practical
 applications of biomedical